Clinical Applications of Magnetoencephalography

Shozo Tobimatsu • Ryusuke Kakigi
Editors

Clinical Applications of Magnetoencephalography

Springer

Editors
Shozo Tobimatsu
Graduate School of Medical Sciences
Department of Clinical Neurophysiology
Fukuoka, Japan

Ryusuke Kakigi
National Institute for Physiological Sciences
Department of Integrative Physiology
Okazaki, Japan

ISBN 978-4-431-55728-9 ISBN 978-4-431-55729-6 (eBook)
DOI 10.1007/978-4-431-55729-6

Library of Congress Control Number: 2016931298

Springer Tokyo Heidelberg New York Dordrecht London

Printed on acid-free paper

Springer Japan KK is part of Springer Science+Business Media (www.springer.com)

Preface

In 1968, David Cohen (*Science* 161:784–786) showed for the first time that it is possible to record magnetic fields from the head, so-called magnetoencephalography (MEG). The extremely weak magnetic fields that are measured are generated by the current flow in the human brain due to electrical processes in the neurons. After the first report on magnetic brain signals, the development of superconducting quantum interference detectors (SQUID) made it possible to make reliable recordings of these weak magnetic fields with appropriate sensitivity. Since the introduction of high-density whole-scalp coverage of MEG in the 1990s, the instrumentation has not changed drastically, yet novel data analyses are advancing the field rapidly by shifting the focus from merely pointing to hotspot activity to seeking stimulus- or task-specific information and to characterizing the dynamics of functional networks. Why did the interest in MEG increase over the last decade in the neuroscience community? There are certainly two major aspects of MEG that make it very suitable for brain imaging: (1) MEG has a high temporal resolution; and (2) the different constituents of the head-like fluid or bone do not affect the spatial distribution of the signals. Over the last decade several different approaches were developed to overcome the limitations of equivalent current dipole modeling and even to reconstruct the activity in deep brain regions. Thus, MEG is nowadays a well-established, non-invasive technique, which can be used over a whole life span and for basic science research and clinical applications.

This handbook is the result of the collective effort by a number of members of the recently formed Japanese Consortium of Clinical MEG. The book has two purposes: to articulate the empirical knowledge gained during the last two decades in the diagnostic use of MEG, and to serve in the clinical training of new users. As the knowledge of the clinical uses of MEG is at present rather limited and in some aspects uncertain, we hope and expect that this book will be augmented and some of its contents will be updated in the future. However, we consider this present volume to be a definite authoritative reference for clinical applications of MEG.

On behalf of all the co-authors, I wish to thank the editorial staff members of Springer Japan for their patience and support in producing this volume.

Fukuoka, Japan Shozo Tobimatsu, M.D., Ph.D.

Contents

Part I

Introduction

Principles of Magnetoencephalography 1

Akira Hashizume and Naruhito Hironaga

Abstract

Magnetoencephalography (MEG) is a neuroimaging tool which obverts functional brain maps or pathognomonic signs with millisecond order, but some computational technique appears to be a black box. To relax such complex matter for general readers, we introduce the principles of MEG computation briefly. This chapter consists of three steps: first, we introduce the basic concept of MEG; next, we describe forward solution; and finally, we address inverse problem for understanding of MEG source localization concept. MEG detects varying magnetic fluxes derived from enormous amount of intracellular currents within pyramidal layer in folded cortices, and estimation of these source activities is essential to the MEG utilization. Here, we introduce typical three source localization algorithms: equivalent current dipole (ECD), minimum norm estimation, and adaptive beam former. To estimate ECDs from measured sensor signals, repeated seeking operation is undertaken so that errors between measured signals and calculated signals are minimized. Fundamental concept of minimum norm estimation has constraint subject to minimizing total currents of all nodes. Minimum variance estimation, the so-called adaptive beam former, has constraint subject to minimizing the variance of time-series current at one node. We hope that this introduction might help readers to understand the basic theory of MEG source localization.

A. Hashizume (✉)
Department of Neurosurgery, Hiroshima University, 1-2-3 Kasumi, Minami-Ku, Hiroshima 734-8551, Japan

Takanobashi Central Hospital, Hiroshima, Japan
e-mail: wj8uc6@bma.biglobe.ne.jp

N. Hironaga
Department of Clinical Neurophysiology, Kyushu University, Fukuoka, Japan

S. Tobimatsu, R. Kakigi (eds.), *Clinical Applications of Magnetoencephalography*, DOI 10.1007/978-4-431-55729-6_1

Keywords
Forward solution • Inverse problem • Sarvas formula • Lead field matrix • Minimum norm estimation • Adaptive beam former

1.1 Principles and Computational Analysis

1.1.1 Origin of Magnetic Fluxes Detected by Neuromagnetometers

Neuromagnetometers detect varying magnetic fields from neuronal electric currents which originate from intracellular currents within pyramidal cells lining cerebral cortices (Fig. 1.1). Only simultaneous depolarization of thousands of pyramidal cells arranged in the same direction can make valid magnetic fields, and it is usually treated as single, some, or numerous equivalent current dipoles (ECDs) [1, 2]. MEG pick-up coils detect only magnetic fluxes crossing the coil surface, i.e., most sensitive to the perpendicular direction. Therefore, many of existing neuromagnetometers detect the magnetic fields generated from currents perpendicular to the surfaces of sulci or fissures. Activities of the following tissues are hard for neuromagnetometers to detect: hippocampus due to folded pyramidal layers, thalamus or basal ganglion due to lack of laminar structure, and cerebellar cortex due to too folded laminar structure of folia. In this chapter, cerebral cortical activities and its numerical modeling are mainly focused.

Notes: Bold letters denote vectors or matrices, and others denote scalar. B (b) (include with subscription) denotes detected magnetic flux density on the MEG sensor array with certain time ranges, b denotes the same as B but for sensor array detection on one time sample or detection on one sensor, and $\hat{b}$ denotes the calculated magnetic flux density using forward simulation. M means the number of the MEG sensors and N means the number of sources, and m means the arbitrary sensor in $1 < m < M$ and n means the arbitrary source (or source point) in $1 < n <$

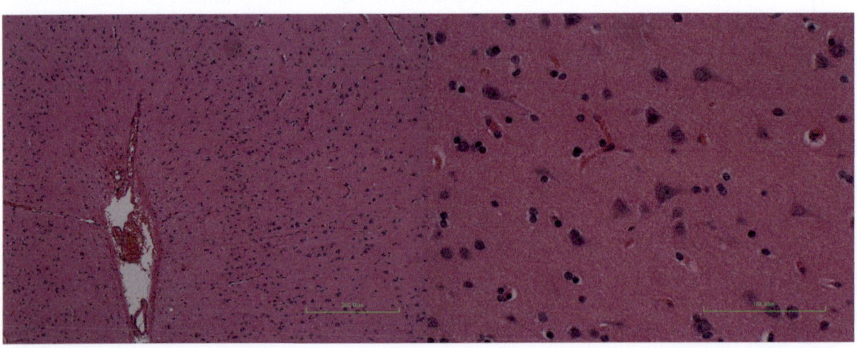

Fig. 1.1 *Left*: Microscopic photo of the human cerebral cortex (H&E stain). *Right*: Microscopic photo of pyramidal cells (H&E stain). MEG detects varying fluxes derived from pyramidal cell layers lining the folded cerebral cortex

N. Q(q) denotes the source activations on the lattice nodes within the brain or cortical polygon vertices and large Q time course expansion as well. $\hat{Q}$ indicates the estimated inverse matrix. Z(z) denotes the normalized statistical z-values with arbitrary unit. The superscript ^{T} denotes transpose matrix, while subscript $_t$ is arbitrary time point. The column number in Figs. 1.8–1.14 is counted from the left side.

1.1.2 Pick-Up Coils

To understand easier, the term "flux" is used for substitution of the magnetic flux density and does not denote physically actual flux throughout this session. Figure 1.2 shows the typical five types of pick-up coils employed in the active working MEG products.

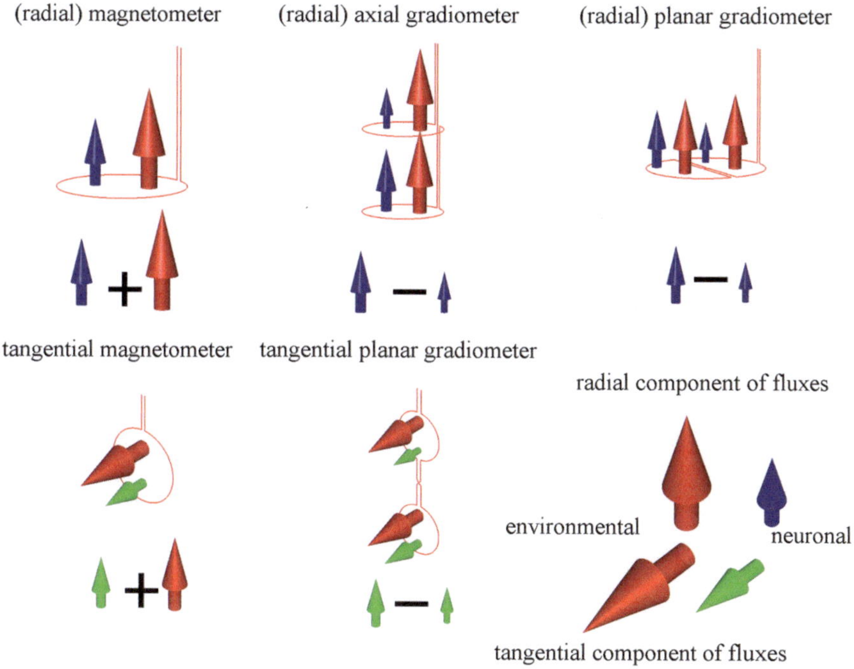

Fig. 1.2 Five types of pick-up coils. The radial magnetometer, the radial axial gradiometer, and the radial planar gradiometer detect radial component of fluxes (*blue arrow*) derived from neuronal currents (*upper row*), while the tangential magnetometer and the tangential planar gradiometer detect tangential component of the fluxes (*green arrow*) (*lower row*). The term "radial" is usually omitted. All pick-up coils detect both neuronal and environmental fluxes (*red arrow*). The gradiometers consisting of two magnetometers which have opposite polarity to each other can cancel the latter. Finally, detected signals are shown below each coil

The terms “radial” and “tangential” indicate the direction of the assumptive spherical conductor. Neuronal or environmental fluxes can be projected into two components, radial and tangential components. The radial magnetometer detects radial component of fluxes and the tangential magnetometer detects tangential component of fluxes. Both radial and tangential magnetometers detect not only neuronal signals but also environmental noises and artifacts from the human body such that generally modeled as

$$\mathrm{b}_{\mathrm{mag}} = f_{\mathrm{mag}} + \mathrm{b}_{\mathrm{noise}}, \tag{1.1}$$

where subscript $_{\mathrm{mag}}$ indicates the magnetometer; thus, $\mathrm{b}_{\mathrm{mag}}$, f_{mag}, and $\mathrm{b}_{\mathrm{noise}}$ are the total fluxes of the magnetometer, the flux from the neuronal sources, and environmental noise, respectively. The gradiometers are designed to cancel out the noises and focus on near-field activity. To achieve these required matters, the gradiometers consist of two magnetometers with opposite polarization. Two magnetometers detect both neuronal signals and noises, but the difference of the signals between the two magnetometers only measures the gradients of the neuronal signals if the noise is constant and can be approximately almost same strength when two magnetometer coils are much closed position. Taking into account the above concept, it is formulized as

$$\begin{aligned}\mathrm{b}_{\mathrm{gra}} &= \mathrm{b}_{\mathrm{mag1}} - \mathrm{b}_{\mathrm{mag2}} = \left(f_{\mathrm{mag1}} + \mathrm{b}_{\mathrm{noise}}\right) - \left(f_{\mathrm{mag2}} + \mathrm{b}_{\mathrm{noise}}\right) \\ &= f_{\mathrm{mag1}} - f_{\mathrm{mag2}},\end{aligned} \tag{1.2}$$

where subscript $_{\mathrm{gra}}$ indicates the gradiometer and $_{\mathrm{mag1}}$ and $_{\mathrm{mag2}}$ indicate the paired two magnetometers which configure the gradiometers. The terms “axial” and “planar” indicate the arranged direction of the two coil planes. The tangential magnetometer and tangential planar gradiometer are equipped in few neuromagnetometers for experimental research usage [3, 4]. Figure 1.3 is the results of calculated magnetic fields with five types of pick-up coils, where the environmental noise does not exist ($\mathrm{b}_{\mathrm{noise}} = 0$). Figure 1.3 lower right is a schema of a longitudinal ECD moving on the surface of a homogeneous spherical conductor with 7 cm radius. The pick-up coil is 10 cm away from the center of the sphere, and the strength of ECDs is 1 n ampere meter [nAm] in longitudinal direction. The maps of the radial magnetometer and the radial axial gradiometer show typical maximum (red) and minimum (blue) pattern; meanwhile, the maps of others show maximum (red) just below the coils, and the feature of the latter resembles bipolar leads of electroencephalography. Note that the red and blue patterns are different from the dipole pattern consisting of efflux and influx.

To increase the accuracy of the coil signal simulation, coil integral is not an ignorable issue. The following integration can be considered for magnetic flux density $\hat{\mathrm{b}}$ [T] on one sensor by equal $\mathrm{N_p}$ divide:

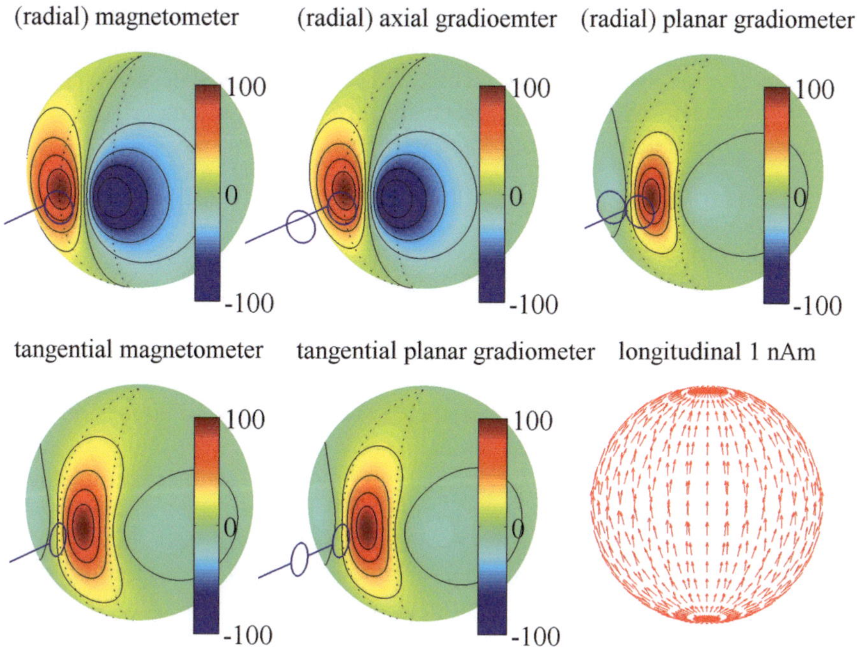

Fig. 1.3 Simulation of sensor signals using five types of pick-up coils described in Fig. 1.2. The coils located on the equator with 10 cm radius detect longitudinal 1 nAm equivalent current dipoles (ECDs) placed at a spherical surface with 7 cm radius (*row left*). Regarding the baseline settings, the radial axial gradiometer, the radial planar gradiometer, and the tangential planar gradiometer are 5 cm, 1.68 cm, and 5 cm, respectively. The counter lines are drawn by 20 fT step. Dotted lines denote 15° (angle of ECD-centroid-sensor). The radial magnetometer and the radial axial gradiometer return maximum or minimum signals when the angle is about 15°. The radial planar gradiometer, tangential magnetometer, and tangential planar gradiometer return maximum or minimum signals when the angle is 0°

$$\hat{\mathrm{b}}_{\mathrm{mag}} = \frac{1}{\mathrm{A}} \sum_{k=1}^{N_{\mathrm{p}}} (w_k \mathrm{A} f_k) = \sum_{k=1}^{N_{\mathrm{p}}} (w_k f_k) \tag{1.3}$$

where N_{p} is the number of the point placed on the coil plane, A is an area of coil, w_k is the weight factor based on the coil area divided by factor N_{p}, and f_k is flux density projected to normal direction at the k-th point. The more increased number of N_{p} is desired, but due to computational burden an approximate number is always used. In the case of magnetometer and $N_{\mathrm{p}} = 4$, so that weight parameter is set to 1/4, the signal of the coil is calculated as

$$\hat{\mathrm{b}}_{\mathrm{mag}} = \frac{1}{\mathrm{A}} \sum_{k=1}^{4} \left(\frac{1}{4} \mathrm{A} f_k \right) = \sum_{k=1}^{4} \left(\frac{1}{4} f_k \right). \tag{1.4}$$

In contrast to magnetometer, polarization should be taken account in gradiometer case. Also, the final output of sensor detection through digital processing depends on each vendor's design. For example, Elekta Neuromag VectorView system is set to $w_k = \pm 1/16.8$ [mm] for radial planar gradiometers aside from coil integration of N_{p} divide. The weights w_k are inverses of the baseline distance of the gradiometer to show that the output is in [T/m] (practically [fT/cm]). For the description of the detail, please consult to the manufacture's specification.

1.1.3 Sensor Array

The previous session states the integration of the magnetic field on the MEG sensor. Next, we describe the calculation of a fixed ECD against hundreds of sensors, namely, sensor array of a neuromagnetometer. Before entering the detail, it is desired to address the exact sensor position and its orientation. Fortunately, some tools are available to obtain such sensor information, especially some MATLAB tools are more prevailed and convenient for simulation study. Figure 1.4 is a schema of the sensor arrays of two representative and two custom-built neuro-magnetometers. The watery square or circle denotes sensor position and blue arrows denote unit vectors normal to the sensor planes. For sensor array, the signal of the m-th sensor $\widehat{\mathrm{b}}_{\mathrm{m}}$ is calculated extending Eq. 1.3 as follows:

$$\left(\hat{\mathrm{b}}_1, \cdots, \hat{\mathrm{b}}_{\mathrm{m}}, \cdots, \hat{\mathrm{b}}_{\mathrm{M}}\right)^{\mathrm{T}} = \left(\sum_{k=1}^{N_1}\left(w_{1,k} f_{1,k}\right), \cdots, \sum_{k=1}^{N_{\mathrm{m}}}\left(w_{\mathrm{m},k} f_{\mathrm{m},k}\right), \cdots, \sum_{k=1}^{N_{\mathrm{M}}}\left(w_{\mathrm{M},k} f_{\mathrm{M},k}\right)\right)^{\mathrm{T}} \tag{1.5}$$

where N_{m}, $w_{\mathrm{m},k}$, and $f_{\mathrm{m},k}$ denote the number of the calculation points, k-th weight, and k-th flux of the m-th sensor, respectively. Details of the actual calculation of the flux $f_{\mathrm{m},k}$ will be stated in Sect. 1.2. In principle, N_{m} and $w_{\mathrm{m},k}$ depend on sensor type and required accuracy level of coil integral. For example, in Elekta Neuromag system selecting four division levels, Eq. 1.4 should be used for the magnetometer, and $w_k = \pm 1/16.8$ should be taken for the planar gradiometers adding to coil integral matter.

1.1.4 Signal Processing and Programming

For understanding of fundamental MEG computational issue, special knowledge of physics and mathematics is not necessary, but you may need basic linear algebra, electromagnetic dynamics, signal processing, computational language, etc. For computation purposes, MATLAB is a widely used tool in this area, for example, signal processing tool implemented in MATLAB, yet as option, is very powerful. Recently, many MEG analysis tools have been provided by several research groups, and almost all MEG data can be converted into MATLAB binary data (i.e., MNE-suite [5], SPM [6], Brainstorm [7], VBMEG [8], FieldTrip [9], etc.). The

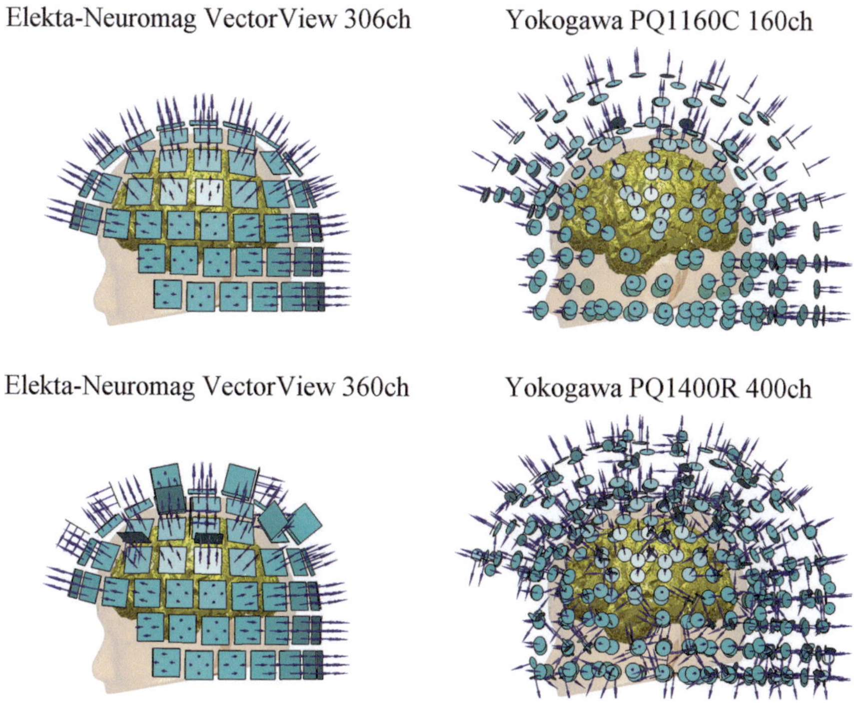

Fig. 1.4 Lateral views of sensor arrays of Elekta Neuromag VectorView (*upper left*), custom-built Elekta Neuromag VectorView with additional 18 tangential coils (*lower left*), Yokogawa's PQ1160C (*upper right*), and custom-built Yokogawa's PQ1400R (*lower right*). The *blue arrows* denote the directions normal to the sensor planes. Ordinary VectorView has 102 sensor units consisting of one radial magnetometer and one pair of orthogonal radial planar gradiometers. PQ1160C has 160 radial axial gradiometers. Note that typical Yokogawa's neuromagnetometer has 160 sensors, but the sensor geometry differs from each other. The PQ1400R has 210 radial axial gradiometers and 190 tangential planar gradiometers

authors recommend for MEG users to challenge programming tools such as MATLAB and its clone freeware, GNU Octave [10], FreeMat [11] or an analog Scilab [12], etc. Although complete compatibility is not guaranteed by these clone software, it should be beneficial at the beginning. If you become familiar with them, you can easily simulate signal processing such as noise cancellation, spatial filter, time-frequency analysis, or so, with few command lines [13, 14].

1.2 Forward Solution

1.2.1 Sarvas Formula

First thing we have to address is the formulization of sensor detection from a specific source generator in the brain which is generally called as "forward solution." To formulize the generated magnetic fields, approximation of the areas where currents flow (i.e., brain) is required. We have to define a conductor head model, in other words, approximate brain volume and conductivity to the computational one. The most prevailed model is a homogeneous spherical conductor model because sphere is mathematically analyzable and simply formulized due to the vanishing of reverse currents and radial component. This formulized spherical model is called as Sarvas formula and has been treated as standard bible in MEG source localization analysis since the 1980s [15]. We address this formula as initial step. Figure 1.5 exhibits a simulated calculation using Sarvas formula in which a spherical head model is integrated with an ECD. This provides circulating fluxes according to the Ampère's right-hand screw rule which shows efflux and influx pattern on a cross-sectioned plane. Due to the simplicity of the formula, the burden of computational

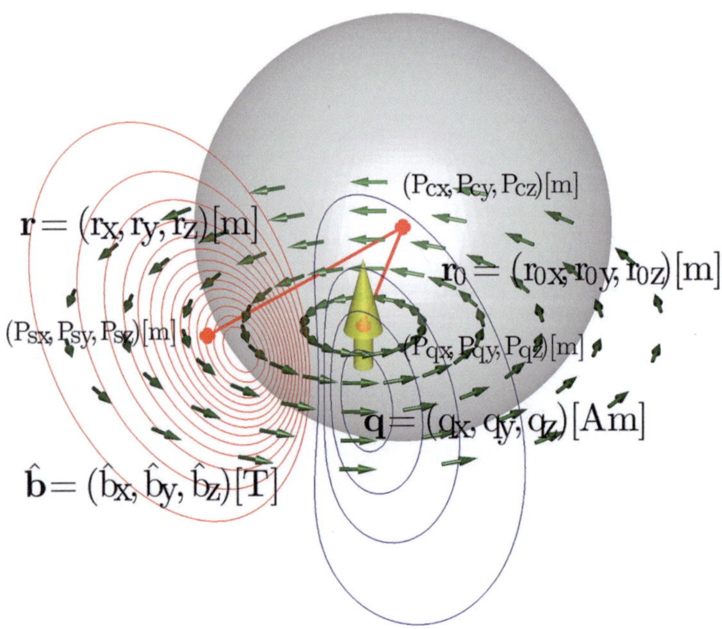

Fig. 1.5 A schematic image of a homogeneous spherical conductor model, Sarvas formula, where an ECD (a *yellow arrow*), circulating fluxes (*green arrows*), and isomagnetic contour map consisting of *red lines* (efflux) and *blue lines* (influx) on an arbitrary plane are illustrated. $\mathbf{r}_0$ is a positional vector from the spherical centroid to the ECD position, $\mathbf{r}$ is a positional vector from the centroid to a sensor position, $\mathbf{q}$ is moment of the ECD, and $\hat{\mathbf{b}}$ is magnetic flux at the sensor position. The coordinate system is Cartesian

cost is much relaxed compare with other complex algorithms. You can calculate magnetic fields from ECDs using the following Sarvas formula:

$$\begin{aligned} &\hat{\mathbf{b}}(\mathbf{r}) = \frac{\mu_0}{4\pi F}[F\mathbf{q} \times \mathbf{r_0} - (\mathbf{q} \times \mathbf{r_0} \cdot \mathbf{r})\nabla F] \\ &\mathbf{a} = \mathbf{r} - \mathbf{r_0} \\ &F = a(ra + r^2 - \mathbf{r_0} \cdot \mathbf{r}) \\ &\nabla F = \left(\frac{a^2}{r} + \frac{\mathbf{a} \cdot \mathbf{r}}{a} + 2a + 2r\right)\mathbf{r} - \left(a + 2r + \frac{\mathbf{a} \cdot \mathbf{r}}{a}\right)\mathbf{r_0} \end{aligned} \tag{1.6}$$

where $\mu_0 = 4\pi \times 10^{-7}\left[N/A^2\right]$ is permeability in vacuum. Sarvas formula needs 3×4 arguments, the position of the spherical centroid $(P_{cx}, P_{cy}, P_{cz})[m]$, the position of the sensor $(P_{sx}, P_{sy}, P_{sz})[m]$, the position of the ECD $(P_{qx}, P_{qy}, P_{qz})[m]$, and the moment of the ECD $\mathbf{q} = \left(q_x, q_y, q_z\right)[Am]$ and outputs the magnetic fields at the sensor position as $\hat{\mathbf{b}}(\mathbf{r}) = \left(\hat{b}_x, \hat{b}_y, \hat{b}_z\right)[T]$. The positional vectors of $\mathbf{r}$, $\mathbf{r}_0$, and $\mathbf{a}$ [m] are calculated as $\mathbf{r} = \left(r_x, r_y, r_z\right) = \left(P_{sx} - P_{cx}, P_{sy} - P_{cy}, P_{sz} - P_{cz}\right)$, $\mathbf{r_0} = \left(r_{0x}, r_{0y}, r_{0z}\right) = \left(P_{qx} - P_{cx}, P_{qy} - P_{cy}, P_{qz} - P_{cz}\right)$, and $\mathbf{a} = \left(a_x, a_y, a_z\right) = \left(r_x - r_{0x}, r_y - r_{0y}, r_z - r_{0z}\right)$, respectively. a and r are length of $\mathbf{a}$ and $\mathbf{r}$, respectively. $\mathbf{q} \times \mathbf{r_0} = \left(q_y r_{0z} - q_z r_{0y}, q_z r_{0x} - q_x r_{0z}, q_x r_{0y} - q_y r_{0x}\right)$ is outer product of $\mathbf{q}$ and $\mathbf{r}_0$, $\mathbf{r_0} \cdot \mathbf{r} = r_{0x}r_x + r_{0y}r_y + r_{0z}r_z$ is inner product of $\mathbf{r}_0$ and $\mathbf{r}$, $\mathbf{q} \times \mathbf{r}_0 \cdot \mathbf{r}$ is dot product of $\mathbf{q} \times \mathbf{r}_0$ and $\mathbf{r}$, and $\nabla F = \left(\frac{a^2}{r} + \frac{\mathbf{a} \cdot \mathbf{r}}{a} + 2a + 2r\right)\mathbf{r} - \left(a + 2r + \frac{\mathbf{a} \cdot \mathbf{r}}{a}\right)\mathbf{r_0}$ is gradient of F. Magnetic field detection on arbitrary sensor points is obtained using inner product of $\hat{\mathbf{b}}(\mathbf{r})$ and the unit normal vector perpendicular to the pick-up coil plane $\mathbf{s} = \left(s_x, s_y, s_z\right)$ and calculated as $\hat{b} = \hat{\mathbf{b}} \cdot \mathbf{s} = \hat{b}_x s_x + \hat{b}_y s_y + \hat{b}_z s_z$. Final output from MEG system should be computed by integrating with Eqs. 1.3, 1.4, and 1.5. Sarvas formula has the term $\mathbf{q} \times \mathbf{r}_0$ and this means the radial component of an ECD produces no fluxes. The parameter of $\mathbf{q}$ can be defined as spherical coordinates $\mathbf{q} = \left(q_r, q_\phi, q_\theta\right)$, where q_r, q_ϕ, and q_θ are the radial, longitudinal tangential, and latitudinal tangential component of an ECD, respectively. Then, conversion equation from the spherical coordinate system to the Cartesian coordinates becomes

$$\begin{aligned} \mathbf{q} &= \left(q_x, q_y, q_z\right) \\ &= \left(-q_\phi \sin\phi - q_\theta \cos\phi \sin\theta,\ q_\phi \cos\phi - q_\theta \sin\phi \sin\theta,\ q_\phi q_\theta \cos\theta\right). \end{aligned} \tag{1.7}$$

Note that q_r is not used in the equation and you can decrease the arguments from $\mathbf{q} = \left(q_x, q_y, q_z\right)$ to $\mathbf{q} = \left(q_\varphi, q_\theta\right)$. This makes sense in ECD estimation to decrease the number of the unknowns. Based on Sarvas formula, you can make programming the function as

$$\widehat{b} = f\left(P_{cx}, P_{cy}, P_{cz}, P_{sx}, P_{sy}, P_{sz}, s_x, s_y, s_z, P_{qx}, P_{qy}, P_{qz}, q_x, q_y, q_z\right). \tag{1.8}$$

Figures 1.3 and 1.5 are made with this function written in MATLAB language.

1.2.2 Lead Field Matrix

We introduce two types of source points. The first one is simple cubic lattice node (Fig. 1.6 left) set within subject's individual brain anatomy, and the second one is vertex (Fig. 1.6 right) set at extracted subject's cortical polygon mesh. Setting source points are requisite preliminary for distributed source analysis. The forward solution is representative using linear matrix and this is generally treated as lead field matrix. Currents at their prefixed points are reconstructed using measured sensor signals. The approach prefixing at vertices is neurophysiologically preferable; however, it includes restriction in subjects with a damaged brain or malformation such as brain tumor, hemorrhage, and lissencephaly, often not available due to failure of extraction of the corticomedullary junction.

Assume an ECD with 1 nAm strength and arbitrary direction at *n*-th point and calculate their sensor array signals as $\hat{\mathbf{b}}_n = \left(\hat{b}_{1,n}, \cdots, \hat{b}_{m,n}, \cdots, \hat{b}_{M,n}\right)^T$. An ECD with q nAm of the same direction at *n*-th point makes $\hat{b}_{m,n}q$ at *m*-th sensor. Using longitudinal (θ), latitudinal (ϕ), and radial (r) 1 nAm currents at each point, the next matrix form is obtained:

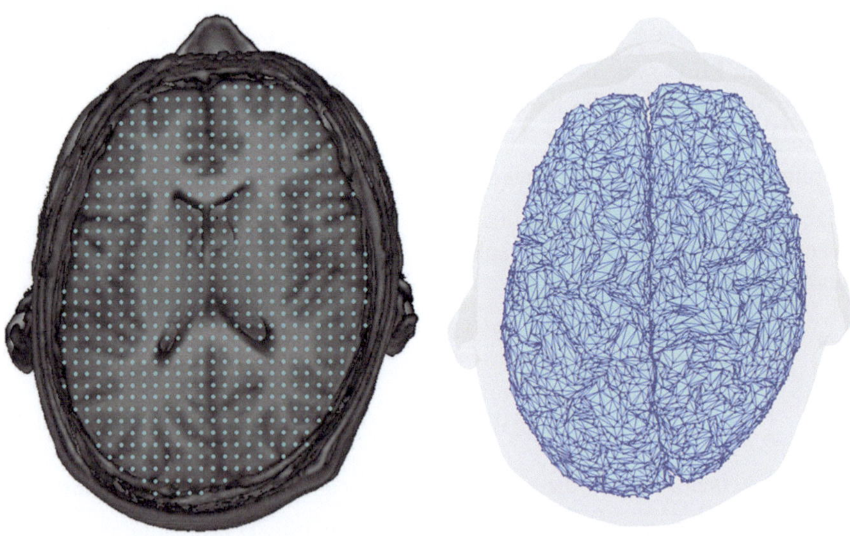

Fig. 1.6 Widely used source points set on simple cubic lattice nodes (aqua points) within brain voxels (*left*) and vertices (*blue points*) of cortical polygon mesh (*right*): those are used for forward calculation, i.e., lead field matrix, and spatial filters

$$\mathbf{L}_{\mathrm{n}} = \begin{pmatrix} \hat{b}_{1,\mathrm{n}\theta}, \cdots, \hat{b}_{\mathrm{m},\mathrm{n}\theta}, \cdots, \hat{b}_{\mathrm{M},\mathrm{n}\theta} \\ \hat{b}_{1,\mathrm{n}\phi}, \cdots, \hat{b}_{\mathrm{m},\mathrm{n}\phi}, \cdots, \hat{b}_{\mathrm{M},\mathrm{n}\phi} \\ \hat{b}_{1,\mathrm{nr}}, \cdots, \hat{b}_{\mathrm{m},\mathrm{nr}}, \cdots, \hat{b}_{\mathrm{M},\mathrm{nr}} \end{pmatrix}^{\mathrm{T}}. \tag{1.9}$$

An ECD with $\mathbf{q}_{\mathrm{n}} = \left(q_{\mathrm{n}\theta}, q_{\mathrm{n}\varphi}, q_{\mathrm{nr}}\right)^{\mathrm{T}}$ at n-th point forms $\hat{\mathbf{b}}_{\mathrm{n}} = \mathbf{L}_{\mathrm{n}} \cdot \mathbf{q}_{\mathrm{n}}$. The signal of the m-th sensor is $\hat{b}_{\mathrm{n,m}} = \hat{b}_{\mathrm{m,n}\theta} q_{\mathrm{n}\theta} + \hat{b}_{\mathrm{m,n}\varphi} q_{\mathrm{n}\varphi} + \hat{b}_{\mathrm{m,nr}} q_{\mathrm{nr}}$. Let us expand $\mathbf{Ln}$ to all N ECDs with three degrees of freedom, and the following matrix is obtained:

$$\mathbf{L} = \begin{pmatrix} \mathbf{L}_1 \\ \vdots \\ \mathbf{L}_{\mathrm{n}} \\ \vdots \\ \mathbf{L}_{\mathrm{N}} \end{pmatrix}^{\mathrm{T}} = \begin{pmatrix} \hat{b}_{1,1\theta} & \cdots & \hat{b}_{\mathrm{m},1\theta} & \cdots & \hat{b}_{\mathrm{M},1\theta} \\ \vdots & \ddots & \vdots & \ddots & \vdots \\ \hat{b}_{1,\mathrm{n}\phi} & \cdots & \hat{b}_{\mathrm{m},\mathrm{n}\phi} & \cdots & \hat{b}_{\mathrm{M},\mathrm{n}\phi} \\ \vdots & \ddots & \vdots & \ddots & \vdots \\ \hat{b}_{1,\mathrm{Nr}} & \cdots & \hat{b}_{\mathrm{m},\mathrm{Nr}} & \cdots & \hat{b}_{\mathrm{M},\mathrm{Nr}} \end{pmatrix}^{\mathrm{T}}. \tag{1.10}$$

Given $\mathbf{q} = (\mathbf{q}_1, \ldots, \mathbf{q}_{\mathrm{n}}, \cdots \mathbf{q}_{\mathrm{N}})^{\mathrm{T}} = \left(q_{1\theta}, \ldots, q_{\mathrm{n}\theta}, q_{\mathrm{n}\phi}, q_{\mathrm{nr}}, \ldots, q_{\mathrm{Nr}}\right)^{\mathrm{T}}$, these sensor signals are $\hat{\mathbf{b}} = \mathbf{L} \cdot \mathbf{q}$, and the m-th sensor signal is

$$\begin{aligned} \hat{b}_{\mathrm{m}} &= \sum_{n=1}^{\mathrm{N}} \left(\hat{b}_{\mathrm{m,n}\theta} q_{\mathrm{n}\theta} + \hat{b}_{\mathrm{m,n}\phi} q_{\mathrm{n}\phi} + \hat{b}_{\mathrm{m,nr}} q_{\mathrm{nr}}\right) = \hat{b}_{\mathrm{m},1\theta} q_{1\theta} \\ &+ \ldots + \hat{b}_{\mathrm{m,n}\phi} q_{\mathrm{n}\phi} + \ldots + \hat{b}_{\mathrm{m,Nr}} q_{\mathrm{Nr}}. \end{aligned} \tag{1.11}$$

Coupling with $\mathbf{q}$ and $\mathbf{L}$ can lead to sensor signals $\overset{\frown}{\mathbf{b}}$, namely, magnetic fields on sensor array; therefore, $\mathbf{L}$ is called "lead field matrix." Once the lead field matrix $\mathbf{L}$ is established, currents $\mathbf{q}$ appear to be solved easily from sensor signals $\overset{\frown}{\mathbf{b}}$; thus, $\mathbf{q} = \frac{\mathbf{b}}{\mathbf{L}}$. One discussion concentrated on how to handle the fraction form of the matrix. Unfortunately, it is not straightforward, but several numerical recipes have been proposed to relax the problem. From Sect. 1.3, we disclose these proposals.

1.2.3 Advanced Study

Due to the restricted space, we only briefly introduce the most fundamental but widely used Sarvas formula via spherical head model. Along to the growth of scanning technology of anatomical image and computational performance, more advanced approaches have been proposed using realistic model [16–20]. We introduce the fraction of these techniques for further study. Actually, the human head is not a sphere: hence, more exact head model might be addressed. The realistic shape using 3D polygon mesh technique is obtained from magnetic resonance image (MRI) in general. High-resolution MRI of the head is now readily performed and routinely included as part of most MEG data analysis. MRI can be used to extract morphological information about individual head shape that can be used to improve the accuracy of the forward model. Under current situation, realistic piecewise homogeneous model obtained from MRI is widely used as accurate model.

Next, we have to consider the influence of conductivity. It is well known that the conductivities in human brain are not homogeneous. Typically, we assume the head consists of a small set of relatively large isotropic homogeneous regions: brain, fluid, skull, and scalp. Simplest piecewise homogeneous model is a concentric homogeneous shells. In contrast to MEG, it is necessary to employ the piecewise homogeneous conductivity model in EEG analysis, because EEG is the technique used to detect the volume current which is much contaminated by the conductivity difference, especially by the skull. However, the main problem in this accurate model is that the conductivities in the living human brain are unknown and they may vary from subject to subject. The skull is typically assumed to be 40–90 times more resistive than the brain and scalp, which are assumed to have similar conductive properties. These values are measured in vitro from postmortem tissue, where conductivity can be significantly altered compared to in vivo values. On the other hand, details of the conductivities force negligible effect on the magnetic field; thus, MEG forward solution is straightforward, and one mono-homogeneous model is commonly used. In general, lead field matrix using the realistic model is solved via the calculus of variations, e.g., boundary element method (BEM) [e.g., 21–24] or finite element methods (FEM) [e.g., 25, 26].

1.3 Inverse Problem

The biomagnetic source localization is performed from values MEG sensor detected. One generalized solution is ECD approach, and the other group is the distributed source analysis. The digitized sensor signals and the current sources can be represented by equations and variables in physics. Due to the overdetermined solution (number of sensors > number of sources, i.e., number of equation > number of variables), the ECD approach has no solution which fulfill the given sensor detection. The distributed source analysis is a technique which uses an amount of prefixed source points in the brain. These source points are marked on the anatomical images, typically using MRI data, and the number of points is generally in the order of 10^3 or more. Thus, in contrast to the ECD approach, the number of variables exceeds the number of equations, and infinite numbers of solutions exist. This is generally called underdetermined solution. The purpose of this approach is to find the best solution out of infinite number of solution via constraining. We try to explain how to find most plausible solution under current technology.

1.3.1 Equivalent Current Dipole (ECD) Estimation

Assume an ECD represents all sensor signals and the net number of unknown is six $(P_{qx}, P_{qy}, P_{qz}, q_{\phi}, q_{\theta}, q_r)$ (see Sect. 1.2.1). For N ECDs, the number of unknown parameters is N × 6, which shall be led through repetitive operation to minimize $\sum_{m=1}^{M} \left(w_m \left(b_m - \hat{b}_m\right)\right)^2$, where w_m is weight of m-th sensor such as baseline of the signal waves and b_m and $\hat{b}_m$ are measured and calculated m-th sensor signals, respectively. $\hat{b}_m$ is the summation of the calculated signal of ECDs, for example, using Sarvas formula (Eq. 1.6). In this case, the number of parameters is arranged to N × 5 since radial component vanishes. For function minimization, many optimized numerical methods have been proposed to search the minimum point (e.g., simplex method, Newton method, Levenberg-Marquardt method [27]). Simplex method is so frequently used due to simplicity of its algorithm, and Elekta Neuromag software employs one of them, Nelder-Mead nonlinear search algorithm. The reliability of the ECD should be statistically evaluated, and the "goodness of fit" (GOF) is commonly used as

$$G = 1 - \frac{\sum_{m=1}^{M} \left(b_m - \hat{b}_m\right)^2}{\sum_{m=1}^{M} b_m \hat{b}_m}. \tag{1.12}$$

Ideally, the number of the ECDs is desired to be the number of the influx and efflux pattern (see Fig. 1.5 for reference) on isomagnetic contour maps. Using many ECDs, high GOF estimation can be easily obtained. Unfortunately there is no decisive statistical reliability index for multiple ECDs and the number of ECDs would be recommended to be as small as possible. Some applications of ECD method have been also proposed. For example, "moving dipole" is one of the popular approaches where the position and the direction of the ECDs vary from moment to moment, but initial position is taken over from results of previous sample point since initial position for each iteration is always disputed [28, 29]. It is called "spatiotemporal dipole" if the positions and the directions of the ECDs are fixed and only the magnitude of the current varies according to the time evolution [30, 31]. The advantage among these proposed methods remain controversial and the selection depends on a user's philosophy. BESA is the representative software of the spatiotemporal dipole [32, 33]. Figure 1.7 is an example of BESA which interprets 15–60 ms time window waveforms from somatosensory evoked fields (SEFs) using spatiotemporal approach with two ECDs (red and blue).

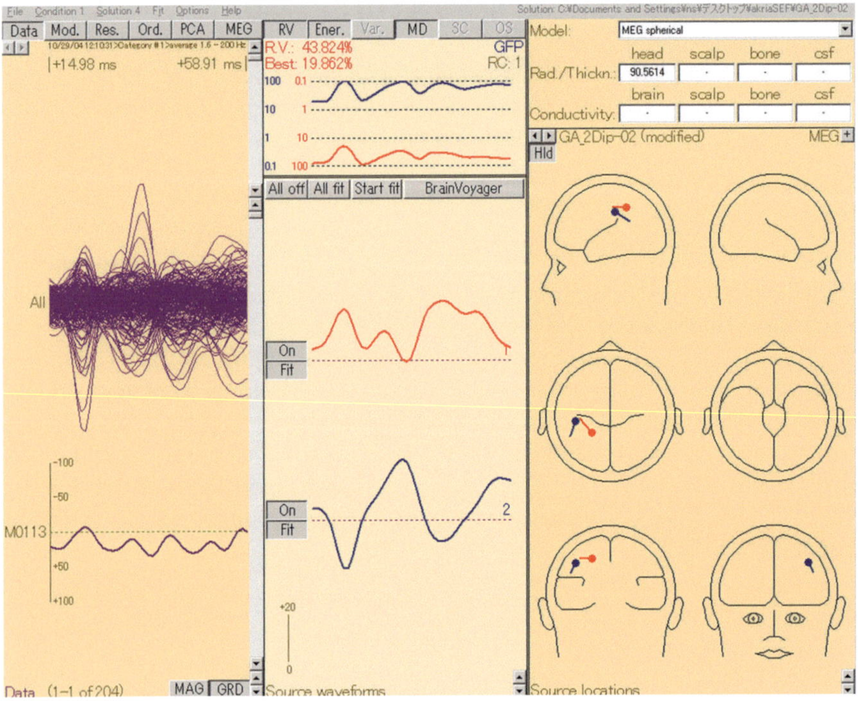

Fig. 1.7 An example of spatiotemporal dipole analysis using BESA software. Somatosensory evoked fields (SEFs) of *right* median nerve stimulation are used. BESA interprets waveforms from 15 to 60 ms as time evolution of two fixed ECDs. The *red line* (*blue line*) curve in the *lower middle* window corresponds to the time series of the *red* (*blue*) ECD

1.3.2 Minimum Norm Estimation

To obtain unique solution, constraint by minimizing total norm of all source points marked in the brain is widely employed. A variety type of minimum norm estimates has been proposed [34–38].

1.3.2.1 What Is Norm?

One of the frequently asked questions from general MEG users is "What is norm?". This is somehow reasonable question because mathematically a definition of norm is a little bit conceptual. Entering in math (linear algebra), a norm is a function that maps from vector space to a length or size. A definition of norm is given in the following formal definition which must satisfy three properties pertaining to scalability and additivity. Given a vector space V over a subfield F of real or complex numbers, a norm on V is a function $\|\| : V \to R$ (R denotes real field) with the following properties:

For all $a \in F$ and all $\mathbf{u}$, $v \in V$,

$$1.\ \|a\mathbf{v}\| = |a|\|\mathbf{v}\|, \text{(absolute scalability)}. \tag{1.13}$$

$$2.\ \|\mathbf{u}+\mathbf{v}\| \le \|\mathbf{u}\| + \|\mathbf{v}\|, \text{(triangle inequality)}. \tag{1.14}$$

$$3.\ \text{If}\ \|\mathbf{v}\| = 0\ \text{then}\ \mathbf{v} = \mathbf{0}, \text{(separates points)}. \tag{1.15}$$

We introduce two representative norms generally used in this area. Given a vector $\mathbf{q} = \left(q_x, q_y, q_z\right)$, L1 norm is defined as

$$\|\mathbf{q}\|_1 = \left(|q_x| + |q_y| + |q_z|\right)^{1/1}, \tag{1.16}$$

and L2 norm is defined as

$$\|\mathbf{q}\|_2 = \left(|q_x|^2 + |q_y|^2 + |q_z|^2\right)^{1/2}. \tag{1.17}$$

The L2 norm is often expressed as squares

$$\|\mathbf{q}\|_2^2 = |q_x|^2 + |q_y|^2 + |q_z|^2. \tag{1.18}$$

For instance, given $\mathbf{q} = (1, -2, 3)$, $\|\mathbf{q}\|_1 = |1|+|-2|+|3| = 6$, and $\|\mathbf{q}\|_2^2 = |1|^2 + |-2|^2 + |3|^2 = 14$. You can easily confirm that L1 and L2 norms fulfill the above three properties.

1.3.2.2 Inverse Matrix

With the distributed source analysis, we shall encounter the problem of the handling of the fraction of the inverse matrix of the lead field matrix $\mathbf{L}$($M \times N$ matrix) (see Sect. 1.2.2). Due to no existence of the unique inverse matrix $\mathbf{L}^{-1}$, the following pseudo-inverse matrices are substituted along with the cases of either underdetermined or overdetermined solution described in Sect. (1.3),

$$\left(\mathbf{L}^T\mathbf{L}\right)^{-1}\mathbf{L}^T \text{if}\ M > N\ \text{and}\ \mathbf{L}^T\left(\mathbf{L}\mathbf{L}^T\right)^{-1} \text{if}\ M < N. \tag{1.19}$$

The second term of Eq. 1.19 indicates the underdetermined solution (i.e., distributed source analysis). General pseudo-inverse matrix can be obtained using singular value decomposition (SVD), and inverse of squared value of very small singular value makes solution instable. Adding tiny unit matrix or noise matrix, truncation to decrease the rank after eigenvalue decomposition and SVD are the methods used to avoid the instability [27], and such methods approximate the original matrix consequently.

1.3.2.3 Minimum L2 Norm Estimation in General

The minimum L2 norm estimation employs the constraint that summation of the square power (Eq. 1.18) of squared root of distributed source currents is minimized. Source activation and sensor detection in one time sample are formulized with lead field matrix **L**:

$$\mathbf{b}_t = \mathbf{L} \cdot \mathbf{q}_t \tag{1.20}$$

s.t. $\sum_{n=1}^{N} \left(q_{n\phi,t}^2 + q_{n\theta,t}^2 + q_{nr,t}^2 \right)$ is minimized. Using inverse matrix $\hat{\mathbf{q}}$ from Eq. 1.19, the solution is given by

$$\mathbf{q}_t = \hat{\mathbf{q}}_t \mathbf{b}_t = \mathbf{L}^T \left(\mathbf{L}\mathbf{L}^T \right)^{-1} \mathbf{b}_t. \tag{1.21}$$

The constraint can be changed, for example, $\sum_{n=1}^{N} \left(q_{n\phi,t}^2 + q_{n\theta,t}^2 \right)$ is minimized for spherical model (Sect. 1.2.1) or $\sum_{n=1}^{N} q_{n,t}^2$ is minimized for the current direction being normal to the cortical surface (orientation constraint). It is thought that the fluxes are not attributed to the currents of the brain stem or cerebellum and source points are not prefixed in general. The constraint of the minimizing L2 norm tends to estimate the larger currents near the sensor and the smaller ones in the deeper brain because fluxes decreases according to the distance between sources and sensors. To avoid this phenomenon, some normalization methods are proposed. We describe two simple approaches as below. Using **Ln** from Eq. 1.9, the solution on each source point n in arbitrary time point t is

$$q_{n,t} = \frac{\mathbf{L}_n^T \left(\mathbf{L}\mathbf{L}^T \right)^{-1}}{\sqrt{\mathbf{L}_n^T \left(\mathbf{L}\mathbf{L}^T \right)^{-2} \mathbf{L}_n}} \mathbf{b}_t. \tag{1.22}$$

And another proposal is

$$q_{n,t} = \frac{\mathbf{L}_n^T \left(\mathbf{L}\mathbf{L}^T \right)^{-1}}{\sqrt{\mathbf{L}_n^T \left(\mathbf{L}\mathbf{L}^T \right)^{-1} \mathbf{L}_n}} \mathbf{b}_t. \tag{1.23}$$

Equation 1.22 is a method based on dynamic statistical parametric mapping (dSPM) proposed by Dale et al. [39] taking sensor noise as white noise, i.e., sensor noise covariance matrix becomes identical, since the inside of the square root of the denominator in Eq. 1.22 for all source points is expressed by $(\hat{\mathbf{q}}\,\hat{\mathbf{q}})_{nn}$, with notification of $()_{nn}$ indicating diagonal operation. Equation 1.23 is based on standardized low-resolution brain electromagnetic tomography (sLORETA) [40, 41] which takes

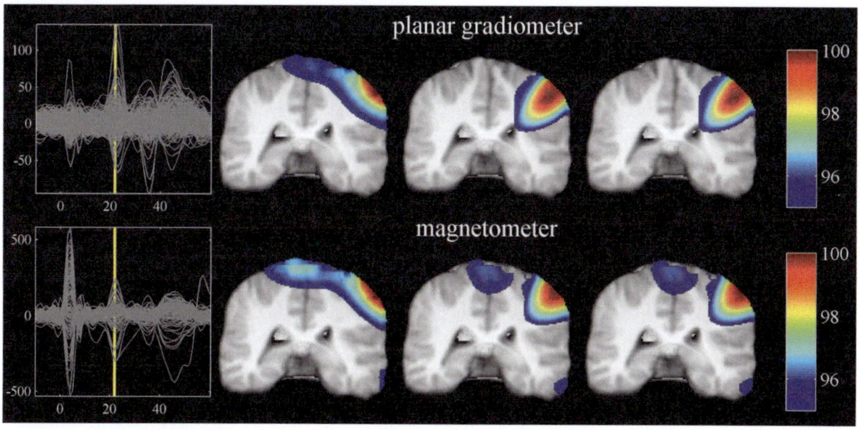

Fig. 1.8 Waveforms of SEF and reconstructed images at 22 ms (*yellow line*) on coronal cross-sectioned brains (first column). Bilateral front poles are removed. *Upper* images are reconstructed from data of 204 planar gradiometers, and lower images are reconstructed from data of 102 magnetometers. Voxels with highest 95–100 % currents are color-coded (scale bar on the *right* in each row). The images on the second column are reconstructed with original minimum L2 norm estimation, Eq. 1.21, and the others are reconstructed with two types of modified methods, Eq. 1.22 for the third column and Eq. 1.23 for fourth column. The results using original methods tend to estimate higher currents on the cortex, i.e., near the sensors, while the improved results tend to move them toward the deeper side of the brain

into account the resolution matrix (inverse matrix × lead field matrix). Indeed, inside of the square root of the denominator in Eq. 1.23 for all source points forms $(\hat{\mathbf{q}}\,\mathbf{L})_{nn}$. See also [42, 43] for the summarization. Figures 1.8, 1.11, 1.12, 1.13, and 1.14 show the reconstructed results from the same SEF at 22 ms or 390 ms, after the right median nerve stimulation. The lattice nodes with maximum 95–100 % currents are color-coded. As seen in Fig. 1.8, due to the effects of additional computation in the denominators, the estimated source currents (third and fourth columns) have moved to deeper side. Figure 1.9 shows reconstructed results from visual evoked fields with blink noise and demonstrated the influence of the artifacts. The original results (second column) interpret the detected signals of basal sensors as large currents in bilateral frontal and temporal lobes near the pterion, which have been moved deeper to the basal side by the normalization approaches of Eq. 1.22 (third column) and Eq. 1.23 (forth column). Thus, improved methods tend to emphasize source activities at the cerebral bases or the deeper brain.

1.3.2.4 Minimum Norm Estimates (MNE)

In previous section, the fundamental concept of minimum L2 norm approach is described. Here, we introduce the widely used application based on minimum L2 norm approach, MNE, developed at the Martinos Center for Biomedical Imaging of the Massachusetts General Hospital by Hämäläinen et al. [1, 19, 25, 35, 36]. This algorithm is a combination of minimum norm approach and least square method based on Tikhonov regularization with addition of the regularization parameter,

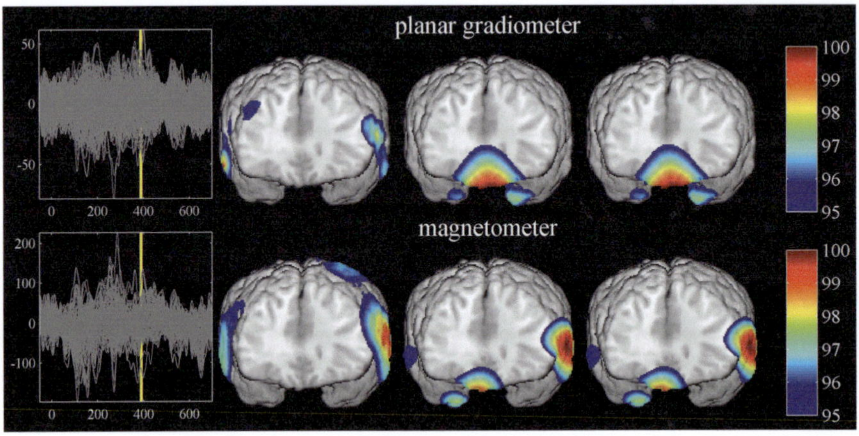

Fig. 1.9 Examples of unexpected results using minimum L2 norm method and its modified methods. Waveforms of visual evoked fields with blink noise and reconstructed images at 390 ms marked with *yellow line* (first column). *Upper* images are reconstructed from planar gradiometers and lower images are reconstructed using magnetometers. Estimated currents in the second column have moved to the deeper side or to the basal side in modified methods as well as shown in Fig. 1.8

noise covariance matrix, and source covariance matrix. Let us expand the vectors b_t, $\mathbf{q}_t$ to matrices **B**, **Q** containing sampled time length and take into account the white noise $\mathbf{B}_{noise}$; Eq. 1.20 can be written as

$$\mathbf{B} = \mathbf{LQ} + \mathbf{B}_{noise}. \tag{1.24}$$

The MNE proposes to generate the unique solution with minimum L2 norm (see Eq. 1.18) among an infinite number of solutions:

$$\hat{\mathbf{Q}} = \mathrm{argmin}_{\mathbf{Q}}\left\{\left\|(\mathbf{LQ}-\mathbf{B})\mathbf{C}^{-1}(\mathbf{LQ}-\mathbf{B})^{\mathrm{T}}\right\|_2^2 + \lambda^2\left\|\mathbf{QR}^{-1}\mathbf{Q}^{\mathrm{T}}\right\|_2^2\right\}, \tag{1.25}$$

where **C** and **R** are the matrix of weighted factor. **C** is a sensor covariance matrix, **R** is a source covariance matrix, and λ^2 is a regularization parameter. The first term in the braces indicates the error noise between the forward solution, **LQ**, and sensor detection, **B**, with weighted factor **C. C** is covariance matrix of certain periods of measured MEG signal. The choices of certain periods will be disputed below. The second term denotes the general L2 norm of given distributed sources with weighted factor **R**. The source covariance matrix **R** is prepared to employ prior information such as the results of fMRI. Without any prior, the sources are assumed to be uncorrelated each other and equal to variance. By means of Lagrange multiplier, the inverse operator matrix $\hat{\mathbf{Q}}$ is given by

$$\hat{\mathbf{Q}} = \mathbf{RL}^{\mathrm{T}}\left(\mathbf{LRL}^{\mathrm{T}} + \lambda^2\mathbf{C}\right)^{-1}. \tag{1.26}$$

The regularization parameter λ^2 is proposed as

$$\lambda^2 = \frac{\mathrm{trace}\left(\left(\mathbf{C}^{-1/2}\mathbf{L}\right)\mathbf{R}\left(\mathbf{C}^{-1/2}\mathbf{L}\right)^{\mathrm{T}}\right)}{\mathrm{M}\cdot\mathrm{SNR}^2} \tag{1.27}$$

where SNR stands for signal-to-noise ratio. For notification of $\mathbf{C}^{-1/2}$, we consider Cholesky factorization of $\mathbf{C} = (\mathbf{C}^{1/2})(\mathbf{C}^{1/2})^{\mathrm{T}}$, and $\mathbf{C}^{-1/2}$ is obtained by an inverse operator of $\mathbf{C}^{-1/2}$. SNR is generally computed as signal divided by noise; here, signal is calculated via forward computation from estimated source activities. The regularization is performed via SNR and in MNE-suite software, default parameter setting for above SNR = 3. Thus, applying the inverse matrix to MEG sensor detection, the solution is

$$\hat{\mathbf{Q}}\,\mathbf{B} = \mathbf{R}\mathbf{L}^{\mathrm{T}}\left(\mathbf{L}\mathbf{R}\mathbf{L}^{\mathrm{T}} + \lambda^2\mathbf{C}\right)^{-1}\mathbf{B}. \tag{1.28}$$

Following to previous section, we consider normalization approach. Dale et al. proposed a noise-normalized estimate, dSPM, [39] as follows:

$$\mathbf{Z}_{\mathrm{dSPM}} = \frac{\hat{\mathbf{Q}}\,\mathbf{B}}{\sqrt{\left(\hat{\mathbf{Q}}\,\mathbf{C}\hat{\mathbf{Q}}^{\mathrm{T}}\right)_{\mathrm{nn}}}}. \tag{1.29}$$

Normalization difference in the denominator from Eq. 1.22 is by adding noise covariance matrix **C**. For evoked response, the default choice for creating covariance matrix **C** is by using a certain pre-trigger range since no evoked responses by stimuli are included in that period. The measurement for epileptic discharge is one of the main topics using MEG, and no trigger stimuli exist in such measurement. Entire raw data is one option to create the covariance matrix **C**. The resting state data analysis has been focused recently, and empty room measurement (or so-called system noise measurement) is required to create noise covariance matrix **C**. We are taking into account the sLORETA [40, 41] as well. Taking into account the regularization parameter λ^2 and source covariance matrix **R** for interpretation of lead field matrix, one computation is ([1] Eqs. 1.8–1.11)

$$\mathbf{Z}_{\mathrm{sLORETA}} = \frac{\hat{\mathbf{Q}}\,\mathbf{B}}{\sqrt{\left(\hat{\mathbf{Q}}\,(\lambda^{-2}\mathbf{R}\mathbf{L}^{\mathrm{T}})\right)_{\mathrm{nn}}}} = \frac{\hat{\mathbf{Q}}\,\mathbf{B}}{\sqrt{\left(\hat{\mathbf{Q}}\,(\mathbf{C} + \lambda^{-2}\mathbf{L}\mathbf{R}\mathbf{L}^{\mathrm{T}})\hat{\mathbf{Q}}^{\mathrm{T}}\right)_{\mathrm{nn}}}}. \tag{1.30}$$

The basic concept of this normalization is the same as Eq. 1.23. The difference is caused by required condition (Eq. 1.25) and additional parameters. As seen inside of the square root of the denominator in the last term, the first term $\left(\hat{\mathbf{Q}}\,\mathbf{C}\hat{\mathbf{Q}}^{\mathrm{T}}\right)_{\mathrm{nn}}$ is equivalent to dSPM (Eq. 1.29). sLORETA takes into account both the variation of measured noise and the variation of the estimated sources [40]. Finally, MNE applies good artifices for practical computation using SVD, and other parameters are also available, for example, depth, weight, orientation constraint, etc. (see [1, 44, 45] for the details). Figure 1.10 shows reconstructed results on the surface

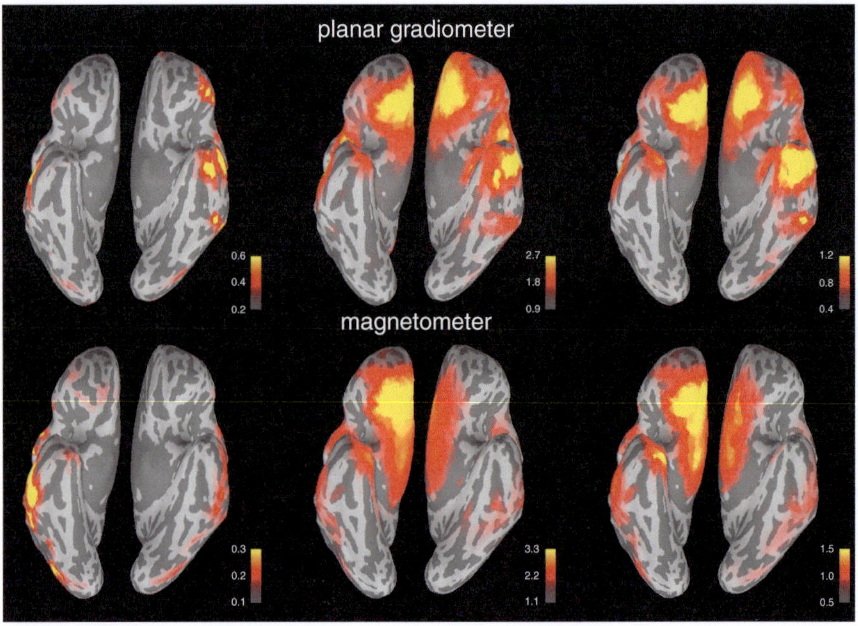

Fig. 1.10 Examples of source activations obtained using MNE-suite software on the inflated cortical vertices using planar gradiometers (*upper*) and magnetometers (*lower*). *Dark gray* pattern denotes folded cortices and *bright gray* pattern denotes unfolded cortices. The first, second, and third columns show source reconstruction estimated by original minimum norm estimates (MNE), dynamic statistical parametric mapping (dSPM), and standardized low-resolution brain electromagnetic tomography (sLORETA), respectively. The same data employed in Fig. 1.9 are used and reconstructed at 390 ms of latency. MNE interprets the blink noise as large currents at bilateral frontal and temporal bases, which are moved to the center of the bilateral frontal lobes through noise normalization methods, dSPM and sLORETA

of the inflated brain from the same blink noise used in Fig. 1.9, obtained from planar gradiometers (upper) and magnetometers (lower) using MNE (first column) and the dSPM (second column) and sLORETA (third column) [39–41]. Relative large sources are estimated mainly at cerebral bases near the sensors using original MNE, and by normalization methods it moves to the deep surface of the cerebral longitudinal fissure. One difference between Figs. 1.9 and 1.10 is the base source point setting. In Fig. 1.9, anatomically based lattice nodes cover nearly the whole cerebrum, while MNE (Fig. 1.10) marks the vertices based on corticomedullary junction and reconstructs the surface left and right hemisphere separately. Intermingling the several issues, results vary even when using the same data. To avoid this phenomenon, noise cancellation is quite important, especially at basal sensors for each application.

1.3.2.5 Minimum L1 Norm Estimation

The constraint of minimum L1 norm is $\mathbf{b}_t = \mathbf{L}\cdot\mathbf{q}_t$ s.t. $\sum_{n=1}^{N}\left(\left|q_{n\phi,t}\right| + \left|q_{n\theta,t}\right| + \left|q_{nr,t}\right|\right)$ is minimized. Large-scale linear programming is required for this solution, which includes sparse submatrix, i.e., the most elements are zero [27]. In general, L1 norm estimation needs time-consuming repetitive computation. This sparsity would be suited to estimate separating plural localized functional areas distinctly, which are interpreted by L2 norm as blurred distributed areas. On the other hand, this sparsity poses a problem when the actual neuronal activities are widely spread, because L1 norm interprets such spread activity as scattered spotty currents. Minimum current estimates (MCEs) are the representative minimum L1 norm spatial filter [38]. Figures 1.11 and 1.12 are examples of MCE using the same data sets employed in Fig. 1.8. The currents estimated at 22 and 80 ms on the lattice nodes are shown in the second column. The currents on the nodes are projected onto the vertices of the cerebral polygon mesh in the third column. Figure 1.11 shows the

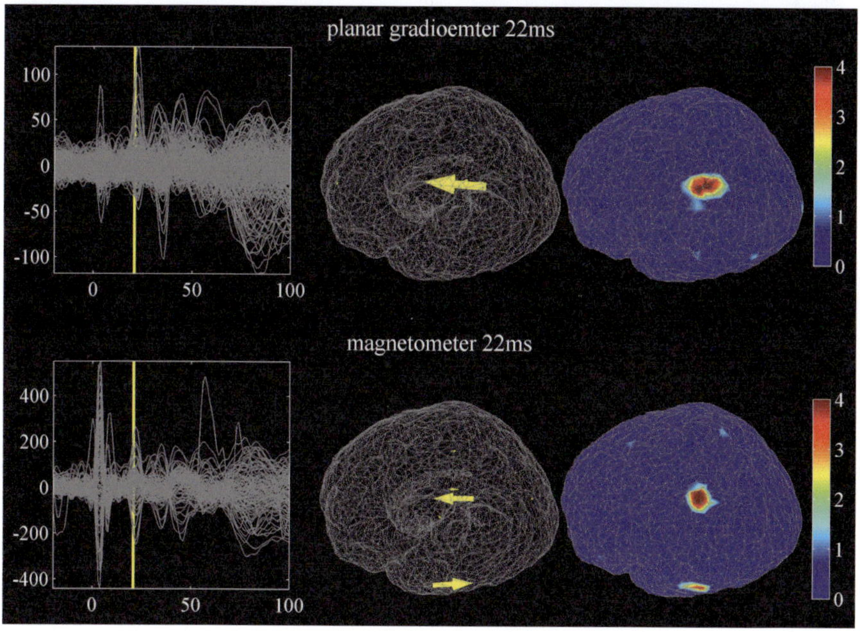

Fig. 1.11 The *left* column shows waveforms of SEF and *yellow lines* indicated the latencies of 22 ms. Minimum L1 norm estimation and minimum current estimates (MCEs) are used for reconstruction. Data of sensor signal and source reconstruction results from planar gradiometers are shown in the *upper row*, and same ones from magnetometers are shown in the *lower row*. MCE estimates the currents on the lattice nodes (*second column*), and these results are projected onto the vertices of the cerebral polygon mesh (*third column*). Comparing with the results using minimum L2 norm estimation shown in Fig. 1.8, MCE shows focalized source activations from the same sensor detection. The unit of the color bar is nAm

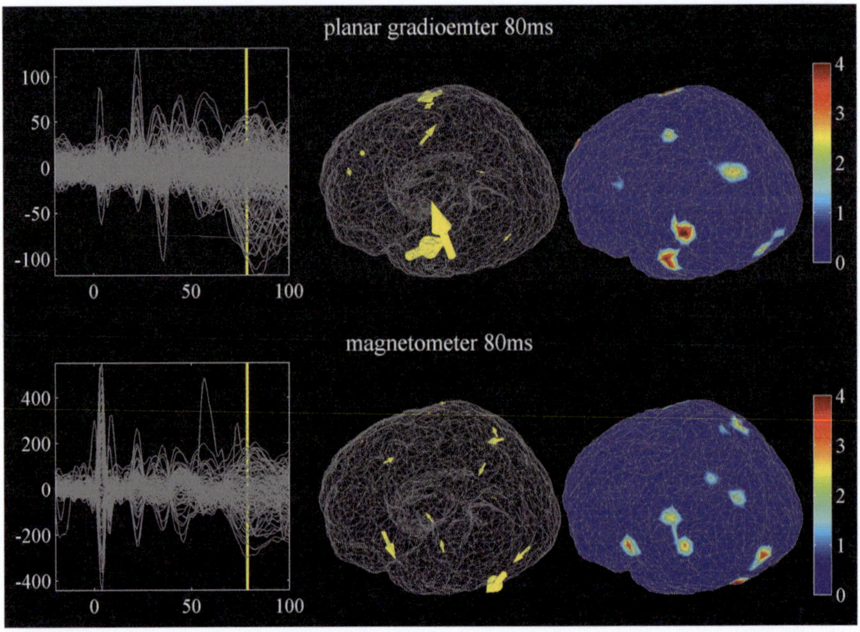

Fig. 1.12 Minimum L1 norm estimation interprets spread activities as scattered spotty sources. Anticlimactic examples of MCE result from somatosensory evoked responses. The same data sets from Fig. 1.11 are employed, but later latency of 80 ms is chosen for this source estimation. The description of each column and row is the same as in Fig. 1.11

better plausible concentrated source localization results, while Fig. 1.12 suggests the anticlimactic spread sources.

1.3.3 Adaptive Beam Former (Minimum Variance Estimation) in General

Beam forming technique has been firstly introduced to optimize TV antenna position and orientation by minimizing the weighted variance of the signal. This technique is also useful in the application of biomedical signals and has been introduced to MEG source localization technique [46]. The constraint of minimum variance estimation forms

$$\mathbf{q}_{\mathrm{n}} = \mathbf{w}_{\mathrm{n}}^{\mathrm{T}} \cdot \mathbf{B} \tag{1.31}$$

under the condition that $\sum_{t=1}^{\mathrm{T_{smp}}} \mathrm{q}_{\mathrm{n,t}}^2$ is minimized, where $\mathrm{T_{smp}}$ denotes total sampled number of the time length of $\mathbf{B}$, $\mathbf{q}_{\mathrm{n}} = \left(\mathrm{q}_{\mathrm{n},1}, \ldots, \mathrm{q}_{\mathrm{n},1}, \ldots, \mathrm{q}_{\mathrm{n},\mathrm{T_{smp}}}\right)$ is time-series current data with an arbitrary direction at n-th source point, and $\mathbf{w}_n^T = (w_{1,n}, \ldots, w_{m,n}, \ldots, w_{M,n})^T$ is weight vector at the same n-th source point. To avoid $\mathbf{w}_n^T = (0, \ldots, 0)^T$, another constraint is added that $\mathbf{w}_n^T \cdot \mathbf{L}_n = \sum_m^M w_{m,n} L_{m,n} = 1$. The name of minimum variance is derived from the constraint that the variance of time-series current at an arbitrary source point is minimum. The relationship between minimum norm and minimum variance is as follows:

$$\mathbf{Q} = \begin{pmatrix} \mathbf{q}_1 \\ \vdots \\ \mathbf{q}_n \\ \vdots \\ \mathbf{q}_N \end{pmatrix} = \begin{pmatrix} q_{1,1} & \cdots & q_{1,t} & \cdots & q_{1,T_{smp}} \\ \vdots & \ddots & \vdots & \ddots & \vdots \\ q_{n,1} & \cdots & q_{n,t} & \cdots & q_{n,T_{smp}} \\ \vdots & \ddots & \vdots & \ddots & \vdots \\ q_{N,1} & \cdots & q_{N,t} & \cdots & q_{N,T_{smp}} \end{pmatrix}$$

minimum variance (row n); minimum norm (column t)

Note that the proposed constrains are not verified neurophysiologically. The minimum norm minimizes $\mathbf{Q}$ in the direction of column (source points) and the minimum variance minimizes $\mathbf{Q}$ in the direction of row (time series). Using Lagrange multiplier, an undetermined coefficient, this problem can be solved, and the answer is (see also [47])

$$\mathbf{q}_n = \hat{\mathbf{q}}_n \cdot \mathbf{B} = \frac{\mathbf{L}_n^T (\mathbf{B}\mathbf{B}^T)^{-1}}{\mathbf{L}_n^T (\mathbf{B}\mathbf{B}^T)^{-1} \mathbf{L}_n} \cdot \mathbf{B}, \tag{1.32}$$

Beam forming technique is called as "adaptive" because $\hat{\mathbf{q}}_n$ depends on $\mathbf{B}$ and consequently the results vary according to the selected time span, frequency band, or selected sensors. Although it has resemblance to arrayed antenna technique, original beam former utilizes the phase information but does not utilize the minimum variance estimation. The neighboring currents on source points are independent and no influence should be taken for the results. By expressing the three degrees of freedom of the current direction (ϕ, θ, r) at an arbitrary source point, the solution is

$$\mathbf{Q}_{\mathrm{n}} = \begin{pmatrix} \mathbf{q}_{\mathrm{n}\varphi} \\ \mathbf{q}_{\mathrm{n}\theta} \\ \mathbf{q}_{\mathrm{nr}} \end{pmatrix} = \begin{pmatrix} \dfrac{\mathbf{L}_{\mathrm{n}\varphi}^{\mathrm{T}}\left(\mathbf{B}\mathbf{B}^{\mathrm{T}}\right)^{-1}}{\mathbf{L}_{\mathrm{n}\varphi}^{\mathrm{T}}\left(\mathbf{B}\mathbf{B}^{\mathrm{T}}\right)^{-1}\mathbf{L}_{\mathrm{n}\varphi}} \\ \dfrac{\mathbf{L}_{\mathrm{n}\theta}^{\mathrm{T}}\left(\mathbf{B}\mathbf{B}^{\mathrm{T}}\right)^{-1}}{\mathbf{L}_{\mathrm{n}\theta}^{\mathrm{T}}\left(\mathbf{B}\mathbf{B}^{\mathrm{T}}\right)^{-1}\mathbf{L}_{\mathrm{n}\theta}} \\ \dfrac{\mathbf{L}_{\mathrm{nr}}^{\mathrm{T}}\left(\mathbf{B}\mathbf{B}^{\mathrm{T}}\right)^{-1}}{\mathbf{L}_{\mathrm{nr}}^{\mathrm{T}}\left(\mathbf{B}\mathbf{B}^{\mathrm{T}}\right)^{-1}\mathbf{L}_{\mathrm{nr}}} \end{pmatrix} \cdot \mathbf{B}. \tag{1.33}$$

If a spherical model is employed, $\mathbf{q}_{\mathrm{nr}}$ is omitted (see Sect. 1.2.1 "Sarvas Formula"). To minimize the variance of the currents, this filter shall allot small weight to the sensor near the currents and large weight to the distant sensor. Consequently, this filter tends to estimate large currents in the deep brain and small currents near the sensor. To avoid this phenomenon, also normalization techniques are proposed as well as minimum L2 norm estimation (see Eqs. 1.22, 1.23, 1.29, and 1.30) [42, 43].

Figure 1.13 shows the results using same data sets in Fig. 1.8 to make comparison between minimum L2 norm estimation and beam forming technique. The time window for beam former is selected from 15 to 30 ms (redly highlighted). The cross-sectioned images of the second column are estimated through original minimum variance (Eq. 1.32); those of the third column are estimated through the modification of

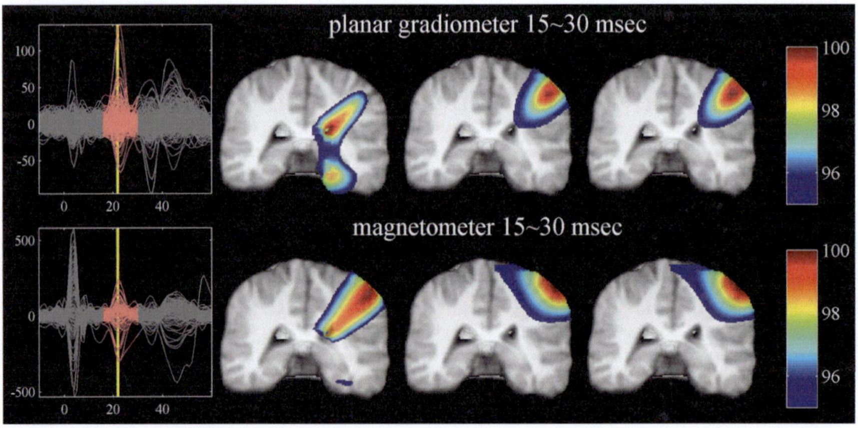

Fig. 1.13 Source reconstruction results obtained using minimum variance methods, so-called beam forming technique. Following Fig. 1.8, SEF waveform at 22 ms latency (*yellow line*) is focused. Due to the demand of time window selection for beam forming technique, 15 ms time window from 15 to 30 ms (*redly* highlighted in the first column) is used for the source reconstruction. Results of original minimum variance, Eq. 1.32, are shown in the second column, and third and fourth columns exhibit the results with improved approach, Eq. 1.33 for third column and Eq. 1.34 for right column. Original minimum variance tends to set large currents in deeper side of the brain, and the normalization methods make them move superficially

$$\mathbf{q}_{\mathrm{n}} = \frac{\widetilde{\mathbf{L}}_{\mathrm{n}}^{\mathrm{T}}\left(\mathbf{B}\mathbf{B}^{\mathrm{T}}\right)^{-1}}{\widetilde{\mathbf{L}}_{\mathrm{n}}^{\mathrm{T}}\left(\mathbf{B}\mathbf{B}^{\mathrm{T}}\right)^{-2}\widetilde{\mathbf{L}}_{\mathrm{n}}}\mathbf{B}, \tag{1.34}$$

where $\widetilde{\mathbf{L}}_{\mathrm{n}} = \frac{\mathbf{L}_{\mathrm{n}}}{\sqrt{\sum_{\mathrm{m}=1}^{\mathrm{M}} \mathrm{L}_{\mathrm{m,n}}^{2}}}$; and those of the fourth column are estimated through the modification of

$$\mathbf{q}_{\mathrm{n}} = \frac{\mathbf{L}_{\mathrm{n}}^{\mathrm{T}}\left(\mathbf{B}\mathbf{B}^{\mathrm{T}}\right)^{-1}}{\sqrt{\mathbf{L}_{\mathrm{n}}^{\mathrm{T}}\left(\mathbf{B}\mathbf{B}^{\mathrm{T}}\right)^{-2}\mathbf{L}_{\mathrm{n}}}}\mathbf{B}. \tag{1.35}$$

As clearly seen, the results from normalization make available to estimate sources superficially. However, selection of time, frequency domain, and sensor of **B** influences the estimated results. Figures 1.13 and 1.14 exhibit the different results using the same data. The difference between Figs. 1.13 and 1.14 is only time domain used for calculation of $\mathbf{B}\mathbf{B}^{\mathrm{T}}$, where the setting for the former is 15–30 ms and the latter is 5–50 ms (redly highlighted). As shown, despite the same centered latency (22 ms), clearly distinct results were shown by different choices of the time window settings. Minimum variance and its normalization methods often employ the ratio of the measurement data between target time domain $\mathbf{B}_{\mathrm{s}}$ and baseline $\mathbf{B}_{\mathrm{c}}$, such as sensor noise or resting state (see also Sect. 1.3.2.4). Relatively long time span (seconds or more) is selected for both $\mathbf{B}_{\mathrm{s}}$ and $\mathbf{B}_{\mathrm{c}}$. For instance, synthetic aperture magnetometry (SAM) employs the following normalization for n-th source point:

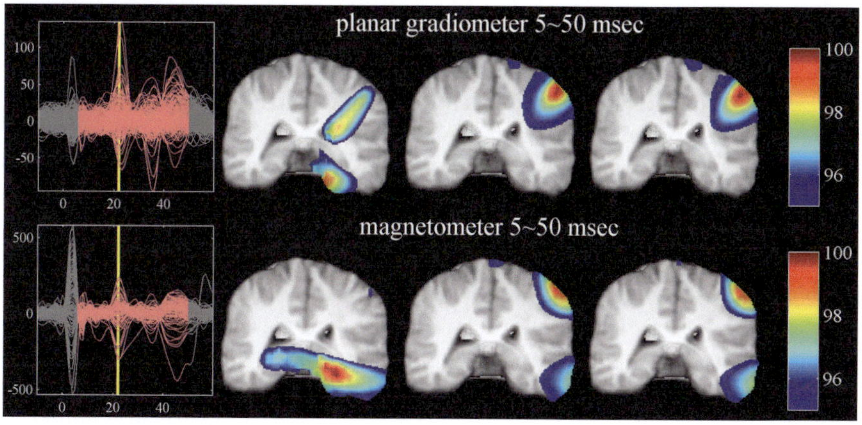

Fig. 1.14 Results under the same condition as in Fig. 1.13. The description of each row and column is the same as the previous figure, except for the time span of 45 ms from 5 to 50 ms (*redly* highlighted). Results of minimum variance, even applying normalization, are affected by selection of time domain

$$Z_n = \sqrt{\frac{\left(\mathbf{L}_n^T\left(\mathbf{B}_s\mathbf{B}_s^T\right)^{-1}\mathbf{L}_n\right)^{-1}}{\left(\mathbf{L}_n^T\left(\mathbf{B}_c\mathbf{B}_c^T\right)^{-1}\mathbf{L}_n\right)^{-1}}}. \tag{1.36}$$

SAM has been developed by Canadian MEG vendor, CTF, and opened a new door to frequency analysis such as event-related synchronization or desynchronization [47–49]. There is no technical problem that other spatial filter algorithm employs resting state data for normalization like SAM.

1.3.4 Limitation of Inverse Solution

We discussed three types of inverse solution (ECD estimation, minimum norm estimation, adaptive beam former). ECD approach provides no solution due to overdetermined solution, while distributed source analysis provides infinite number of solutions. In either case, this biomedical inverse solution is generally called mathematically ill-posed solution. Therefore, no warrantable solution is given such that some penalty should be taken account. In ECD analysis, initial position and local minima are also disputed. MNE provides stable solution, but the problem with this solution is its less sensitivity to focal activity, because minimizing norm always suppresses such focal activity and tends to provide spreader distribution. In general, the reconstructed current data $\mathbf{q}_n$ at n-th source point is expressed mathematically as $\mathbf{q}_n = \mathbf{W}_n B + \mathbf{n}_n$, where $\mathbf{W}_n$ is a mixing matrix and $\mathbf{n}_n$ is noise vector at the same n-th source point. The spatial filter acts as if making blurred images consisting of thousands of or more pixels from hundreds of pixel data because $\mathbf{q}_n$ remains merely a mixture from $\mathbf{B}$. It is unlikely that spatial filter extracts hidden signals on the $\mathbf{q}_n$ which is not seen in measured signal in $\mathbf{B}$. Therefore, one of the expected functions of spatial filter becomes signal separation of $\mathbf{B}$. Independent component analysis (ICA) [50] and factorial analysis are other tools specific to signal separation, and they do not require conductor models for signal separation purpose. The $\mathbf{W}_n$ always includes approximation of inverse matrix which deteriorates original information of $\mathbf{B}$. If the sensor is radial planar gradiometers, there is a simple and convenient method to guess cortical activities. The way is projection of sensor signals onto underlying cortex. Figure 1.15 is an example, where waveforms of the same SEF employed in Figs. 1.7 and 1.8 are projected onto the cortex [51, 52]. Sensor projection method cannot express brain activities in the deep brain or on the cerebral bases; however, fluxes from such regions are usually weak for MEG sensor to detect. This utility might be useful for confirmation of lateral aspect of cerebrum as initial step.

In this chapter we discussed the algorithm of MEG forward and inverse solution for source localization purpose. The inverse solution has been much paid attention by researchers and engineers to achieve accurate and stable source localization. However, other factors are also intermingled for accurate source localization. Among others, the following two factors underlying source analysis are fundamental:

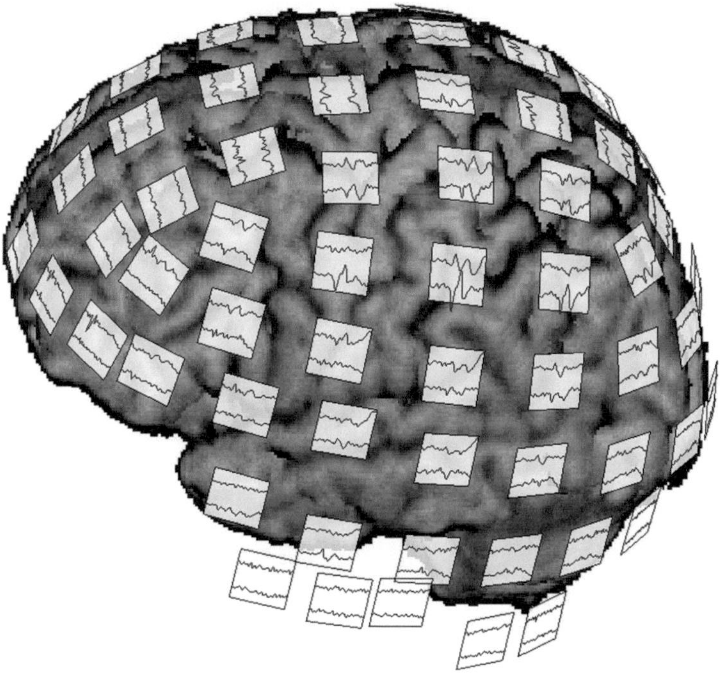

Fig. 1.15 Schema of sensor projection onto the brain surface. Waveforms of SEF obtained from planar gradiometers are projected onto the brain surface just below the sensors directed from sensor position and along with perpendicular orientation. The waveforms with highest amplitude are moved near the right postcentral sulcus

(1) preprocessing of the MEG signal and (2) precise alignment of the brain position to the anatomical image obtained by MRI. Even the sophisticated inverse algorithm will be unsuccessful if factors (1) and (2) are not correctly performed in advance. For example, recently ICA has been focused to improve the SNR [50, 53–55] related with item (1). With respect to item (2), the magnetic digitizer (Polhemus) has been mainly used for MEG-MRI co-registration. Recently, much accurate and rapid measurement has been proposed using laser scanning technique [56]. These techniques also improve the source localization results.

References

1. Fernando H, da Silva L. Electrophysiological basis of MEG signals. In: Hansen PC, Kringelbach ML, Salmelin R, editors. MEG -an introduction to methods. New York: Oxford University Press; 2010. p. 1–23.
2. Okada Y. Toward understanding the physiological origins of neuromagnetic signals. In: Lu ZL, Kaufman L, editors. Magnetic source imaging of the human brain. New York: Psychology Press; 2008. p. 43–76.

3. Nurminen J, Taulu S, Nenonen J, Helle L, Simola J, Ahonen A. Improving MEG performance with additional tangential sensors. IEEE Trans Biomed Eng. 2013;60:2559–66. doi:10.1109/TBME.2013.2260541.
4. Sato MA, Yoshioka T, Kajihara S, Toyama K, Goda N, Doya K, Kawato M. Hierarchical Bayesian estimation for MEG inverse problem. NeuroImage. 2004;23:806–26.
5. MNE Home – Martinos Center for Biomedical Imaging http://martinos.org/mne/stable/index.html. Accessed 7 Jan 2015.
6. SPM – Statistical Parametric Mapping. http://www.fil.ion.ucl.ac.uk/spm/. Accessed 7 Jan 2015.
7. Brainstorm. http://neuroimage.usc.edu/brainstorm/. Accessed 7 Jan 2015.
8. VBMEG – Variational Bayesian Multimodal EncephaloGraphy. http://vbmeg.atr.jp/?lang=en. Accessed 7 Jan 2015.
9. FieldTrip. http://fieldtrip.fcdonders.nl/. Accessed 7 Jan 2015.
10. GNU Octave. https://www.gnu.org/software/octave/. Accessed 7 Jan 2015.
11. FreeMat. http://freemat.sourceforge.net/. Accessed 7 Jan 2015.
12. Scilab. http://www.scilab.org/. Accessed 7 Jan 2015.
13. EEGLAB. http://sccn.ucsd.edu/eeglab/. Accessed 7 Jan 2015.
14. Iversen JR, Makeig S. MEG/EEG data analysis using EEGLAB. In: Supek S, Aine CJ, editors. Magnetoencephalography. From signals to dynamic cortical networks. Heidelberg: Springer; 2014. p. 199–05. doi:10.1007/978-3-642-33045-2.
15. Sarvas J. Basic mathematical and electromagnetic concepts of the biomagnetic inverse problem. Phys Med Biol. 1987;32:11–22.
16. Heller L, Volegov P. Electric and magnetic fields of the brain. In: Supek S, Aine CJ, editors. Magnetoencephalography. From signals to dynamic cortical networks. Heidelberg: Springer; 2014. p. 73–105. doi:10.1007/978-3-642-33045-2.
17. Haueisen J, Knösche TR. Forward modeling and tissue conductivities. In: Supek S, Aine CJ, editors. Magnetoencephalography. From signals to dynamic cortical networks. Heidelberg: Springer; 2014. p. 107–27. doi:10.1007/978-3-642-33045-2.
18. Gençer NG, Acar CE, Tanzer IO. Forward problem solution of magnetic source imaging. In: Lu ZL, Kaufman L, editors. Magnetic source imaging of the human brain. New York: Psychology Press; 2008. p. 77–100.
19. Hämäläinen M, Hari R, Ilmoniemi RJ, Knuutila J, Lounasmaa OV. Magnetoencephalography – theory, instrumentation, and applications to noninvasive studies of the working human brain. Rev Mod Phys. 1993;65:413–97.
20. Baillet S. The dowser in the fields: searching for MEG sources. In: Hansen PC, Kringelbach ML, Salmelin R, editors. MEG -an introduction to methods. New York: Oxford University Press; 2010. p. 83–123.
21. Uranker L. Common compact analytical formulas for computation of geometry integrals on a basic Cartesian sub-domain in boundary and volume integral methods. Eng Anal Bound Elem. 1990;7:124–9.
22. Van't ED, de Munck JC, Kaas AL. A fast method to derive realistic BEM models foe E/MEG source reconstruction. IEEE Trans Biomed Eng. 2001;48:1434–43.
23. de Munck JC. A linear discretization of the volume conductor boundary integral equation using analytically integrated elements. IEEE Trans Biomed Eng. 1992;39:669–71.
24. Fuchs M, Drenckhahn R, Wischmann HA, Wagner M. An improved boundary element method for realistic volume-conductor modeling. IEEE Trans Biomed Eng. 1998;45:980–97.
25. Haueisen J, Ramon C, Eiselt M, Brauer H, Nowak H. Influence of tissue resistivities on neuromagnetic fields and electric potentials studied with a finite element model of the head. IEEE Trans Biomed Eng. 1997;44:727–35.
26. Wolters CH, Grasedyck L, Hackbusch W. Efficient computation of lead field bases and influence matrix for the FEM-based EEG and MEG inverse problem. Inverse Probl. 2004;20:1099–116.
27. Press WH, Teukolsky SA, Vetterling WT, Flannery BP. Numerical recipes -the art of scientific computing. 3rd ed. Cambridge: Cambridge University Press; 2007.

28. Cuffin BN. A comparison moving dipole inverse solutions using EEG's and MEG's. IEEE Trans Biomed Eng. 1985;32:905–10.
29. Cuffin BN. Effects of measurement errors and noise on MEG moving dipole inverse solutions. IEEE Trans Biomed Eng. 1986;33:854–61.
30. Scherg M, Voncramon D. Two bilateral sources of the late AEP as identified by a spatio-temporal dipole model. Electroencephalogr Clin Neurophysiol. 1985;62:32–44.
31. Scherg M, Voncramon D. A new interpretation of the generators of BAEP waves-I-V: results of a spatio-temporal model. Electroencephalogr Clin Neurophysiol. 1985;62:290–9.
32. Salmelin R. Multi-dipole modeling in MEG. In: Hansen PC, Kringelbach ML, Salmelin R, editors. MEG -an introduction to methods. New York: Oxford University Press; 2010. p. 124–55.
33. Scherg M, Buchner H. Somatosensory evoked potentials and magnetic fields: separation of multiple source activities. Physiol Meas. 1993;14:35–9.
34. Jensen O, Hesse C. Estimating distributed representations of evoked responses and oscillatory brain activity. In: Hansen PC, Kringelbach ML, Salmelin R, editors. MEG -an introduction to methods. New York: Oxford University Press; 2010. p. 156–85.
35. Hämäläinen MS, Ilmoniemi RJ. Interpreting magnetic fields of the brain: minimum norm estimates. Med Biol Eng Comput. 1994;32:35–42.
36. Hämäläinen MS, Lin FH, Mosher JC. Anatomically and functionally constrained minimum-norm estimates. In: Hansen PC, Kringelbach ML, Salmelin R, editors. MEG -an introduction to methods. New York: Oxford University Press; 2010. p. 186–215.
37. Wang JZ, Kaufman L. Magnetic source imaging: search for inverse solutions. In: Lu ZL, Kaufman L, editors. Magnetic source imaging of the human brain. New York: Psychology Press; 2008. p. 101–33.
38. Uutela K, Hämäläinen M, Somersalo E. Visualization of magnetoencephalographic data using minimum current estimates. NeuroImage. 1999;10:173–80.
39. Dale AM, Liu AK, Fischl BR, Buckner RL, Belliveau JW, Lewine JD, et al. Dynamic statistical parametric mapping: combining fMRI and MEG for high-resolution imaging of cortical activity. Neuron. 2000;26:55–67.
40. Pascual-Marqui RD. Standardized low-resolution brain electromagnetic tomography (sLORETA): technical details. Methods Find Exp Clin Pharmacol. 2002;24:5–12.
41. LORETA. http://www.uzh.ch/keyinst/loreta.htm. Accessed 8 Jan 2015.
42. Sekihara K1, Sahani M, Nagarajan SS. A simple nonparametric statistical thresholding for MEG spatial-filter source reconstruction images. NeuroImage. 2005;27:368–76.
43. Sekihara K, Sahani M, Nagarajan SS. Localization bias and spatial resolution of adaptive and non-adaptive spatial filters for MEG source reconstruction. NeuroImage. 2005;25:1056–67.
44. Lin FH, Belliveau JW, Dale AM, Hämäläinen MS. Distributed current estimates using cortical orientation constraints. Hum Brain Mapp. 2006;27:1–13.
45. Molins A, Stufflebeam SM, Brown EN, Hämäläinen MS. Quantification of the benefit from integrating MEG and EEG data in minimum l2-norm estimation. NeuroImage. 2008;42: 1069–77.
46. van Veen BD, van Drongelen W, Yuchman M, Suzuki A. Localization of brain electrical activity via linearly constrained minimum variance filtering. IEEE Trans Biomed Eng. 1997; 44:867–80.
47. Robinson SE, Vuba J. Functional neuroimaging by synthetic aperture magnetometry (SAM). In: Yoshimoto T, Kotani M, Kuriki S, Karibe H, Nakasato N, editors. Recent advances in biomagnetism -Proceeding of the 11th international congress of biomagnetism. Tohoku University Press; 1999. p. 302–5.
48. Hirata M1, Kato A, Taniguchi M, Ninomiya H, Cheyne D, Robinson SE, Maruno M, Kumura E, Ishii R, Hirabuki N, Nakamura H, Yoshimine T. Frequency-dependent spatial distribution of human somatosensory evoked neuromagnetic fields. Neurosci Lett. 2002;318:73–6.

49. Herdman AT, Cheyne D. A practical guide for MEG and beamforming. In: Handy TC, editor. Brain signal analysis -advances in neuroelectric and neuromagnetic methods. Cambridge: Massachusetts Institute of Technology; 2009. p. 99–140.
50. Hyvärinen A, Juha Karhunen J, Oja E. Independent component analysis. New York: Wiley-Interscience; 2001.
51. Hashizume A, Kurisu K, Iida K, Arita K, Akimitsu T, Nagasaki N. Development of a freeware for analysis of neuromagnetic epileptic discharges. Hiroshima J Med Sci. 2010;59:21–5.
52. hns_meg. http://meg.aalip.jp/freeware/freewareE.html. Accessed 8 Jan 2015.
53. Tang AC, Pearlmutter BA, Malaszenko NA, Phung DB, Reeb BC. Independent components of magnetoencephalography: localization. Neural Comput. 2002;14:1827–58.
54. Hironaga N, Ioannides AA. Localization of individual area neuronal activity. NeuroImage. 2007;34:1519–34.
55. Tang AC, Pearlmutter BA. Independent components of magnetoencephalography: localization and single-trial response onset detection. In: Lu ZL, Kaufman L, editors. Magnetic source imaging of the human brain. New York: Psychology Press; 2008. p. 159–201.
56. Hironaga N, Hagiwara K, Ogata K, Hayamizu M, Urakawa T, Tobimatsu S. Proposal for a new MEG–MRI co-registration: a 3D laser scanner system. Clin Neurophysiol. 2014;125: 2404–12.

Part II

Motor System

Basic Functions and Clinical Applications

2

Takashi Nagamine and Masao Matsuhashi

Abstract

Investigation for motor system by magnetoencephalographic (MEG) recordings has focused on detecting the primary motor area. In order to collect data with high S/N ratio, selection of a motor paradigm and an analysis method to assure events time-locked to movement onset is important. Self-paced movement paradigm has been widely employed, because the same paradigm in electroencephalographic (EEG) and subdural recordings has provided bunch of physiological data. The motor field (MF) component in movement-related cortical magnetic fields (MRCFs) is believed to reflect the final stage of motor execution in MI. Frequency analysis on the background rhythm related to movement can be also utilized to estimate motor-related areas.

Considering the difficulties in performing the self-paced movement, easier motor paradigms such as isometric contraction or cyclic repetitive movements have been recently introduced. Although estimated sources from these paradigms have provided sources of reasonable locations, physiological significance should await for further study.

Keywords

Movement-related cortical magnetic fields (MRCFs) • Self-paced movement • Frequency analysis • Event-related desynchronization (ERD)/event-related synchronization (ERS) • Coherence

T. Nagamine (✉)
Department of Systems Neuroscience, School of Medicine, Sapporo Medical University, South 1, West 17, Chuo-ku, Sapporo 060-8556, Japan
e-mail: nagamine@sapmed.ac.jp

M. Matsuhashi
Human Brain Research Center, Graduate School of Medicine, Kyoto University, 54 Kawahara-cho, Syogoin, Sakyo-ku, Kyoto 606-8507, Japan

S. Tobimatsu, R. Kakigi (eds.), *Clinical Applications of Magnetoencephalography*, DOI 10.1007/978-4-431-55729-6_2

2.1 Basic Principles and Physiology (Fig. 2.1)

2.1.1 Movement and Posture

The whole range of motor functions from voluntarily initiated precise grasping to reflexive reactions can be mainly characterized through a description of the temporal change of coordinates in space for each part of the body or the whole body. Each bodily motion is governed by muscle contraction, even if the body part keeps still in space, the placement of which is ruled by the motor system. These muscle contractions are without exception regulated by motor neurons situated in the anterior horn in the spinal cord or brainstem motor nucleus. The route originating from the motor neurons to a muscle is named the final common pathway, which receives various kinds of inputs from central sources.

Movement and posture are the two main features of motor function, whose space coordinates change and remain unchanged, respectively. As individuals, we can be vividly aware of our own movement, but we are rarely conscious of our posture.

2.1.2 Movement

Movement is categorized into three kinds of bodily motions: voluntary movement, automatic movement, and reflex.

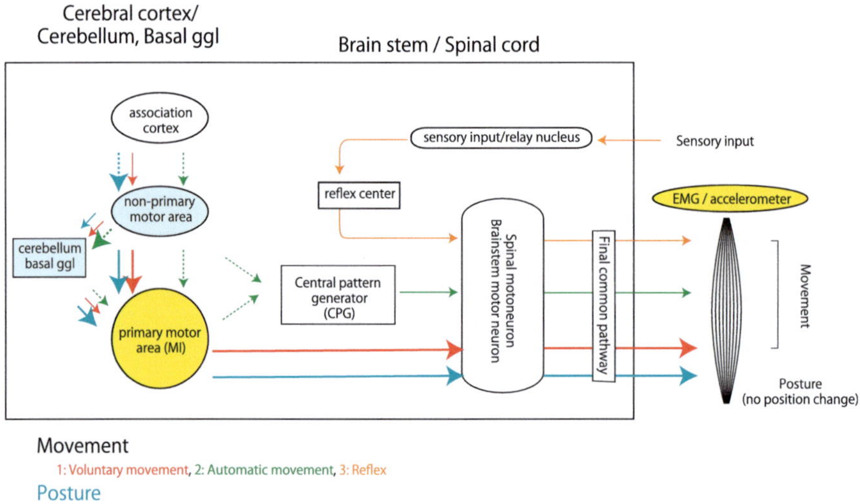

Fig. 2.1 Scheme for motor systems and targets for magnetoencephalographic (MEG) measurement. *Yellow shading* indicates the main targets used for MEG analysis and *light blue* shows possible sites

Voluntary movement is a goal-directed behavior launched by volition and is generated by the cerebral cortex with associated activation of basal ganglia and the cerebellum.

Automatic movement is characterized by a stereotyped rhythmic movement with recursive elements, such as swallowing and walking. It is believed that a neural circuit called the "central pattern generator" exists in the spinal cord or brainstem to generate automatic movement.

Reflex is a stereotyped movement occurring unconsciously and caused by a specific stimulus. The relevant reflex center sits in the spinal cord or brainstem.

2.1.3 Voluntary Movement

Voluntary movement is initiated by volition and requires several steps, including, preparation, initiation, maintenance, and termination. Among these, the initiation step is represented by an abrupt change from the preceding resting position, which is usually caused by a drastic increase in activity in the primary motor area (MI). This change is projected to the periphery through the corticospinal or corticobulbar tract.

2.1.4 Involuntary Movements

Unwanted movements unsuitable for the conditions of body parts are a manifestation of disorders of motor control and categorized as involuntary movements. The site responsible in the nervous system spreads over various sites from peripheral to central areas, including the cerebral cortex.

2.2 Clinical Applications

2.2.1 Physiological Investigation of Control of Movement and Posture

For detecting electrical activity in cortical areas, self-paced movement has been commonly recorded using an EEG [1, 2]. Waveforms obtained using this approach are called movement-related cortical potentials (MRCPs). Various generators have been clarified by investigations with scalp recordings [3, 4] and subdural recordings [5, 6]. Following on from these recordings, the same approach has been adopted since the early era of magnetoencephalographic (MEG) recordings [7–12] to acquire movement-related cortical magnetic fields (MRCFs).

Compared with the investigation of sensory systems, the temporal sequence of neural activities related to motor function is not as well correlated with events visible to observers. In order to achieve the best possible MEG signals, short brisk movements are utilized to evoke alterations in brain activity correlated with movement onset.

Empirically, neural activities in the lower levels are more closely correlated with the onset of movement than those in higher levels. However, the detection of motor-related activities of spinal motor neurons or brainstem motor neurons is difficult because they do not produce large magnetic fields owing to their anatomical alignment, although they are tightly linked to the electromyography (EMG) activities in the temporal domain. Therefore, cortical activities in the MI, which are tightly linked to EMG activities after spinal motor neurons, are considered to be the best targets for MEG investigation. This tight connection between cortex and muscle activity is strengthened if we focus on the initiation of brisk movement, because rapid firing of pyramidal cells in the MI area becomes apparent.

At the termination stage of voluntary movement, a similar but opposite phenomenon must be happening. However, the tail-off of muscle activity is not so abrupt as to allow alignment of times across trials; therefore, few studies have explored this stage [13]. During the maintenance stage of voluntary movement, although there may be some continuing neural activities so as to maintain connection between cortical and muscular activities, they cannot be detected because they are only sustained for a rather short period in the case of voluntary movement.

If this maintenance stage of voluntary movement becomes elongated, it can be regarded as posture maintenance, which enables us to obtain a clear signal as for the correlation between cortex and muscle using coherence analysis (see Sect. 2.6.3).

2.2.2 Generator Sources for Involuntary Movement

In addition to physiological surveys of motor control, we can detect sources for some types of involuntary movements, if the generator responsible is situated in a limited area of the cerebral cortex. Because some types of cortical myoclonus are characterized by brisk contraction of the small body part, it is likely that the lesion responsible should be located in the cerebral cortex, which has been proven by several reports using jerk-locked back averaging [14–18]. There are reports dealing with non-cortical lesions in patients with tremor [19] and widespread lesions responsible in a patient with subacute sclerosing panencephalitis [20], although their relevant cortical lesions were not clearly demonstrated.

2.3 Principal Analysis Methods (Fig. 2.2)

The self-paced movement method has been widely adopted. Subjects are requested to move their finger, hand, or foot on one side with intervals exceeding several seconds. Subjects should make a brisk and short contraction and keep still in between the movements. In order to delineate somatotopic organization, a distal part of a limb is often used.

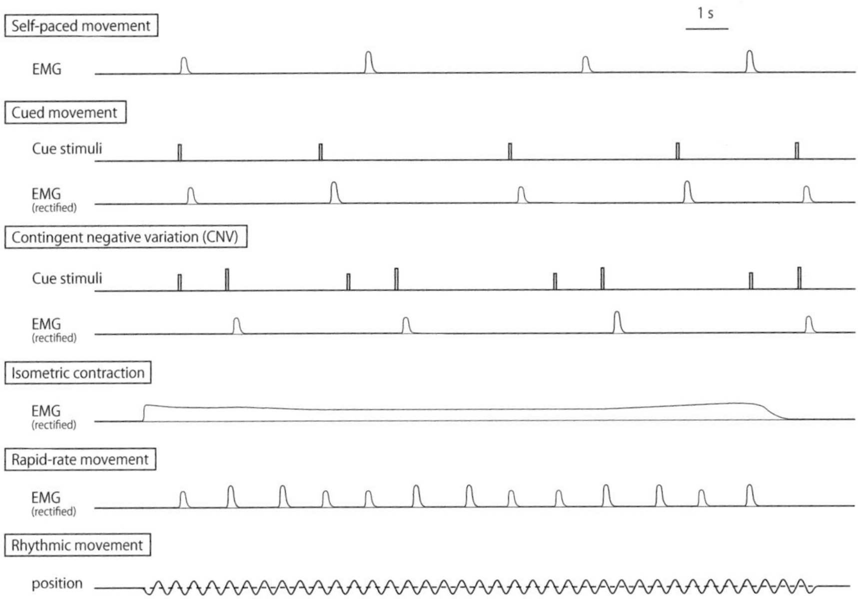

Fig. 2.2 Approaches used for the investigation of motor systems

Since some subjects experience difficulties in deciding when they have to move because of the long interval requested for this self-paced movement, some centers use assistance to indicate when movements should start. Delivery of cue signals such as a tone or light with intervals of several seconds can persuade subjects to start movement. Intervals can be either fixed or randomly assigned, and subjects do not have to be so strict in performing rapid responses.

Contingent negative variation (CNV) is a famous and well-utilized task in the investigation of motor control. A pair of stimuli (S1 and S2) with an interval of around 2 s serve as a warning and imperative stimuli, and they are repeated with an interval of 3–5 s between the stimulus pair. A time interval of 2 s between S1 and S2 is appropriate for subjects to predict the timing of S2, so subjects have to respond as soon as possible.

Isometric contraction maintained for longer than several seconds can be regarded as a task focusing the maintenance stage of voluntary movement. In order to verify the constant condition of muscle contraction, it is useful to employ some visual feedback or give guidance to adjust contraction.

Because conventional self-paced movements with a long interval are not easy for some subjects, rapid rate or short interval movement can be utilized. Repetitions of brisk movement up to 1–2-Hz can leave a rather silent period of muscle activity in between the repetitions.

Rhythmic and cyclic movement is similar but different from this rapid rate movement. Back-and-forth motion of 2–4 Hz does not leave an EMG silent period, allowing an assumption that it can mimic a sinusoidal profile, and coherence analysis between EMG and brain activity can be applied assuming brain activities also share a cyclic temporal sequence.

2.4 Recording and Analysis of Self-Paced Movement

2.4.1 Preparation

Besides usual preparations for MEG recording, it is important to ensure detection of alterations in motor output of the target body part. Selection of appropriate muscles and movements and methods varies across experiments. Surface EMG over the target muscle with cup electrodes is the most commonly used method. Electrodes can be placed on any part of the body and can be applied not only for recordings of movement but also for posture. Button pressing with a natural return designed without causing magnetic noise can be adopted. Light-emitting diodes are also utilized to check displacement of body parts. Adjustment of the allowance from the operating point is critical to define the time point to determine the initiation of movement.

Equally important is confirming the stillness of other body parts at other times except for the intended movement. For this purpose, EMG electrodes are placed on some body parts other than the target muscle, including the contralateral side.

Examiners should check the magnetic environment before actual measurement starts. When we find slow magnetic fluctuation, asking the subject to hold their breath is of help to check. Sometimes buttons on clothes or dental implants can cause magnetic noise. Screening conditions during the performance of an actual assessment is essential. Magnetized materials often cause large magnetic noise even if the movement is tiny. Putting a cushion under the return of a moving platform is useful to avoid noise caused by touchdown by the moving object. EMG cables should be fixed at several points so that they do not swing.

Eye movement can significantly distort the magnetic field related to movement, because its temporal profile is similar to movement. It is useful to set a visual target to which subjects are requested to fix their vision in order to avoid eye movement. Electrooculogram (EOG) monitoring is strongly advised.

2.4.2 Recording Procedure

Before introducing subjects into a magnetically shielded room, recording baseline condition of magnetic fields without the subject is recommended (empty-room measurement). This data can be used for confirmation of background noise, essential data for noise reduction filters, and uncovering overlooked magnetic materials attached to the subject. Data acquisition parameters for MEG signals require a sampling rate of at least 500 Hz, with a band-pass filter set between 0.1 and 100 Hz. Depending on the necessity for detecting slow magnetic fields preceding movement, a high-pass filter can be widened to 0.03 Hz or narrowed to 1 Hz. When putting subjects adjusted to the MEG sensor, the examiner should check that subject's head fits well around the sensorimotor area.

As for data collection, if an examiner sets an electronic pulse for averaging, it is possible to get only averaged data utilizing an automatic noise cancelation system to avoid large signals, including eye movement. It would also be useful to accumulate repetition-time segmented waveforms during the predetermined analysis time window. However, with these methods, we cannot properly exclude trials including unwanted noise segments. Therefore, it would be preferable to save all the continuous data including EMG, EOG, and trigger pulse.

Fifty to 100 repetitions of a movement are required to obtain averaged waveforms with a reasonable S/N ratio. Considering artifact rejection to exclude trials with contaminating noises, two or three sessions each with 30–50 movements are advised. Averaging procedures can be done either online or off-line. In case of recording continuous data and leaving final analysis to be done off-line, it would be preferable to obtain online averages to check recording conditions.

Because the task is rather simple and repetitive, adequate rest periods should be provided once the movement has been repeated 30–50 times. After a break, recording the head position with respect to the dewar should be done, because subjects tend to change their head position during the rest period.

2.4.3 Analysis

2.4.3.1 Trigger Point

Because MRCF aims to explore temporal changes in brain activity, an average across a certain number of repetitions of the movement is the first step. If continuous data is going to be processed, EMG onset should be visually tagged from continuous data. If the background EMG level serving as the resting condition becomes noisy, the triggering point may fluctuate. In this case, exclusion of data from the time period with increased EMG background from averaging could be considered.

2.4.3.2 Averaging Window

For off-line analysis, the time window for averaging can be adjusted by taking the shortest interval between two successive movements within the session. If the interval is short compared with the intended time window, the trials with a short interval should be excluded from the average. Depending on the purpose, settings for premovement and postmovement time periods vary from 2–4 s to 1–2 s, respectively. This assignment should be determined carefully, as the first 10–20 % of the total analysis window is usually assigned for the baseline period for amplitude calculation. Confirmation of reproducibility by superimposition of two different sessions can provide insight for contamination by noise elements. If these sets of averages show similarities, the sessions can be averaged to obtain group average waveforms to be used for further analysis.

2.4.3.3 Components of MRCF: (Figs. 2.3 and 2.4)

The components of MRCF have usually been named in accordance with Kristeva et al. [21]. The readiness field (RF) shown over the central area contralateral to the moving side corresponds to the first slow increasing magnetic activity prior to movement onset. The onset of RF is approximately several hundred milliseconds before movement onset, and it is usually later than the initiation of the slow EEG component starting 2–3 s before onset. This EEG component, Bereitschaftspotential (BP), is bilaterally distributed, and is followed by NS' in the EEG starting several hundred milliseconds before movement onset, which is dominant over the contralateral side [3]. Taken together, it is most likely that the RF component is a counterpart of NS' or BP2 rather than BP or BP1. This RF activity shows a peak around the onset of movement, and this is called the motor field (MF).

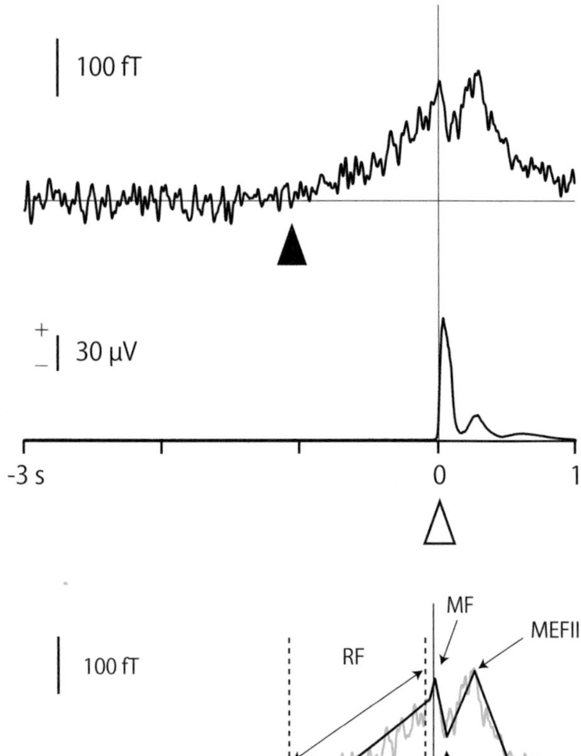

Fig. 2.3 Sample movement-related cortical magnetic fields (MRCFs) after self-paced extension of the right middle finger. *Upper panel*: Magnetometer waveform over the left central area, low-pass filter set at 30 Hz and high-pass filter at 0.03 Hz, with a sampling rate of 603 Hz. *Lower panel*: Average of high-pass filter (1 Hz) and rectified electromyography (EMG) of the right extensor digitorum communis muscle. *Open triangle*: EMG onset. *Solid triangle*: readiness field (RF) onset

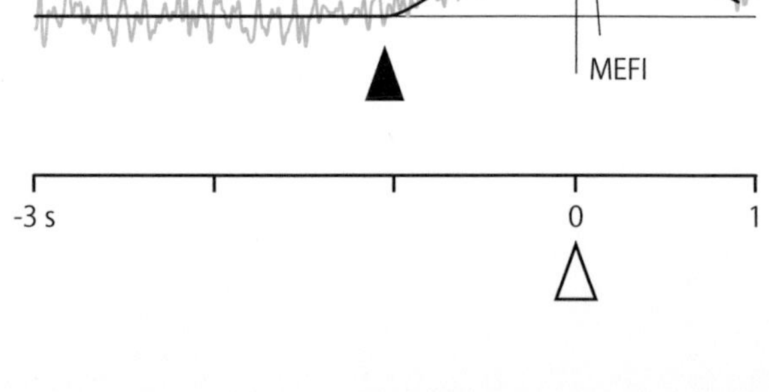

Fig. 2.4 Schematic description of each component of the movement-related cortical magnetic field. *RF* readiness field, *MF* motor field, *MEFI* movement-evoked field I, *MEFII* movement-evoked field II

It is considered that this is equivalent to motor potential (MP) in EEG recordings, which is sometimes clearly differentiated from the preceding slow cortical potential of NS' or BP2 [22]. However, it is rare to encounter a prominent MF component emerging on top of the preceding RF. Hence, it is difficult to use these two terms differently, especially for the analysis for generator sources. Because RF is characterized by its onset time, it is advised to use in the temporal domain; how early this component appears etc. On the other hand, MF is better for use in cases of source localization. Following MF, a peak in the opposite direction at approximately 100 ms after EMG onset is noted as movement-evoked field I (MEFI) and the following peak at about 200 ms corresponds to movement-evoked field II (MEFII).

2.4.3.4 Source Analysis with an Equivalent Current Dipole Model

For resolving the sources for intracranial sources for each component of MRCF, equivalent current dipole (ECD) assumption is widely used. At the MF, around 30 sensors covering the field extrema of opposing influx and efflux are sufficient to localize the ECD, whereas a smaller number of sensor locations with planar gradiometers centering on the field extrema are required for analysis. Compared with the prominent peak of N20m in somatosensory evoked fields, accepting a threshold of parameters indicating a correlation between observed and those estimated hypothetically can be set to a lower value for reliable ECDs. Correlation coefficients of 0.90–0.95 or goodness of fit of 0.80–0.90 are acceptable (Fig. 2.5).

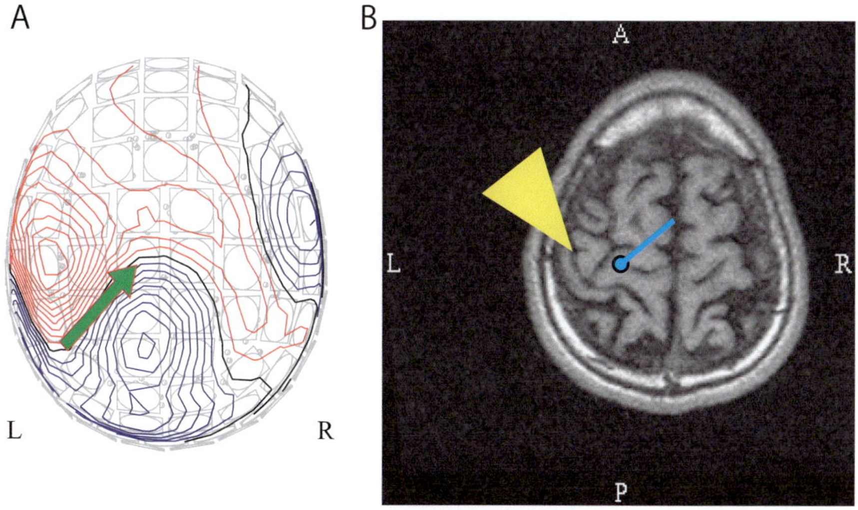

Fig. 2.5 Source location estimated by the assumption of a single dipole source, at 6 ms before the movement onset taken from the same data as Fig. 2.3. (**a**) Contour map and approximate source location. *Red* indicates an outward magnetic field and *blue* indicates an inward field. (**b**) Estimated current dipole location projected onto the subject's own magnetic resonance imaging (MRI); the *circle* indicates the dipole location and the *bar* indicates the dipole orientation. A *yellow triangle* indicates the *left* central sulcus

Because slow magnetic artifacts largely distort the background baseline level, it can be reasonable to localize the MF component by setting a high-pass filter around 0.1 Hz at the expense of detecting the RF component [23]. Exploration with multiple generator sources has been performed using BESA [23, 24]. The MEFI component was located posterior to the central sulcus and may be different from N20m of the SEF [25, 26].

2.4.3.5 Spatial Filter

In modeling the source of MEG signals, the motor system can be a good target to introduce a spatial filter assuming sources are distributed within a certain volume or on a surface. Utilization of L1 minimum norm (minimum current estimate [MCE] [27]) or L2 minimum norm estimates (MNEs [28]) can be adopted, because we can start by making a minimum of assumptions. An example using MNE is shown in Fig. 2.6. These analytical methods do not require a priori information concerning the source structures. Spatial filters optimized to pass activity from a specific brain areas suppressing activity from other areas are called beamformers and can present narrower source distribution [29].

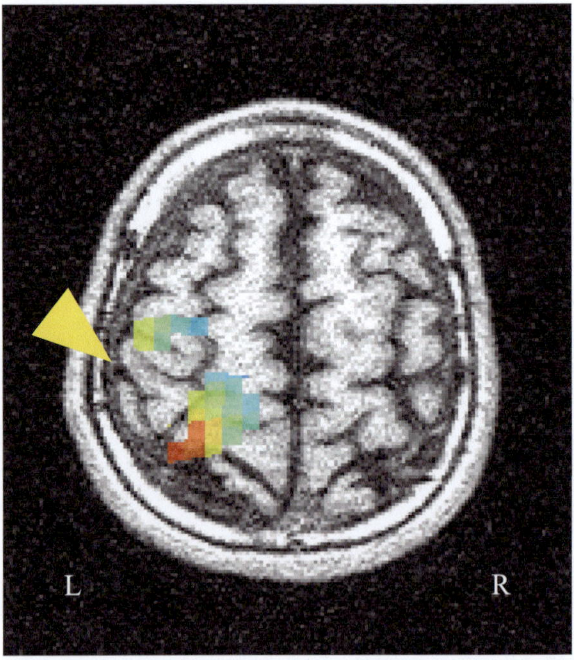

Fig. 2.6 Source distribution by spatial filter (minimum norm estimate) projected onto the subject's own MRI, 6 ms before the movement onset, taken from the same subject as Figs. 2.3 and 2.5. Pixels with an F-value threshold of 40 calculated from the baseline power are shown. *Red* areas indicates higher power than *blue* areas. A *yellow triangle* indicates the *left* central sulcus

2.5 Contingent Negative Variation

In EEG investigations, CNV is another well-known approach for surveying motor control, especially sensorimotor integration [30]. Using the same method, several MEG attempts have been reported. They demonstrated MI activity following the S2 time period [31–35]. As for the researcher recording using subdural electrodes, few reports have dealt with this sensorimotor integration [36]. This report demonstrated the involvement of areas other than the MI, whereas reports of conventional self-paced movement demonstrated activation of a rather restricted area of the MI. Therefore, source analysis as for CNV is required, and single ECD assumption may fail to provide an accurate location for the MI.

2.6 Advanced Techniques

2.6.1 Event-Related Desynchronization/Event-Related Synchronization

In the activities related to motor function, mu rhythm has been widely observed in routine EEG recordings [37]. Pfurtscheller's group developed a method for the study of temporal changes in power in the EEG background oscillatory rhythm [38, 39]. Power decrease in alpha or beta bands in association with movement onset or a certain task was named event-related desynchronization (ERD) and was considered to indicate increased cortical excitability. The opposite phenomenon of a power increase was termed event-related synchronization (ERS), and this was believed to be related to decreased excitability (Fig. 2.7). Salmelin introduced this method to MEG by keeping the analysis time bin the same as that of the sampling rate, by adopting the name of temporal spatial evolution [40].

This analysis revealed other aspects of brain activity other than slow components. Usually, the RF of a slow magnetic shift precedes the movement by several hundred milliseconds, and ERD already starts 3 s before the onset [41]. This clear contrast between the slow and background activities indicate that these two components manifest completely different physiological events.

2.6.2 Rapid Rate Movement

Self-paced movement methods have been regarded as the standard approach to detect the MI. However, because this approach requires several task regulations such as control of eye movement, brisk movement, and keeping relaxed muscles between the movements, some subjects cannot perform the task properly. In order to overcome this situation, methods with self-paced movement with shorter intervals have been introduced [42]. Because only a short time period of around 400 ms can be left as a baseline for amplitude measurement in the case of 2-Hz repetitive movements, a time period of 500 ms was introduced for baseline

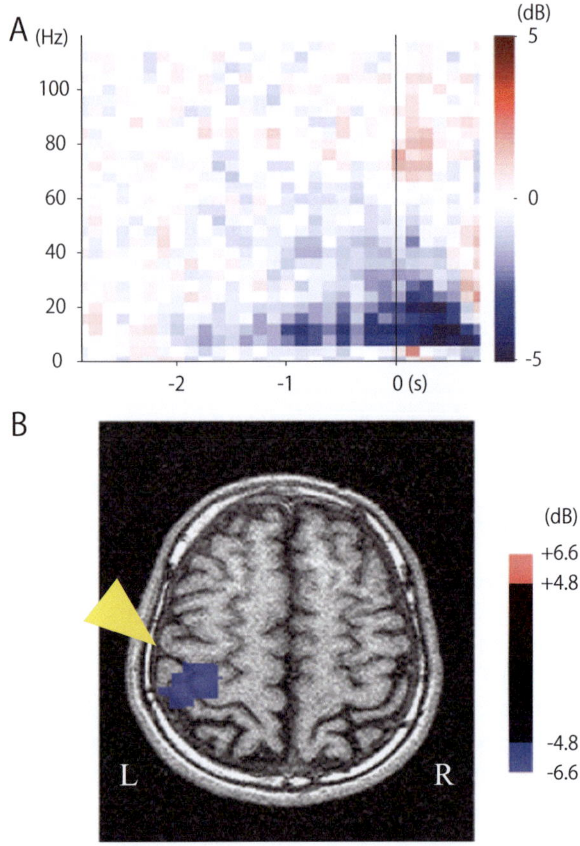

Fig. 2.7 Time–frequency analysis. (**a**) Time–frequency representation of gradiometer activity over the left central area related to self-paced extension of the right middle finger (same data source as in Figs. 2.3, 2.5, and 2.6). Note event-related desynchronization (ERD) in the alpha to beta band starts around 2 s before movement onset, whereas beta frequency event-related synchronization (ERS) occurs 6–700 ms after movement onset. (**b**) Projection of the ERD source distribution at 12 Hz, 100 ms before movement onset, using a minimum norm estimate spatial filter. A *yellow triangle* indicates the *left* central sulcus

measurement. Because of its steadiness, this baseline period can be achieved by more subjects.

2.6.3 Corticomuscular Coherence (CMC)

Recent EEG studies of coherence between EEG and EMG showed corticomuscular oscillatory relations in humans [43–47] as well as in monkeys [48]. In a human MEG study, Salenius et al. demonstrated increased coherence in the frequency range of 15 ± 33 Hz between MEG signals from the primary sensorimotor cortex [44]. This activity was suppressed temporarily by transient ischemic sensory deafferentation [49] but regained its strength after the end of ischemia, indicating that sensory feedback is not essential for coherence.

This procedure enables researchers to examine speech [50] and tongue movement [51]. Because this method is easier compared with conventional self-paced movement especially for patients, this can be used extensively in clinical situations.

The source-level analysis of CMC is achieved either by using cross-correlogram [44] or by DICS [52] (Fig. 2.8).

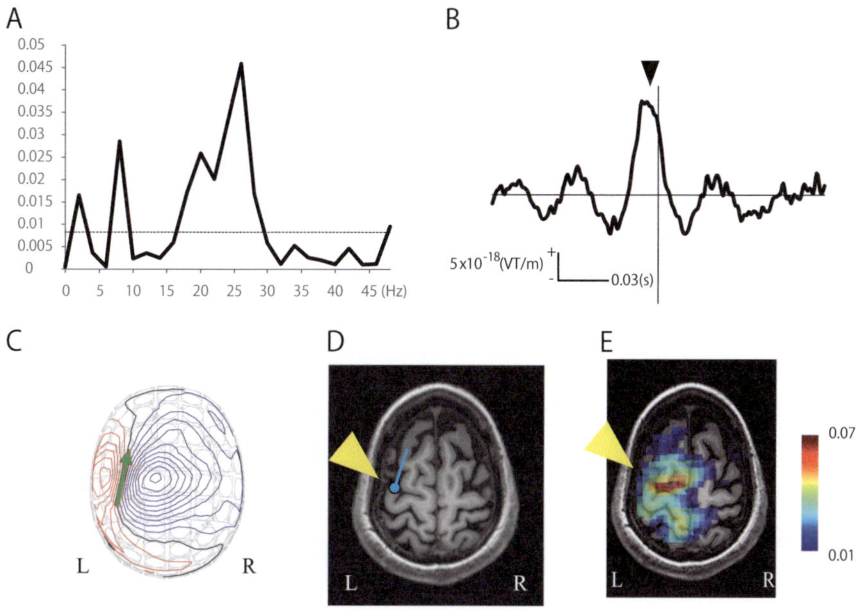

Fig. 2.8 Corticomuscular coherence analysis of right middle finger tonic extension. (**a**) Amplitude squared coherence between high-pass filtered (1 Hz) and rectified electromyography (EMG) and gradiometer over the *left* central area. A *horizontal broken line* indicates the 95 % confidence level. (**b**) Correlogram calculated by inverse Fourier transformation of the cross-spectra of the same data. A *solid triangle* indicates gradiometer activity precedes EMG by about 5 ms. (**c**) Contour map and approximate source location of correlogram activity at 5 ms time lead. (**d**) Estimated current dipole location projected onto the subject's own MRI. (**e**) Source distribution of the coherent source by spatial filter (minimum norm estimate) projected onto the subject's own MRI, 26 Hz, threshold set at the 95 % confidence level. A *yellow triangle* indicates the *left* central sulcus

2.6.4 Corticokinematic Coherence (CKC) [53]

Following the method of rapid rate movements, which still leaves silent period for EMG, repetitions of flexion-extensions of a body part produce steady-state movements without showing silent EMG activity. Bourguignon's group developed a new method using a three-axis accelerometer that attaches to a finger, and subjects were instructed to make cyclic movements at a pace of 3 Hz. Cross-correlogram and coherence spectra demonstrated increased coherence at a driving frequency of 3 Hz and its harmonic frequency. The coherence was statistically significant with sources in the MI area close to the area representing the corresponding body part. Because this method is easy and straightforward for recording, it will become an easy and robust approach for MI surveys (Fig. 2.9).

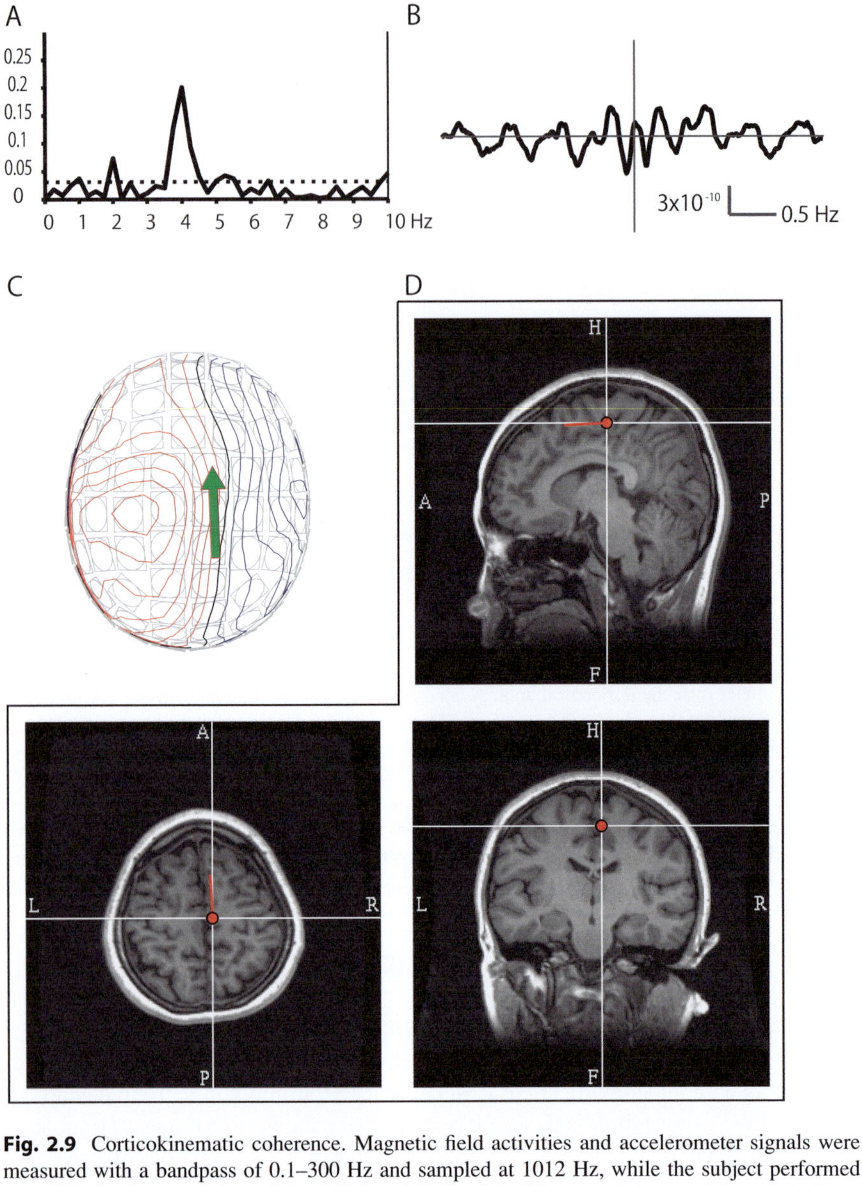

Fig. 2.9 Corticokinematic coherence. Magnetic field activities and accelerometer signals were measured with a bandpass of 0.1–300 Hz and sampled at 1012 Hz, while the subject performed repetitive flexion-extension of the *left toe* at 2 Hz. (**a**) Coherence between the central planar gradiometer sensor and accelerogram. *Horizontal axis*: frequency from 0 to 10 Hz, *Vertical axis*: coherence. There are two coherence peaks of 0.07 at 2 Hz and 0.2 at 4 Hz. (**b**) Cross-correlogram (**c**) Source estimated from the cross-correlogram at a peak of −74 ms (Courtesy of Dr Hidekazu Saito)

2.7 Future Directions

Without doubt, the most important target for clinical MEG of the motor system is to delineate the MI. Using these new techniques that utilize coherence, success rates for localizing the MI area have improved. However, compared with investigations for other systems, such as somatosensory systems, sampling a sufficient number of averages is not so easy owing to the difficulty to executing motor analyses, especially in patients. Physiological meaning for solutions obtained from these new techniques is to be clarified also.

For physiological investigations, analysis of the separation between the MI and non-primary motor areas is awaited. This development may be accomplished not only by analytical procedures, but these require advancements in hardware capable of higher spatial resolution.

References

1. Bates JA. Electrical activity of the cortex accompanying movement. J Physiol. 1951;113 (2–3):240–57.
2. Kornhuber HH, Deecke L. Changes in the brain potential in voluntary movements and passive movements in man: readiness potential and reafferent potentials. Pflugers Arch Gesamten Physiol Menschen Tiere. 1965;284:1–17.
3. Shibasaki H, Barrett G, Halliday E, Halliday AM. Components of the movement-related cortical potential and their scalp topography. Electroencephalogr Clin Neurophysiol. 1980;49(3–4):213–26.
4. Vaughan Jr HG, Costa LD, Ritter W. Topography of the human motor potential. Electroencephalogr Clin Neurophysiol. 1968;25(1):1–10.
5. Ikeda A, Luders HO, Burgess RC, Shibasaki H. Movement-related potentials recorded from supplementary motor area and primary motor area. Role of supplementary motor area in voluntary movements. Brain: J Neurol. 1992;115(Pt 4):1017–43.
6. Neshige R, Luders H, Shibasaki H. Recording of movement-related potentials from scalp and cortex in man. Brain: J Neurol. 1988;111(Pt 3):719–36.
7. Deecke L, Boschert J, Weinberg H, Brickett P. Magnetic fields of the human brain (Bereitschaftsmagnetfeld) preceding voluntary foot and toe movements. Exp Brain Res. 1983;52(1):81–6.
8. Antervo A, Hari R, Katila T, Poutanen T, Seppanen M, Tuomisto T. Cerebral magnetic fields preceding self-paced plantar flexions of the foot. Acta Neurol Scand. 1983;68(4):213–7.
9. Hari R, Antervo A, Salmi T. Slow EEG potentials preceding self-paced plantar flexions of hand and foot. Acta Physiol Scand. 1983;119(1):55–9. doi:10.1111/j.1748-1716.1983.tb07305.x.
10. Cheyne D, Weinberg H. Neuromagnetic fields accompanying unilateral finger movements: pre-movement and movement-evoked fields. Exp Brain Res. 1989;78(3):604–12.
11. Kristeva-Feige R, Walter H, Lutkenhoner B, Hampson S, Ross B, Knorr U, Steinmetz H, Cheyne D. A neuromagnetic study of the functional organization of the sensorimotor cortex. Eur J Neurosci. 1994;6(4):632–9.
12. Nagamine T, Toro C, Balish M, Deuschl G, Wang B, Sato S, Shibasaki H, Hallett M. Cortical magnetic and electric fields associated with voluntary finger movements. Brain Topogr. 1994;6(3):175–83.
13. Toma K, Nagamine T, Yazawa S, Terada K, Ikeda A, Honda M, Oga T, Shibasaki H. Desynchronization and synchronization of central 20-Hz rhythms associated with voluntary

muscle relaxation: a magnetoencephalographic study. Exp Brain Res. 2000;134(4):417–25. doi:10.1007/s002210000483.

14. Ugawa Y, Uesaka Y, Terao Y, Yumoto M, Hnajima R, Sakai K. Pathophysiology of sensorimotor cortex in cortical myoclonus. Clin Neurosci. 1995;3(4):198–202.
15. Mima T, Nagamine T, Ikeda A, Yazawa S, Kimura J, Shibasaki H. Pathogenesis of cortical myoclonus studied by magnetoencephalography. Ann Neurol. 1998;43(5):598–607. doi:10.1002/ana.410430507.
16. Timmermann L, Gross J, Schmitz F, Freund HJ, Schnitzler A. Involvement of the motor cortex in pseudochoreoathetosis. Mov Disord: Off J Mov Disord Soc. 2001;16(5):876–81.
17. Sudmeyer M, Pollok B, Hefter H, Gross J, Wojtecki L, Butz M, Timmermann L, Schnitzler A. Postural tremor in Wilson's disease: a magnetoencephalographic study. Mov Disord: Off J Mov Disord Soc. 2004;19(12):1476–82. doi:10.1002/mds.20240.
18. Butz M, Timmermann L, Gross J, Pollok B, Sudmeyer M, Kircheis G, Haussinger D, Schnitzler A. Cortical activation associated with asterixis in manifest hepatic encephalopathy. Acta Neurol Scand. 2014;130(4):260–7. doi:10.1111/ane.12217.
19. Volkmann J, Joliot M, Mogilner A, Ioannides AA, Lado F, Fazzini E, Ribary U, Llinas R. Central motor loop oscillations in parkinsonian resting tremor revealed by magnetoencephalography. Neurology. 1996;46(5):1359–70.
20. Oga T, Ikeda A, Nagamine T, Sumi E, Matsumoto R, Akiguchi I, Kimura J, Shibasaki H. Implication of sensorimotor integration in the generation of periodic dystonic myoclonus in subacute sclerosing panencephalitis (SSPE). Mov Disord: Off J Mov Disord Soc. 2000;15(6):1173–83.
21. Kristeva R, Cheyne D, Deecke L. Neuromagnetic fields accompanying unilateral and bilateral voluntary movements: topography and analysis of cortical sources. Electroencephalogr Clin Neurophysiol. 1991;81(4):284–98.
22. Shibasaki H, Hallett M. What is the Bereitschaftspotential? Clin Neurophysiol: Off J Int Fed Clin Neurophysiol. 2006;117(11):2341–56. doi:10.1016/j.clinph.2006.04.025.
23. Hoshiyama M, Kakigi R, Berg P, Koyama S, Kitamura Y, Shimojo M, Watanabe S, Nakamura A. Identification of motor and sensory brain activities during unilateral finger movement: spatiotemporal source analysis of movement-associated magnetic fields. Exp Brain Res. 1997;115(1):6–14.
24. Onishi H, Sugawara K, Yamashiro K, Sato D, Suzuki M, Kirimoto H, Tamaki H, Murakami H, Kameyama S. Neuromagnetic activation following active and passive finger movements. Brain Behav. 2013;3(2):178–92. doi:10.1002/brb3.126.
25. Niimi M, Ohira T, Akiyama T, Hiraga K, Kaneko Y, Ochiai M, Fukunaga A, Kobayashi M, Kawase T. Source analysis of the magnetic field evoked during self-paced finger movements. Neurol Res. 2008;30(3):239–43. doi:10.1179/016164107X230801.
26. Weinberg H, Cheyne D, Crisp D. Electroencephalographic and magnetoencephalographic studies of motor function. Adv Neurol. 1990;54:193–205.
27. Uutela K, Hamalainen M, Somersalo E. Visualization of magnetoencephalographic data using minimum current estimates. NeuroImage. 1999;10(2):173–80. doi:10.1006/nimg.1999.0454.
28. Hamalainen MS, Ilmoniemi RJ. Interpreting magnetic fields of the brain: minimum norm estimates. Med Biol Eng Comput. 1994;32(1):35–42.
29. Hillebrand A, Barnes GR. Beamformer analysis of MEG data. Int Rev Neurobiol. 2005;68:149–71. doi:10.1016/S0074-7742(05)68006-3.
30. Walter WG. The contingent negative variation: an electro-cortical sign of sensori-motor reflex association in man. Prog Brain Res. 1968;22:364–77. doi:10.1016/S0079-6123(08)63519-0.
31. Basile LF, Rogers RL, Bourbon WT, Papanicolaou AC. Slow magnetic flux from human frontal cortex. Electroencephalogr Clin Neurophysiol. 1994;90(2):157–65.
32. Elbert T, Rockstroh B, Hampson S, Pantev C, Hoke M. The magnetic counterpart of the contingent negative variation. Electroencephalogr Clin Neurophysiol. 1994;92(3):262–72.

33. Hultin L, Rossini P, Romani GL, Hogstedt P, Tecchio F, Pizzella V. Neuromagnetic localization of the late component of the contingent negative variation. Electroencephalogr Clin Neurophysiol. 1996;98(6):435–48.
34. Gomez CM, Fernandez A, Maestu F, Amo C, Gonzalez-Rosa JJ, Vaquero E, Ortiz T. Task-specific sensory and motor preparatory activation revealed by contingent magnetic variation. Brain Res Cogn Brain Res. 2004;21(1):59–68. doi:10.1016/j.cogbrainres.2004.05.005.
35. Babiloni C, Brancucci A, Pizzella V, Romani GL, Tecchio F, Torquati K, Zappasodi F, Arendt-Nielsen L, Chen AC, Rossini PM. Contingent negative variation in the parasylvian cortex increases during expectancy of painful sensorimotor events: a magnetoencephalographic study. Behav Neurosci. 2005;119(2):491–502. doi:10.1037/0735-7044.119.2.491.
36. Matsumoto R, Ikeda A, Ohara S, Matsuhashi M, Baba K, Yamane F, Hori T, Mihara T, Nagamine T, Shibasaki H. Motor-related functional subdivisions of human lateral premotor cortex: epicortical recording in conditional visuomotor task. Clin Neurophysiol: Off J Int Fed Clin Neurophysiol. 2003;114(6):1102–15.
37. Chatrian GE, Petersen MC, Lazarte JA. The blocking of the rolandic wicket rhythm and some central changes related to movement. Electroencephalogr Clin Neurophysiol. 1959;11 (3):497–510.
38. Pfurtscheller G, Aranibar A. Event-related cortical desynchronization detected by power measurements of scalp EEG. Electroencephalogr Clin Neurophysiol. 1977;42(6):817–26.
39. Pfurtscheller G, Lopes da Silva FH. Event-related EEG/MEG synchronization and desynchronization: basic principles. Clin Neurophysiol: Off J Int Fed Clin Neurophysiol. 1999;110 (11):1842–57.
40. Salmelin R, Hari R. Spatiotemporal characteristics of sensorimotor neuromagnetic rhythms related to thumb movement. Neuroscience. 1994;60(2):537–50.
41. Nagamine T, Kajola M, Salmelin R, Shibasaki H, Hari R. Movement-related slow cortical magnetic fields and changes of spontaneous MEG- and EEG-brain rhythms. Electroencephalogr Clin Neurophysiol. 1996;99(3):274–86.
42. Gerloff C, Uenishi N, Nagamine T, Kunieda T, Hallett M, Shibasaki H. Cortical activation during fast repetitive finger movements in humans: steady-state movement-related magnetic fields and their cortical generators. Electroencephalogr Clin Neurophysiol. 1998;109 (5):444–53.
43. Conway BA, Halliday DM, Farmer SF, Shahani U, Maas P, Weir AI, Rosenberg JR. Synchronization between motor cortex and spinal motoneuronal pool during the performance of a maintained motor task in man. J Physiol. 1995;489(Pt 3):917–24.
44. Salenius S, Portin K, Kajola M, Salmelin R, Hari R. Cortical control of human motoneuron firing during isometric contraction. J Neurophysiol. 1997;77(6):3401–5.
45. Salenius S, Salmelin R, Neuper C, Pfurtscheller G, Hari R. Human cortical 40 Hz rhythm is closely related to EMG rhythmicity. Neurosci Lett. 1996;213(2):75–8.
46. Brown P, Salenius S, Rothwell JC, Hari R. Cortical correlate of the Piper rhythm in humans. J Neurophysiol. 1998;80(6):2911–7.
47. Halliday DM, Conway BA, Farmer SF, Rosenberg JR. Using electroencephalography to study functional coupling between cortical activity and electromyograms during voluntary contractions in humans. Neurosci Lett. 1998;241(1):5–8.
48. Baker SN, Olivier E, Lemon RN. Coherent oscillations in monkey motor cortex and hand muscle EMG show task-dependent modulation. J Physiol. 1997;501(Pt 1):225–41.
49. Pohja M, Salenius S. Modulation of cortex-muscle oscillatory interaction by ischaemia-induced deafferentation. Neuroreport. 2003;14(3):321–4. doi:10.1097/01.wnr.0000058518.74643.96.
50. Ruspantini I, Saarinen T, Belardinelli P, Jalava A, Parviainen T, Kujala J, Salmelin R. Corticomuscular coherence is tuned to the spontaneous rhythmicity of speech at 2–3 Hz. J Neurosci: Off J Soc Neurosci. 2012;32(11):3786–90. doi:10.1523/JNEUROSCI.3191-11.2012.

51. Maezawa H, Mima T, Yazawa S, Matsuhashi M, Shiraishi H, Hirai Y, Funahashi M. Contralateral dominance of corticomuscular coherence for both sides of the tongue during human tongue protrusion: an MEG study. NeuroImage. 2014;101:245–55. doi:10.1016/j.neuroimage.2014.07.018.
52. Gross J, Kujala J, Hamalainen M, Timmermann L, Schnitzler A, Salmelin R. Dynamic imaging of coherent sources: studying neural interactions in the human brain. Proc Natl Acad Sci U S A. 2001;98(2):694–9. doi:10.1073/pnas.98.2.694.
53. Bourguignon M, De Tiege X, Op de Beeck M, Pirotte B, Van Bogaert P, Goldman S, Hari R, Jousmaki V. Functional motor-cortex mapping using corticokinematic coherence. NeuroImage. 2011;55(4):1475–9. doi:10.1016/j.neuroimage.2011.01.031.

Part III

Somatosensory System

Basic Function

3

Kei Nakagawa, Koji Inui, and Ryusuke Kakigi

Abstract

Magnetoencephalography (MEG) is a powerful noninvasive tool for examining the temporal and spatial aspects of sensory processing in the brain. In this chapter, MEG responses elicited by innocuous and noxious somatic stimuli were reviewed. The development of devices for the selective stimulation of each peripheral receptor (i.e., A-beta mechanoreceptors, A-delta nociceptors, pain-related C nociceptors, and itch-related C nociceptors) enables us to record cortical responses to the activities of a given receptor and compare these responses among somatosensory submodalities. MEG responses to the stimulation of cutaneous A-beta mechanoreceptors suggest a serial mode of processing through the primary and secondary somatosensory cortices. By using the intraepidermal electrical stimulation (IES) method or an electrical itch stimulus, similar serial processing was found for pain and itch systems; however, some differences were observed (e.g., the area 3b sources following IES were absent and the itch stimulus induced a specific activity in the precuneus). These findings can be helpful in further basic studies on the physiology of the somatosensory system and clinical studies to elucidate the pathophysiology of somatosensory dysfunctions.

K. Nakagawa, Ph.D. (✉)
Department of Integrative Physiology, National Institute for Physiological Sciences, 38 Nishigonaka Myodaiji, Okazaki, Aichi 444-8585, Japan
e-mail: keinakag@hiroshima-u.ac.jp

K. Inui • R. Kakigi
Department of Integrative Physiology, National Institute for Physiological Sciences, 38 Nishigonaka Myodaiji, Okazaki, Aichi 444-8585, Japan

SOKENDAI, The Graduate University for Advanced Studies, Hayama, Kanagawa 240-0193, Japan

S. Tobimatsu, R. Kakigi (eds.), *Clinical Applications of Magnetoencephalography*,
DOI 10.1007/978-4-431-55729-6_3

Keywords
Somatosensory-evoked magnetic field (SEF) • Temporal processing • Intraepidermal electrical stimulation (IES) • Itching

3.1 Introduction

Magnetoencephalography (MEG) is a powerful tool for investigating sensory processing in the brain. It has excellent spatial and temporal resolution, in the order of mm and ms, respectively. The spatial resolution of MEG is similar to that of functional magnetic resonance imaging (fMRI) and positron emission tomography (PET), whereas its temporal resolution is far superior [1, 2]. Therefore, it is useful for evaluating the hierarchical organization of cortical processing of somatic information.

MEG has a number of advantages over somatosensory-evoked potentials using electroencephalography (EEG). (1) It is specialized to detect cortical activities because it is unaffected by the various tissue layers of the head, whereas EEG activities are widely distributed over the scalp by volume conductors [3]. (2) It can easily detect activities arising in the central sulcus, i.e., area 3b or area 4, because it is sensitive to currents tangential to the skull [2]. (3) It has the clinical advantage of detecting the reference point of the central sulcus without using invasive techniques. However, MEG also has some disadvantages. For example, it is not always easy to solve neuromagnetic inverse solutions, and it is not very sensitive to activities in the radially oriented or very deep sources [2]. Therefore, recording somatosensory-evoked magnetic fields (SEFs) provides insights in both basic and clinical studies if researchers understand the principals and significance of the elicited responses.

In this chapter, we reviewed several findings of previous SEF studies and considered possibilities for further clinical applications. We discussed differences and similarities in hierarchical processing between innocuous and noxious somatosensory information.

3.2 The Tactile System

Neuromagnetic responses induced by innocuous electrical stimuli were first demonstrated approximately 40 years ago [4]. Researchers have since obtained vast amounts of information from basic and clinical studies. For example, one study succeeded in representing a large part of the somatosensory homunculus map using the spatial advantage of MEG [5]. Another study indicated that SEFs were informative for assessing the functional state of the somatosensory system of patients with somatosensory disorders [6].

3.2.1 Physiological Significance of SEF Components

SEFs are generally elicited by electrically stimulating a peripheral nerve or cutaneous receptor percutaneously. Median nerve stimulation is commonly used because it elicits three clear SEF deflections, N20m-P35m(P30m)-P60m, which arise in and around the primary somatosensory cortex (SI) contralateral to the stimulated side, as shown in Fig. 3.1. However, the physiological significance of each component is still being debated.

The earliest deflection pointing toward the cortical surface, N20m, is generated in area 3b of the contralateral SI [2, 7–10]. It is considered to reflect the excitatory postsynaptic potentials (EPSP) produced by the synchronous activity of cortical pyramidal neurons in layers III and IV. Thus, the N20m source is used as a reference point for activation of the SI hand area [11].

On the other hand, the generator of P35m currently remains unknown. Previous studies indicated that it was generated in some areas of the SI [2, 9]. The

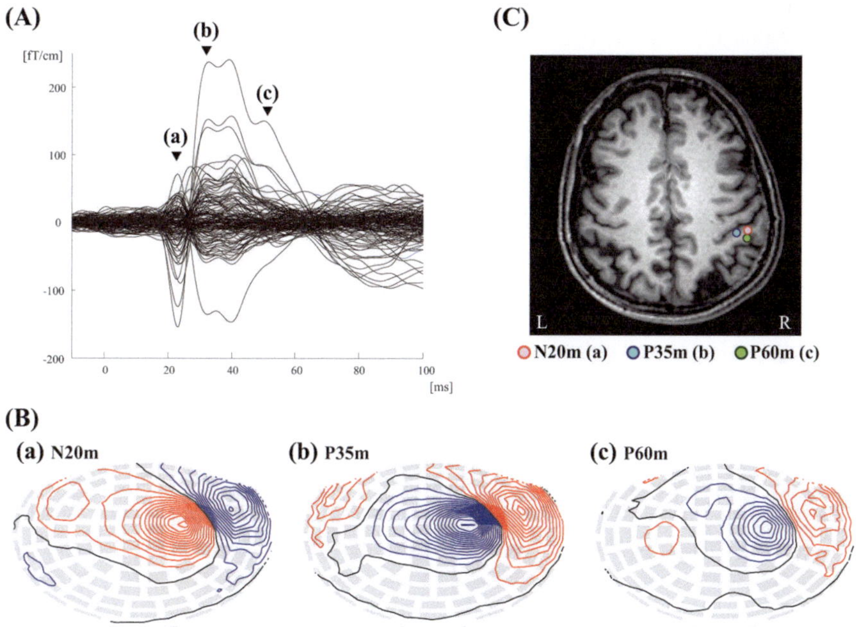

Fig. 3.1 Somatosensory-evoked magnetic fields following *left* median nerve stimulation. The stimulus was a constant-current square-wave pulse that was 0.2 ms in duration at the motor threshold. (**a**) Waveforms from 204 gradiometers with a filter of 0.5–100 Hz. Three clear components (**a**–**c**) were detected in the early phase after the stimulus. (**b**) Isocontour maps (40 ft/step) of the three components (N20m, P35m, and P60m). Areas surrounded by *red* and *blue* contour lines in the isocontour map show the outflux and influx of magnetic fields, respectively. (**c**) Locations of the equivalent current dipoles (ECD) of the three components. The P35m ECD was located medial to the N20m ECD, and the P60m ECD was located posterior to the N20m ECD

intracellular current of P35m is directed toward the cortical depth, which indicates an active current source in deep cortical layers. Therefore, it has been suggested to reflect the behavior of inhibitory postsynaptic potentials (IPSP) generated by GABAergic neurons in the deep layers [12–14]. The functional difference between N20m and P35m is also evident from findings of previous studies using repetitive stimuli [13] or recovery function [15, 16]. The P35m component is also known to be more sensitive to the movement gating effect than N20m [17–20].

P35m and P60m are considered to be derived from a similar generator because they have similar orientations and some common behaviors, such as they are both decreased by increments in the stimulus rate [13]. However, they behave differently under certain conditions. The ECD location of P60m was found to be posterior to that of P35m [11, 21], and the recovery function of P60m clearly differed from that of P35m [12]. Only P60m was attenuated toward the end of a lengthy stimulus series [22]. Patients with the pediatric degenerative disease, the CLN5 form of neuronal ceroid lipofuscinosis, have been shown to have greatly enlarged N20m and P35m, but intact P60m [23]. Moreover, only P60m was inhibited after the application of transcranial direct current stimulation over SI [21]. These findings indicate that P60m reflects the activity of an extended source, possibly involving areas 1 and 2.

3.2.2 Serial Processing of A-Beta Mechanoreceptor Signals

Tactile information from A-beta mechanoreceptors is transmitted to the brain via thalamocortical afferents. MEG can precisely detect the timing of the arrival of signals to multiple cortical areas. By using this advantage of MEG, we revealed the temporal relationship of cortical activities including SI, the parietal cortex, and secondary somatosensory cortex (SII) following a tactile stimulus using multi-dipole analysis software (brain electric source analysis, BESA) on the contralateral hemisphere to the stimulated side [24].

The temporal waveforms of the equivalent current dipoles (ECDs) induced by a tactile stimulus to the dorsum of the left hand are shown in Fig. 3.2. The first source for activity that peaked at 21 and 30 ms (the onset latency, 14.4 ms) was located in the posterior bank of the central sulcus, corresponding to area 3b. The second source for activity in SI that peaked at 25 and 34 ms (the onset latency, 18.0 ms) was in a location that was more medial, superior, and posterior to the area 3b source, suggesting that it originated from area 1. The third source in the posterior parietal cortex (PPC) that peaked at 28 and 37 ms (the onset latency, 22.4 ms) was in its caudal region around the intraparietal sulcus, possibly corresponding to area 5. Although we could not detect the activity of area 2, which is located in the crown of the postcentral gyrus, because MEG cannot detect activities from radial dipoles, serial processing elicited from area 3b to areas 1 and 5 was congruent with the anatomical findings suggesting hierarchical somatic processing via cortico-cortical connections in the postcentral gyrus [25]. The onset latency of the area 4 source (14.5 ms) was not different from that of the area 3b source, indicating that

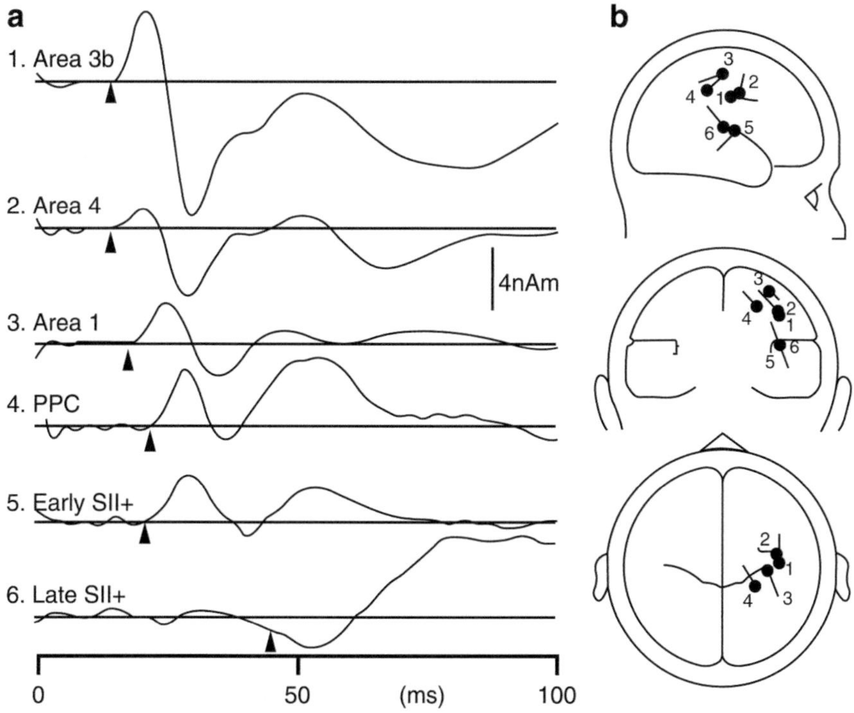

Fig. 3.2 Cortical activities following stimulation of the dorsum of the *left hand*. Grand-averaged waveforms of the time course of each equivalent current dipole (ECD) (**a**) and the mean location and orientation of each source (**b**). *Arrowheads* indicate the mean onset latency (This figure has been reproduced from Inui et al. [24] with the permission of the Oxford University Press)

the initial activation of area 4 was independent of the activities of SI, which was also consistent with anatomical findings that projections from area 3b to area 4 were absent or very weak [26, 27]. The area 4 source may have been driven by thalamic inputs, whereas the later activities of the source may come from area 1 or 2.

Regarding the lateral sulcus region, we detected two sources (early and late SII). The peak latencies of the early SII source were 30 and 40 ms, while those of the late SII source were 56 and 90 ms, which were very similar to the two separable components, N30op and N60/90, in intracranial EEG studies [28, 29]. The onset latency of the early SII (21.7 ms) was longer than the area 3b source, suggesting serial mode processing in SI and SII. This serial processing was consistent with previous findings [30, 31]. On the other hand, the hierarchical equivalence of SI and SII was detected in animal [32, 33] and human studies [34]. However, we might miss the detection of this parallel processing because the initial activity of SII was very weak.

3.2.3 Roles of Bilateral SII Regions

In addition to the contralateral temporal region (mainly SII), the ipsilateral region was activated by one-side somatic stimulation. These SII activities were more sensitive to higher-order functions such as attention, decision making, and objective recognition than SI activity [35–41]. SII activity was enhanced in magnitude by centrifugal regulation (motor-related neural activities before the afferent inputs of muscles), whereas SI activity (mainly P35m) was reduced [42]. This finding indicated that SII plays a different functional role than SI in somatosensory-motor regulation.

We recently recorded automatic responses by sudden somatic changes and revealed that (1) change-related somatic responses were detected in SI and SII, and the magnitude of these responses was affected by the status of the sensory store and comparisons between preceding and new events [43]; (2) the somatosensory offset response was also clearly detected in bilateral SII, but it was less apparent in SI [44]; and (3) the administration of nicotine, which is known to have enhancing effects on attention and cognition, enhanced change-related ipsilateral SII activity [45]. These results indicated that SII has more important roles as a detector of sensory changes.

3.2.4 Possibilities for Further Applications

SEFs have provided us with further insights into cortical function. For example, the phenomenon in which cortical responses to a sensory stimulus are attenuated when the stimulus is repeated represents a valuable tool for evaluating function. Paired-pulse suppression, quantified by the ratio of the second response amplitude to the first, is often used as a marker of recovery function in the nervous system [15, 16]. The conventional theory of suppression was determined by the number of synapses or tactile discrimination skills because suppression was found to be reduced in the elderly [46–48]. We consider that suppression indicates not only a passive attenuation process, such as adaptation, fatigue, or a refractory period, but also the presence of an active inhibitory process because suppression was obtained even if the first stimulus was very weak [49, 50]. This is supported by our recent study in which the protective effects of the first stimulus were found to have functional consequences beyond simple suppression to the second response [51]. A certain group of patients such as those with schizophrenia have defective sensory inhibitory processes; therefore, these measurements may be useful as an evaluation of cortical inhibitory processes.

Recording the background rhythmic activities of high-frequency bands also enables us to provide new and important insights into the brain network related to somatosensory function. For example, a previous study demonstrated temporal binding between SI and SII by analyzing the oscillatory gamma-band activities of SEFs [52]. Other studies reported that even higher-frequency oscillation (>300 Hz) activities were observed over the contralateral SI and suggested that they may reflect a postsynaptic neural network in areas 3b and 1 [53, 54].

3.3 The Pain System

Difficulties are associated with evaluating the degree to which subjects feel pain sensations because they are derived from internal experiences. In the past 25 years, the development of noninvasive techniques to measure brain function has enabled us to directly examine the temporal processing of pain sensations. The nociceptive system can be activated in various ways, including chemical, thermal, electrical, and mechanical stimuli. Since each method has its own advantages and disadvantages, researchers have to select an appropriate device. A laser stimulus is one of the useful means to activate the nociceptive system and is often used in research and clinical testing [55]. However, the equipment needed is expensive, and laser beams cannot be applied to the same area repetitively. While an electrical stimulus is easy to use, it cannot selectively activate the nociceptive system.

To overcome these issues, we have developed a new device, an intraepidermal electrical stimulation (IES) device, using a concentric bipolar needle electrode (Fig. 3.3), and succeeded in stimulating cutaneous A-delta nociceptors without concomitantly activating A-beta mechanoreceptors [56–58]. The advantages of IES are the following: (1) nociceptive fiber terminals located in the epidermis and superficial layer of the dermis can be selectively stimulated; (2) insertion of the needle electrode causes no bleeding or visible damage to the skin; (3) it provides

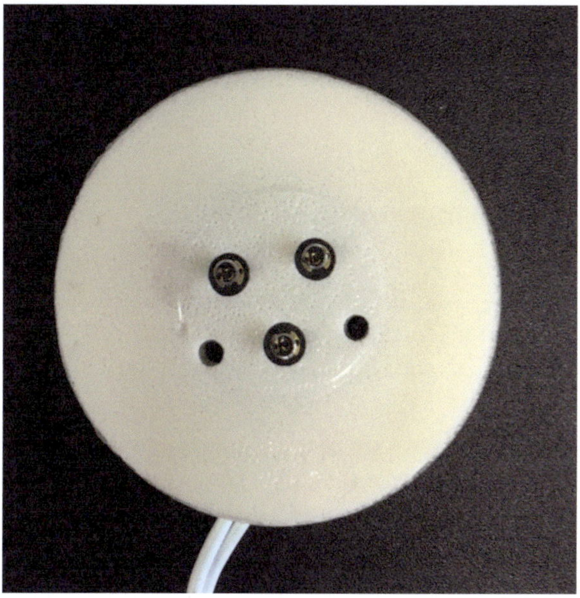

Fig. 3.3 An IES electrode (NM-980 W, Nihon Kohden Inc., Tokyo, Japan). It consists of three concentric bipolar electrodes with an outer ring, 1.3 mm in diameter, and an inner needle that protrudes 0.02 mm from the outer ring. A weak current passed through this electrode elicits a sharp pricking sensation mediated by signals ascending through A-delta fibers without concomitant tactile sensations

precisely time-locked stimuli and is easy to use, requiring no specialized skills; and (4) subjects feel less discomfort because it can evoke clear and selective cortical responses with a weak current [56–59]. In this section, we reviewed our previous studies using IES.

3.3.1 Serial Processing of A-Delta Nociceptor Signals

We investigated differences in temporal processing between IES and transcutaneous stimulation (TS) within SI and SII using MEG, as depicted in Fig. 3.4 [60, 61]. Similar to the previously discussed study (Sect. 3.2.2, [24]), TS elicited two distinct SI sources that originated from area 3b (peaking at 20 and 30 ms) and area 1 (peaking

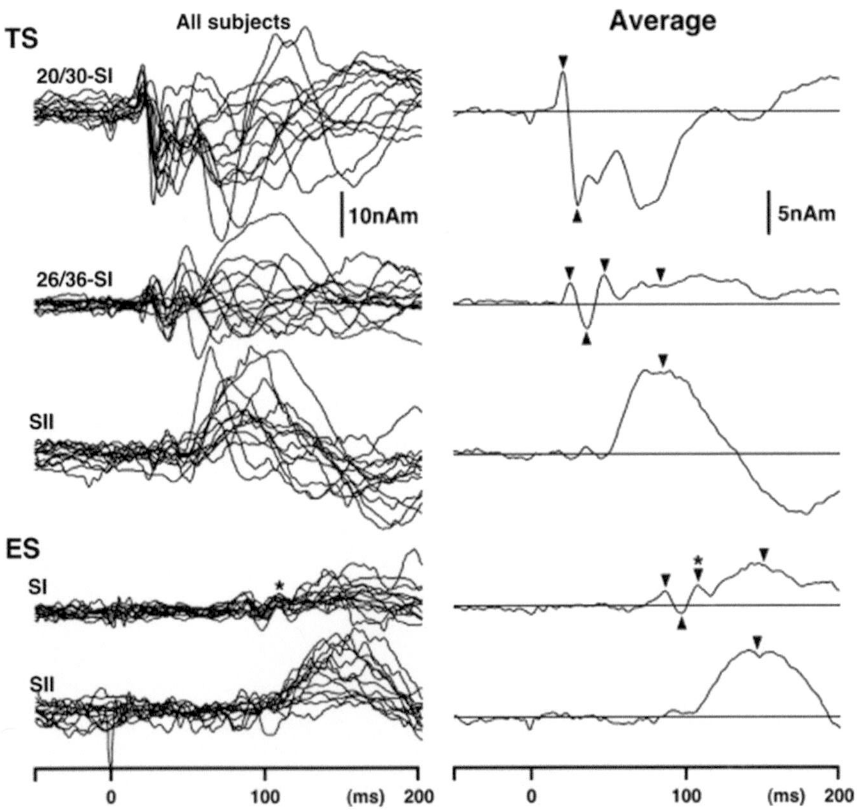

Fig. 3.4 Comparison of cortical activities between noxious (IES) and innocuous (TS) stimulations. Waveforms show the time course of the source strength of each dipole. TS activated area 3b (20/30-SI), area 1 (26/36-SI), and SII, whereas the activation of area 3b was absent for IES. *Arrowheads* indicate the mean peak latency of each component. Waveforms of each subject (*left*) and the grand-averaged ones (*right*) are shown (This figure has been reproduced from Inui et al. [61] with the permission of the John Wiley and Sons)

at 26, 36, and 49 ms) prior to SII activity (peaking at 87 ms). On the other hand, we found one weak SI (area 1) response (peaking at 88, 98, and 109 ms) and strong SII activity (peaking at 148 ms) following IES. This result implied that the sequential processing pattern in area 1 to SII was similar between the tactile and pain systems. Furthermore, the conduction time from SI to SII was similar, which supported both signals reaching SII from SI via similar pathways. Although previous studies failed to identify initial SI activity following noxious stimuli because the components were very weak and easily overlooked [56, 62], we demonstrated for the first time serial pain processing from SI to SII using IES. However, activity in area 3b following IES was absent, which may have due to a functional difference in SI between the two systems. For example, the main role of SI in pain processing is to localize the stimulated site, whereas tactile processing requires the execution of elaborated capacities.

3.3.2 Serial Processing of C Nociceptor Signals

It is more difficult to selectively stimulate C nociceptors than cutaneous A-beta mechanoreceptors and A-delta nociceptors. This is mainly due to its high electrical threshold, which inevitably leads to the concomitant activation of thicker fibers. Therefore, the selective activation of C nociceptors by an electrical stimulus was previously deemed impossible. However, we succeeded by using IES [63] based on previous findings: an anodal current was theoretically effective at stimulating cutaneous fibers running vertical to the skin surface [64], a pulse with a longer duration was necessary to stimulate thinner fibers [65], and the marked summation of C fiber impulses was required to produce a painful sensation [66]. Our IES parameters for stimulating C nociceptors were as follows: (1) the anode was the inner needle, and the cathode was the outer ring (opposite setting to the stimulus for A-delta nociceptors); (2) a train of three sharp triangular electrical pulses, with a short interstimulus interval (20 ms), were applied; and (3) three electrodes separated by 10 mm were used for augmentation [63]. By applying this method to the hand and forearm, we observed evoked potentials with the mean P1 latencies of 1,007 and 783 ms, respectively. The peripheral conduction velocity was 1.5 ± 0.7 m/s, which was consistent with previous findings using a laser stimulus [67]. An MEG study revealed activation in the contralateral SI and bilateral opercular regions (mainly SII) ([59]; see in Fig. 3.5).

3.3.3 Further Possibilities of Applying IES

As described above, IES is advantageous in terms of cost, noninvasiveness, convenience, and selectivity. By using these advantages, we believe that IES will provide new insights into the physiology of the nociceptive system. For example, we

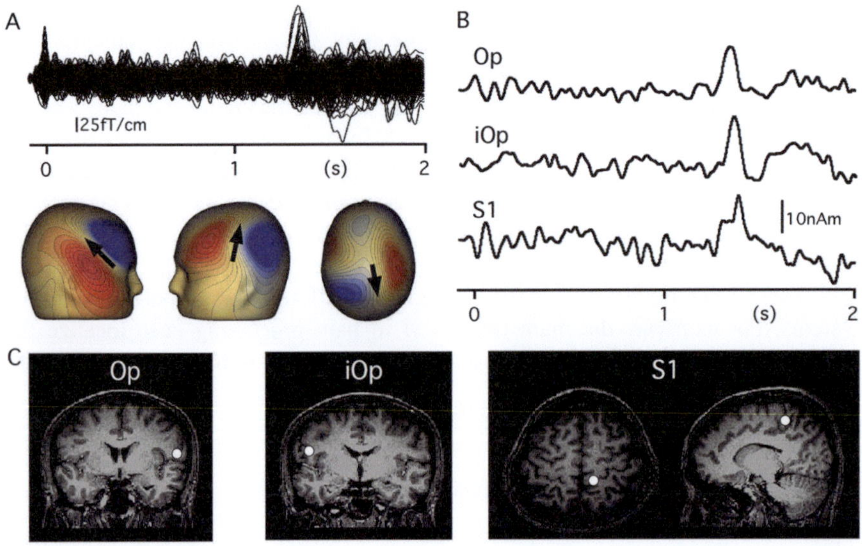

Fig. 3.5 Typical magnetic responses elicited by the stimulation of C nociceptors. (**a**) Superimposed waveforms evoked by IES on the dorsum of the *left foot* and isocontour maps at three selected latencies. Areas surrounded by *red* and *blue* contour lines in the isocontour map show the outflux and influx of magnetic fields. (**b**) Source strength waveforms of ECDs estimated in the opercular region (Op) and SI. (**c**) The location of each ECD superimposed on the subject's own MR images (This figure has been reproduced from Motogi et al. [59] with the permission of Elsevier)

confirmed a somatotopic map in SI in the nociceptive system by applying IES to various body surface areas including the limbs, neck, and face [68]. IES can be easily applied to any surface of the body. This finding indicates that humans have a high capacity for localizing noxious stimuli.

In clinical studies, IES may be suitable for evaluating somatic dysfunctions such as small fiber neuropathy. We previously demonstrated the utility of IES for assessing peripheral nerve function by using a lidocaine-induced neuropathy model [69, 70]. Decrements in the evoked potential and subjective pain rating score were observed following the transdermal application of lidocaine in stimulations of A-delta and C nociceptors. These findings indicate that IES has the potential to be used as a clinical tool for elucidating the pathophysiology of patients with small fiber neuropathy.

3.4 The Itch System

Itching is a sensation induced by pruritogens such as histamine. The signals responsible for this sensation are transmitted to the brain through C fibers and the spinothalamic tract [71]. However, the subtype of C fibers is different from that for pain sensations in that it is insensitive to mechanical and heat stimuli [72]. Therefore, itching is not a weak pain.

In the last 20 years, researchers have attempted to clarify which brain regions are involved in itch sensations and identified broad brain regions, such as the insula, cingulate cortex, prefrontal cortex, inferior parietal cortex, (pre)supplementary motor cortex, somatosensory cortex, basal ganglia, and cerebellum [73–78]. However, the temporal aspects of processing itching in the brain have not been elucidated because histamine induces a long-lasting and dull itch sensation, which is not appropriate for measuring stimulus-locked brain activity. To overcome this issue, we have developed a device to induce itch sensations with an electrical stimulus of constant-current square pulses (20 pulses; duration, 2 ms; frequency, 50 Hz; intensity, 0.1–0.7 mA) to the skin [79].

This device enabled us to elicit clear MEG responses, as shown in Fig. 3.6 [80]. Typical magnetic responses were obtained in three distinct regions: the right/left front-temporal and posterior part of the centro-parietal regions. The sources of these responses were estimated to be located in the bilateral SII/insula and precuneus. Precuneus activity was unique for itch sensations not reported in other pain and tactile studies, implying a difference from the pain system. The peak latency of ipsilateral SII/insula (785 ± 76 ms) was significantly longer than that for the contralateral source (740 ± 76 ms). This difference was associated with callosal transmission. The latency of the precuneus did not significantly differ from those of the two SII/insula sources (peak latency, 783 ± 76 ms). Considering this result and anatomical findings that SII/insula receives projections from the lateral thalamus [71] while the precuneus receives them from the medial thalamus [81], itching appears to have two independent neural circuits associated with the precuneus and SI-SII processing.

3.5 Conclusion and Future Prospects

We reviewed the findings of neuromagnetic responses elicited by A-beta, A-delta, and C fiber-mediated somatic sensations. The development of stimulating devices in the last two decades has provided us with prospective insights into the serial and parallel processing of sensory information. Since temporal and spatial resolution is superior in MEG, it allows us to understand the physiology of the somatosensory system and elucidate the pathology of somatosensory dysfunctions. We believe that the hierarchical processing introduced here will be the foundation of further basic and clinical studies.

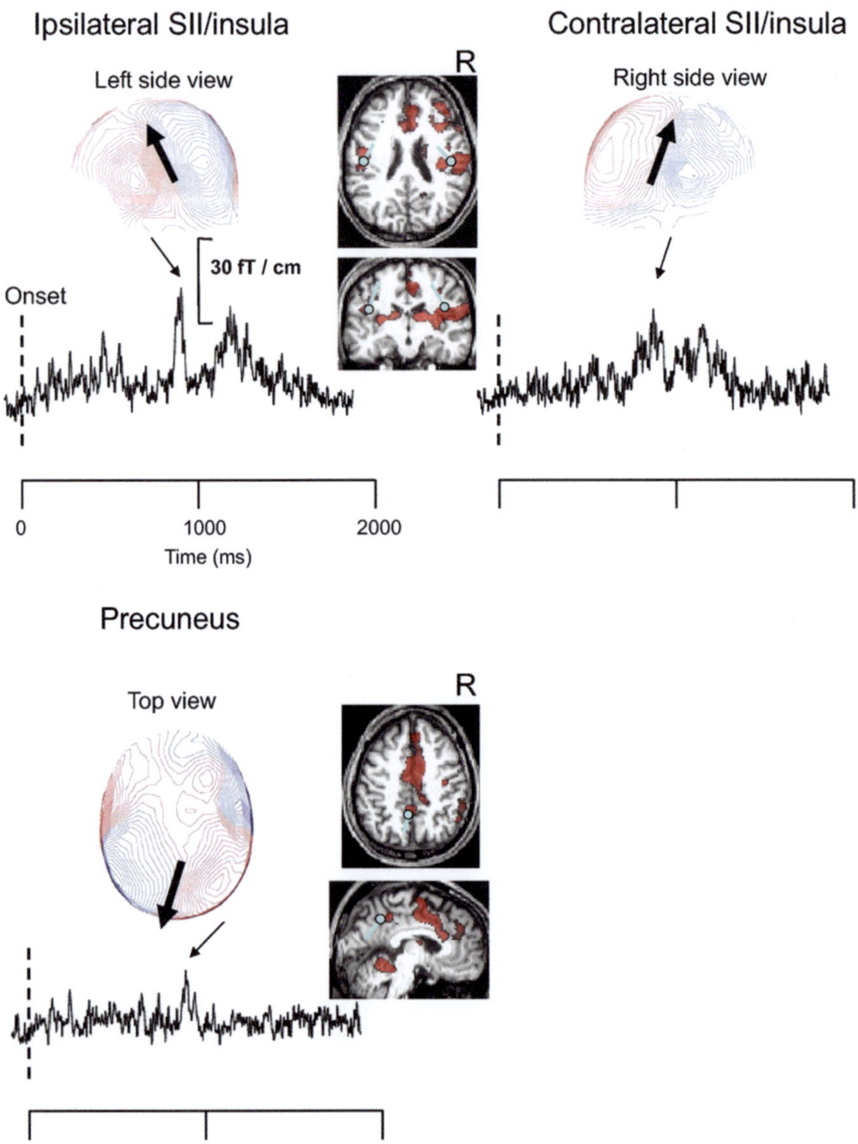

Fig. 3.6 MEG responses elicited by an electrical itch stimulus to the *left hand*. Typical images of the three source locations (*blue circles and bars*), isocontour maps of magnetic fields, and typical area mean signals are shown. *Red regions* in MR images are areas that were significantly activated by the itch stimulus in the functional MRI experiment. Areas surrounded by *red* and *blue* contour lines in the isocontour map show the outflux and influx of magnetic fields. *Thin arrows* indicate the latency point at which isocontour maps and dipoles were obtained (This figure has been reproduced from Mochizuki et al. [80])

Acknowledgments We are very grateful to Dr. Hideki Mochizuki for his suggestions and comments on itch study. This study is supported by COI STREAM of MEXT, Japan.

References

1. Hämäläinen MS. Magnetoencephalography: a tool for functional brain imaging. Brain Topogr. 1992;5:95–102.
2. Kakigi R, Hoshiyama M, Shimojo M, Naka D, Yamasaki H, Watanabe S, Xiang J, Maeda K, Lam K, Itomi K, Nakamura A. The somatosensory evoked magnetic fields. Prog Neurobiol. 2000;61:495–523. doi:10.1016/S0301-0082(99)00063-5.
3. Ishibashi H, Tobimatsu S, Shigeto H, Morioka T, Yamamoto T, Fukui M. Differential interaction of somatosensory inputs in the human primary sensory cortex: a magnetoencephalographic study. Clin Neurophysiol. 2000;111:1095–102. doi:10.1016/S1388-2457(00)00266-2.
4. Brenner D, Lipton J, Kaufman L, Williamson SJ. Somatically evoked magnetic fields of the human brain. Science. 1978;199:81–3.
5. Nakamura A, Yamada T, Goto A, Kato T, Ito K, Abe Y, Kachi T, Kakigi R. Somatosensory homunculus as drawn by MEG. Neuroimage. 1998;7:377–86. doi:10.1006/nimg.1998.0332.
6. Wikström H, Roine RO, Salonen O, Lund KB, Salli E, Ilmoniemi RJ, Aronen HJ, Huttunen J. Somatosensory evoked magnetic fields from the primary somatosensory cortex (SI) in acute stroke. Clin Neurophysiol. 1999;110:916–23. doi:10.1016/S1388-2457(99)00026-7.
7. Wood CC, Cohen D, Cuffin BN, Yarita M, Allison T. Electrical sources in human somatosensory cortex: identification by combined magnetic and potential recordings. Science. 1985; 227:1051–3. doi:10.1126/science.3975600.
8. Kawamura T, Nakasato N, Seki K, Kanno A, Fujita S, Fujiwara S, Yoshimoto T. Neuromagnetic evidence of pre- and post-central cortical sources of somatosensory evoked responses. Electroencephalogr Clin Neurophysiol. 1996;100:44–50.
9. Hari R, Forss N. Magnetoencephalography in the study of human somatosensory cortical processing. Philos Trans R Soc Lond B Biol Sci. 1999;354:1145–54. doi:10.1098/rstb.1999.0470.
10. Huttunen J, Tolvanen H, Heinonen E, Voipio J, Wikström H, Ilmoniemi RJ, Hari R, Kaila K. Effects of voluntary hyperventilation on cortical sensory responses. Electroencephalographic and magnetoencephalographic studies. Exp Brain Res. 1999;125:248–54.
11. Huttunen J, Komssi S, Lauronen L. Spatial dynamics of population activities at S1 after median and ulnar nerve stimulation revisited: an MEG study. Neuroimage. 2006;32: 1024–31. doi:10.1016/j.neuroimage.2006.04.196.
12. Huttunen J, Pekkonen E, Kivisaari R, Autti T, Kahkonen S. Modulation of somatosensory evoked fields from SI and SII by acute GABA A-agonism and paired-pulse stimulation. Neuroimage. 2008;40:427–34. doi:10.1016/j.neuroimage.2007.12.024.
13. Wikström H, Huttunen J, Korvenoja A, Virtanen J, Salonen O, Aronen H, Ilmoniemi RJ. Effects of interstimulus interval on somatosensory evoked magnetic fields (SEFs): a hypothesis concerning SEF generation at the primary sensorimotor cortex. Electroencephalogr Clin Neurophysiol. 1996;100:479–87.
14. Lim M, Kim JS, Chung CK. Modulation of somatosensory evoked magnetic fields by intensity of interfering stimuli in human somatosensory cortex: an MEG study. Neuroimage. 2012;61: 660–9. doi:10.1016/j.neuroimage.2012.04.003.
15. Hoshiyama M, Kakigi R. Two evoked responses with different recovery functions in the primary somatosensory cortex in humans. Clin Neurophysiol. 2001;112:1334–42.
16. Hoshiyama M, Kakigi R. New concept for the recovery function of short-latency somatosensory evoked cortical potentials following median nerve stimulation. Clin Neurophysiol. 2002;113:535–41.

17. Cohen LG, Starr A. Localization, timing and specificity of gating of somatosensory evoked potentials during active movement in man. Brain. 1987;110:451–67.
18. Kakigi R, Koyama S, Hoshiyama M, Watanabe S, Shimojo M, Kitamura Y. Gating of somatosensory evoked responses during active finger movements magnetoencephalographic studies. J Neurol Sci. 1995;128:195–204.
19. Forss N, Jousmaki V. Sensorimotor integration in human primary and secondary somatosensory cortices. Brain Res. 1998;781:259–67.
20. Murase N, Kaji R, Shimazu H, Katayama-Hirota M, Ikeda A, Kohara N, Kimura J, Shibasaki H, Rothwell JC. Abnormal premovement gating of somatosensory input in writer's cramp. Brain. 2000;123:1813–29.
21. Sugawara K, Onishi H, Yamashiro K, Kojima S, Miyaguchi S, Kirimoto H, Tsubaki A, Tamaki H, Shirozu H, Kameyama S. The effect of anodal transcranial direct current stimulation over the primary motor or somatosensory cortices on somatosensory evoked magnetic fields. Clin Neurophysiol. 2015;126:60–7. doi:10.1016/j.clinph.2014.04.014.
22. Huttunen J, Jaaskelainen IP, Hirvonen J, Kaakkola S, Ilmoniemi RJ, Pekkonen E. Scopolamine reduces the P35m and P60m deflections of the human somatosensory evoked magnetic fields. Neuroreport. 2001;12:619–23.
23. Lauronen L, Huttunen J, Kirveskari E, Wikström H, Sainio K, Autti T, Santavuori P. Enlarged SI and SII somatosensory evoked responses in the CLN5 form of neuronal ceroid lipofuscinosis. Clin Neurophysiol. 2002;113:1491–500.
24. Inui K, Wang X, Tamura Y, Kaneoke Y, Kakigi R. Serial processing in the human somatosensory system. Cereb Cortex. 2004;14:851–7. doi:10.1093/cercor/bhh043.
25. Iwamura Y. Hierarchical somatosensory processing. Curr Opin Neurobiol. 1998;8:522–8.
26. Vogt BA, Pandya DN. Cortico-cortical connections of somatic sensory cortex (areas 3, 1 and 2) in the rhesus monkey. J Comp Neurol. 1978;177:179–91. doi:10.1002/cne.901770202.
27. Burton H, Fabri M. Ipsilateral intracortical connections of physiologically defined cutaneous representations in areas 3b and 1 of macaque monkeys: projections in the vicinity of the central sulcus. J Comp Neurol. 1995;355:508–38. doi:10.1002/cne.903550404.
28. Frot M, Mauguiere F. Timing and spatial distribution of somatosensory responses recorded in the upper bank of the sylvian fissure (SII area) in humans. Cereb Cortex. 1999;9:854–63.
29. Barba C, Frot M, Mauguiere F. Early secondary somatosensory area (SII) SEPs. Data from intracerebral recordings in humans. Clin Neurophysiol. 2002;113:1778–86.
30. Hari R, Reinikainen K, Kaukoranta E, Hämäläinen M, Ilmoniemi R, Penttinen A, Salminen J, Teszner D. Somatosensory evoked cerebral magnetic fields from SI and SII in man. Electroencephalogr Clin Neurophysiol. 1984;57:254–63.
31. Disbrow E, Roberts T, Poeppel D, Krubitzer L. Evidence for interhemispheric processing of inputs from the hands in human S2 and PV. J Neurophysiol. 2001;85:2236–44.
32. Zhang HQ, Murray GM, Coleman GT, Turman AB, Zhang SP, Rowe MJ. Functional characteristics of the parallel SI- and SII-projecting neurons of the thalamic ventral posterior nucleus in the marmoset. J Neurophysiol. 2001;85:1805–22.
33. Zhang HQ, Zachariah MK, Coleman GT, Rowe MJ. Hierarchical equivalence of somatosensory areas I and II for tactile processing in the cerebral cortex of the marmoset monkey. J Neurophysiol. 2001;85:1823–35.
34. Karhu J, Tesche CD. Simultaneous early processing of sensory input in human primary (SI) and secondary (SII) somatosensory cortices. J Neurophysiol. 1999;81:2017–25.
35. Mima T, Nagamine T, Nakamura K, Shibasaki H. Attention modulates both primary and second somatosensory cortical activities in humans: a magnetoencephalographic study. J Neurophysiol. 1998;80:2215–21.
36. Burton H, Abend NS, MacLeod AM, Sinclair RJ, Snyder AZ, Raichle ME. Tactile attention tasks enhance activation in somatosensory regions of parietal cortex: a positron emission tomography study. Cereb Cortex. 1999;9:662–74.

37. Steinmetz PN, Roy A, Fitzgerald PJ, Hsiao SS, Johnson KO, Niebur E. Attention modulates synchronized neuronal firing in primate somatosensory cortex. Nature. 2000;404:187–90. doi:10.1038/35004588.
38. Pruett Jr JR, Sinclair RJ, Burton H. Neural correlates for roughness choice in monkey second somatosensory cortex (SII). J Neurophysiol. 2001;86:2069–80.
39. Kida T, Nishihira Y, Wasaka T, Nakata H, Sakamoto M. Differential modulation of temporal and frontal components of the somatosensory N140 and the effect of interstimulus interval in a selective attention task. Brain Res Cogn Brain Res. 2004;19:33–9. doi:10.1016/j.cogbrainres.2003.10.016.
40. Nakata H, Inui K, Nishihira Y, Hatta A, Sakamoto M, Kida T, Wasaka T, Kakigi R. Effects of a go/nogo task on event-related potentials following somatosensory stimulation. Clin Neurophysiol. 2004;115:361–8. doi:10.1016/j.clinph.2003.09.013.
41. Chapman CE, el Meftah M. Independent controls of attentional influences in primary and secondary somatosensory cortex. J Neurophysiol. 2005;94:4094–107. doi:10.1152/jn.00303.2005.
42. Kida T, Wasaka T, Inui K, Akatsuka K, Nakata H, Kakigi R. Centrifugal regulation of human cortical responses to a task-relevant somatosensory signal triggering voluntary movement. Neuroimage. 2006;32:1355–64. doi:10.1016/j.neuroimage.2006.05.015.
43. Otsuru N, Inui K, Yamashiro K, Urakawa T, Keceli S, Kakigi R. Effects of prior sustained tactile stimulation on the somatosensory response to the sudden change of intensity in humans: an magnetoencephalography study. Neuroscience. 2011;182:115–24. doi:10.1016/j.neuro science.2011.03.019.
44. Yamashiro K, Inui K, Otsuru N, Kida T, Kakigi R. Somatosensory off-response in humans: an MEG study. Neuroimage. 2009;44:1363–8. doi:10.1016/j.neuroimage.2008.11.003.
45. Kodaira M, Wasaka T, Motomura E, Tanii H, Inui K, Kakigi R. Effects of acute nicotine on somatosensory change-related cortical responses. Neuroscience. 2013;229:20–6. doi:10.1016/j.neuroscience.2012.10.060.
46. Lenz M, Tegenthoff M, Kohlhaas K, Stude P, Höffken O, Gatica Tossi MA, Kalisch T, Kowalewski R, Dinse HR. Increased excitability of somatosensory cortex in aged humans is associated with impaired tactile acuity. J Neurosci. 2012;32:1811–6. doi:10.1523/JNEUROSCI.2722-11.2012.
47. Cheng CH, Lin YY. Aging-related decline in somatosensory inhibition of the human cerebral cortex. Exp Brain Res. 2013;226:145–52. doi:10.1007/s00221-013-3420-9.
48. Goto S, Fujisawa Y, Uemura JI, Yamada S, Hoshiyama M, Hirayama M. Disinhibitory shift of recovery curve of somatosensory-evoked response in elderly: a magnetoencephalographic study. Clin Neurophysiol. 2014. doi:10.1016/j.clinph.2014.09.018.
49. Wühle A, Mertiens L, Rüter J, Ostwald D, Braun C. Cortical processing of near-threshold tactile stimuli: an MEG study. Psychophysiology. 2010;47:523–34. doi:10.1111/j.1469-8986.2010.00964.x.
50. Wühle A, Preissl H, Braun C. Cortical processing of near-threshold tactile stimuli in a paired-stimulus paradigm – an MEG study. Eur J Neurosci. 2011;34:641–51. doi:10.1111/j.1460-9568.2011.07770.x.
51. Nakagawa K, Inui K, Yuge L, Kakigi R. Inhibition of somatosensory-evoked cortical responses by a weak leading stimulus. Neuroimage. 2014;101:416–24. doi:10.1016/j.neuroimage.2014.07.035.
52. Hagiwara K, Okamoto T, Shigeto H, Ogata K, Somehara Y, Matsushita T, Kira J, Tobimatsu S. Oscillatory gamma synchronization binds the primary and secondary somatosensory areas in humans. Neuroimage. 2010;51:412–20. doi:10.1016/j.neuroimage.2010.02.001.
53. Hashimoto I, Mashiko T, Imada T. Somatic evoked high-frequency magnetic oscillations reflect activity of inhibitory interneurons in the human somatosensory cortex. Electroencephalogr Clin Neurophysiol. 1996;100:189–203.

54. Ozaki I, Hashimoto I. Exploring the physiology and function of high-frequency oscillations (HFOs) from the somatosensory cortex. Clin Neurophysiol. 2011;122:1908–23. doi:10.1016/j.clinph.2011.05.023.
55. Treede RD, Lorenz J, Baumgärtner U. Clinical usefulness of laser-evoked potentials. Neurophysiol Clin. 2003;33:303–14. doi:10.1016/j.neucli.2003.10.009.
56. Inui K, Tran TD, Qiu Y, Wang X, Hoshiyama M, Kakigi R. Pain-related magnetic fields evoked by intra-epidermal electrical stimulation in humans. Clin Neurophysiol. 2002;113:298–304.
57. Inui K, Tran TD, Hoshiyama M, Kakigi R. Preferential stimulation of Adelta fibers by intra-epidermal needle electrode in humans. Pain. 2002;96:247–52.
58. Inui K, Kakigi R. Pain perception in humans: use of intraepidermal electrical stimulation. J Neurol Neurosurg Psychiatry. 2012;83:551–6. doi:10.1136/jnnp-2011-301484.
59. Motogi J, Kodaira M, Muragaki Y, Inui K, Kakigi R. Cortical responses to C-fiber stimulation by intra-epidermal electrical stimulation: an MEG study. Neurosci Lett. 2014;570:69–74. doi:10.1016/j.neulet.2014.04.001.
60. Inui K, Tran TD, Qiu Y, Wang X, Hoshiyama M, Kakigi R. A comparative magnetoencephalographic study of cortical activations evoked by noxious and innocuous somatosensory stimulations. Neuroscience. 2003;120:235–48.
61. Inui K, Wang X, Qiu Y, Nguyen BT, Ojima S, Tamura Y, Nakata H, Wasaka T, Tran TD, Kakigi R. Pain processing within the primary somatosensory cortex in humans. Eur J Neurosci. 2003;18:2859–66.
62. Kanda M, Nagamine T, Ikeda A, Ohara S, Kunieda T, Fujiwara N, Yazawa S, Sawamoto N, Matsumoto R, Taki W, Shibasaki H. Primary somatosensory cortex is actively involved in pain processing in human. Brain Res. 2000;853:282–9.
63. Otsuru N, Inui K, Yamashiro K, Miyazaki T, Ohsawa I, Takeshima Y, Kakigi R. Selective stimulation of C fibers by an Intra-Epidermal needle electrode in humans. Open Pain J. 2009;2:53–6.
64. Rettay F. Electrical nerve stimulation. New York: Springer; 1990.
65. Grill Jr WM, Mortimer JT. The effect of stimulus pulse duration on selectivity of neural stimulation. IEEE Trans Biomed Eng. 1996;43:161–6. doi:10.1109/10.481985.
66. Adriaensen H, Gybels J, Handwerker HO, Van Hees J. Suppression of C-fibre discharges upon repeated heat stimulation may explain characteristics of concomitant pain sensations. Brain Res. 1984;302:203–11.
67. Qiu Y, Fu Q, Wang X, Tran TD, Inui K, Iwase S, Kakigi R. Microneurographic study of C fiber discharges induced by CO2 laser stimulation in humans. Neuroscience Lett. 2003;353:25–8.
68. Omori S, Isose S, Otsuru N, Nishihara M, Kuwabara S, Inui K, Kakigi R. Somatotopic representation of pain in the primary somatosensory cortex (S1) in humans. Clin Neurophysiol. 2013;124:1422–30. doi:10.1016/j.clinph.2013.01.006.
69. Otsuru N, Inui K, Yamashiro K, Miyazaki T, Takeshima Y, Kakigi R. Assessing Adelta fiber function with lidocaine using intraepidermal electrical stimulation. J Pain. 2010;11:621–7. doi:10.1016/j.jpain.2009.10.001.
70. Kodaira M, Inui K, Kakigi R. Evaluation of nociceptive Adelta- and C-fiber dysfunction with lidocaine using intraepidermal electrical stimulation. Clin Neurophysiol. 2014;125:1870–7. doi:10.1016/j.clinph.2014.01.009.
71. Andrew D, Craig AD. Spinothalamic lamina I neurons selectively sensitive to histamine: a central neural pathway for itch. Nat Neurosci. 2001;4:72–7. doi:10.1038/82924.
72. Schmelz M, Schmidt R, Bickel A, Handwerker HO, Torebjork HE. Specific C-receptors for itch in human skin. J Neurosci. 1997;17:8003–8.
73. Hsieh JC, Hagermark O, Stahle-Backdahl M, Ericson K, Eriksson L, Stone-Elander S, Ingvar M. Urge to scratch represented in the human cerebral cortex during itch. J Neurophysiol. 1994;72:3004–8.
74. Darsow U, Drzezga A, Frisch M, Munz F, Weilke F, Bartenstein P, Schwaiger M, Ring J. Processing of histamine-induced itch in the human cerebral cortex: a correlation analysis with

dermal reactions. J Invest Dermatol. 2000;115:1029–33. doi:10.1046/j.1523-1747.2000.00193.x.

75. Drzezga A, Darsow U, Treede RD, Siebner H, Frisch M, Munz F, Weilke F, Ring J, Schwaiger M, Bartenstein P. Central activation by histamine-induced itch: analogies to pain processing: a correlational analysis of O-15 H2O positron emission tomography studies. Pain. 2001;92:295–305.
76. Mochizuki H, Sadato N, Saito DN, Toyoda H, Tashiro M, Okamura N, Yanai K. Neural correlates of perceptual difference between itching and pain: a human fMRI study. Neuroimage. 2007;36:706–17. doi:10.1016/j.neuroimage.2007.04.003.
77. Leknes SG, Bantick S, Willis CM, Wilkinson JD, Wise RG, Tracey I. Itch and motivation to scratch: an investigation of the central and peripheral correlates of allergen- and histamine-induced itch in humans. J Neurophysiol. 2007;97:415–22. doi:10.1152/jn.00070.2006.
78. Schneider G, Stander S, Burgmer M, Driesch G, Heuft G, Weckesser M. Significant differences in central imaging of histamine-induced itch between atopic dermatitis and healthy subjects. Eur J Pain. 2008;12:834–41. doi:10.1016/j.ejpain.2007.12.003.
79. Mochizuki H, Inui K, Yamashiro K, Ootsuru N, Kakigi R. Itching-related somatosensory evoked potentials. Pain. 2008;138:598–603. doi:10.1016/j.pain.2008.02.017.
80. Mochizuki H, Inui K, Tanabe HC, Akiyama LF, Otsuru N, Yamashiro K, Sasaki A, Nakata H, Sadato N, Kakigi R. Time course of activity in itch-related brain regions: a combined MEG-fMRI study. J Neurophysiol. 2009;102:2657–66. doi:10.1152/jn.00460.2009.
81. Schmahmann JD, Pandya DN. Anatomical investigation of projections from thalamus to posterior parietal cortex in the rhesus monkey: a WGA-HRP and fluorescent tracer study. J Comp Neurol. 1990;295:299–326. doi:10.1002/cne.902950212.

Clinical Applications

4

Hideaki Onishi and Shigeki Kameyama

Abstract

Magnetoencephalography is sensitive primarily to current sources tangential to the skull. Therefore, currents generated in area 3b of primly somatosensory cortex (S1) and area 4 of primary motor cortex located on the posterior and anterior banks of the central sulcus, respectively, are easily detected. The SEFs generated by peripheral mixed nerve stimulation (e.g., median nerve) have been widely used to investigate the physiology of normal somatosensory cortical processing. In this chapter, we describe various SEF components elicited by median nerve simulation and their modulation by stimulus intensity and frequency, interfering stimulation, movement, age, medication, and disease. In addition, we describe the characteristics of SEFs generated by stimulation of various nerves and body parts and of those generated by different stimulus modalities, for example, stimulation by transcutaneous electrical currents, pneumatics, brushes, pins driven by piezoelectric actuators, lasers, intraepidermal electrical currents, intramuscular motor point, and passive movement.

Keywords

Magnetoencephalography • Somatosensory evoked magnetic field • Primary somatosensory cortex • Somatosensory cortical processing

H. Onishi (✉)
Institute for Human Movement and Medical Sciences, Niigata University of Health and Welfare, 1398 Shimami-cho, Kita-ku, Niigata-City 950-3198, Japan
e-mail: onishi@nuhw.ac.jp

S. Kameyama
Nishi-Niigata Chuo National Hospital, Niigata-City, Japan

S. Tobimatsu, R. Kakigi (eds.), *Clinical Applications of Magnetoencephalography*,
DOI 10.1007/978-4-431-55729-6_4

4.1 SEF Following Median Nerve Stimulation

Somatosensory evoked magnetic field (SEF) is clearly observed over the sensorimotor cortex contralateral to the stimulated side. It is generally accepted that the peak latencies of prominent deflections occur at approximately 20 ms (N20m), 35 ms (P35m), and 60 ms (P60m) after median nerve stimulation at the wrist [1–10] (Fig. 4.1).

4.1.1 N20m

The earliest deflection of the SEF waveform peaks approximately at 20 ms (N20m) after the stimulation and is generally believed to reflect early excitatory postsynaptic potentials generated by activation of thalamocortical axons synapsing onto pyramidal neurons in area 3b of S1.

4.1.2 P35m

The second deflection of the SEF waveform peaks approximately at 35 ms (P35m) after stimulation. Unlike N20m, the source location of P35m is still disputed. Huttunen et al. (1987) concluded that both N20m and P35m were generated in the S1 hand area, but they suggested that P35m may be partly generated in the motor cortex [5]. Subsequently, Tiihonen et al. (1989) proposed the contribution of area 4 in the precentral cortex to P35m [11]. Alternatively, Fross et al. [7] and Kakigi et al. [6] indicated that the source locations of both N20m and P35m were in the S1 hand area, and Kakigi suggested that the short-latency deflections less than 40 ms were hybrid signals from activity in areas 3 and 1 of S1. Wikstrom et al. (1996) proposed that P35m reflects inhibitory postsynaptic potentials (IPSPs) in the postcentral cortex based on the similar recovery cycles [12]. On the other hand, Kawamura et al. [13] suggested a contribution of area 4 to P35m and so did Tiihonen et al. [11]. Huang et al. (2000) used a two-dipole model to estimate the source location of P35m and indicated that the source of P35m was located in both anterior and posterior banks of the central sulcus [10]. While the exact source location is unknown, the source of P35m is certainly more medial and anterior to that of N20m, with the current pointing posteriorly [5, 14–16].

4.1.3 P60m

The third deflection of the SEF waveform peaks approximately at 60 ms (P60m) after stimulation; the precise source location of P60m is uncertain. Huttunen et al. (2006) suggested that the source of P60m is located significantly posterior, and P60m may arise in part from the postcentral gyrus and sulcus, possibility areas 1 and 2 in S1 [14].

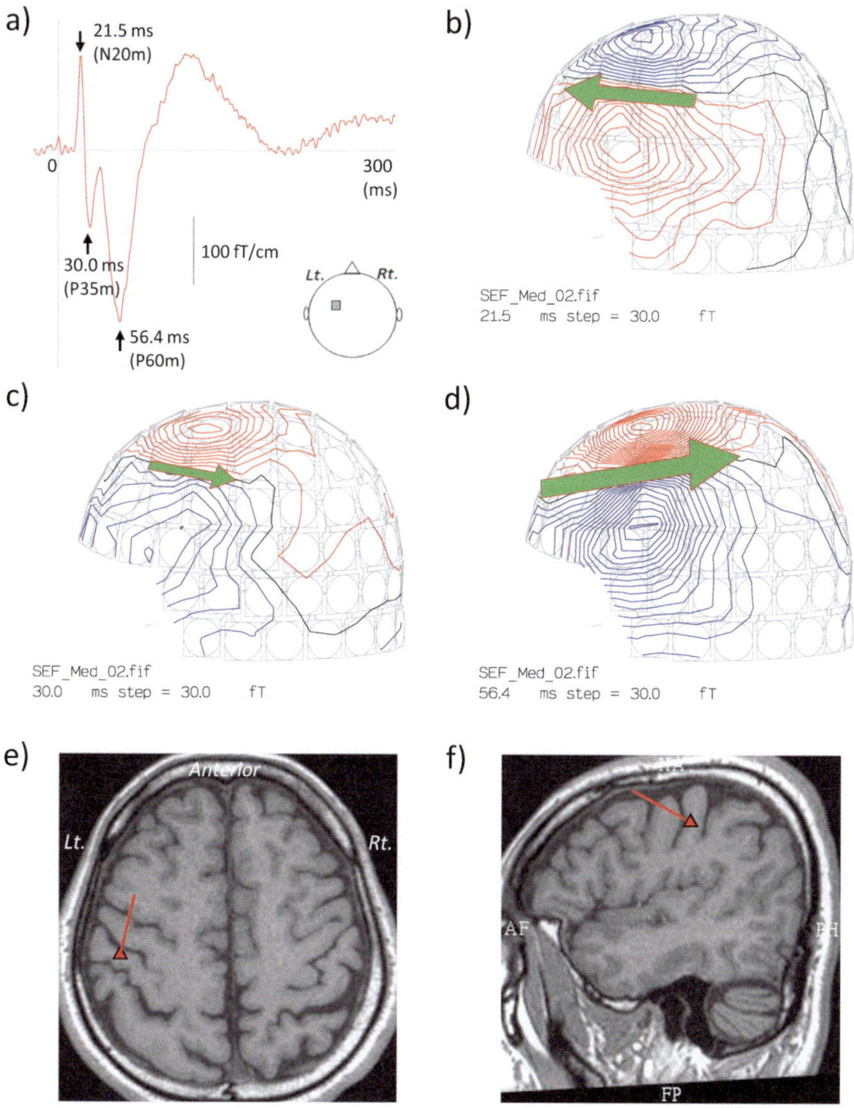

Fig. 4.1 SEF waveform, isocontour map, and equivalent current dipole (*ECD*) following right median nerve stimulation at the wrist. (**a**) SEF waveform obtained over the *left* sensorimotor cortex. (**b–d**) Isocontour map at the peak of (**b**) the earliest (21.5 ms), (**c**) the second (30.0 ms), and (**d**) the third (56.4 ms) deflections after stimulation. (**e, f**) ECD estimated at the peak of the earliest deflection (N20m) overlapping the axial (**e**) and sagittal planes (**f**) of the subject's MRI

4.1.4 Middle-Latency Deflections

Deflections longer than 70 ms after stimulation reflect activity of the secondary somatosensory cortex (S2) located in the upper wall of the sylvian fissure. Responses are observed bilaterally, with peak at approximately 80 ms after stimulation, but typically slightly earlier and stronger in the contralateral than the ipsilateral hemisphere [1, 2, 6–9, 17]. Middle-latency activity is also observed on the contralateral posterior parietal cortex (PPC), with peak at 70–140 ms after stimulation [7–9, 17].

4.2 SEF Following Transcutaneous Electrical Stimulation

Inui et al. (2004) examined the spatiotemporal properties of human somatosensory processing by collecting 800–900 artifact-free SEF responses to transcutaneous electrical stimulation (TS) at the dorsum of the hand using a 37-channel magneto-encephalography (MEG) system and a multidipole model for source localization [18]. They found activation in areas 3b, 4, 1, 5, and S2 of the contralateral hemisphere, with peaks of source activity at 20.8/29.8 ms (first peak/second peak) in area 3b, 21.6/30.4 ms in area 4, 24.5/33.5 ms in area 1, 28.1/36.2 in PPC, 29.6/40.0 ms in S2 (early response), and 55.8/89.5 ms in S2 (late response) after stimulation (Fig. 4.2).

4.3 Modulation of SEF Components

These SEF deflections are modulated by stimulus intensity, stimulus frequency, interfering stimuli, age, and several medications targeting synaptic transmission.

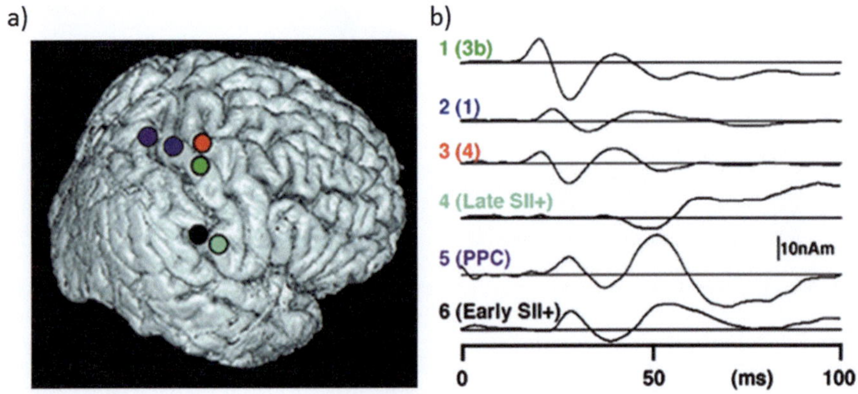

Fig. 4.2 Locations and time courses of source activities following transcutaneous electrical stimulation. (**a**) Magnetic source image: *Green circle*, area 3b; *dark blue circle*, area 1; *orange circle*, area 4; *light blue*, late secondary somatosensory cortex (S2); *purple circle*, PPC; *black circle*, early S2. (**b**) Time course of each source activity (Adapted from Inui et al. [18])

4.3.1 Effects of Stimulus Intensity

Some investigators have reported that the amplitude of SEF deflections following electrical stimulation is influenced by stimulus intensity, with both S1 and S2 responses increasing in amplitude with greater stimulus intensity [16, 19–24]. No cortical activity was observed with stimulus intensity lower than the subjective sensory threshold. Above the sensory threshold, S1 activity at N20m, P35m, and P60m increased with stimulus intensity up to strong motor threshold [20] or three times sensory threshold [24] and then saturated. The S2 response is saturated at lower stimulus intensity than the S1 response, reaching a maximum at motor threshold [20] or two times sensory threshold [24].

4.3.2 Effects of Stimulus Frequency and Repetitive Pulse

SEF waveforms are affected by stimulus frequency [interstimulus interval (ISI)] and number of stimuli. P35m and P60m are significantly reduced with decreasing ISI [4, 11, 12, 25]. Moreover, Hamada et al. [26] reported that the recovery of human S1 and S2 SEF signals generated by electrical stimulation of the median nerve (with low-pass filtering of 50 Hz) was differentially affected by ISI from 100 to 500 ms, with S1 P60m attenuation most marked at 100 ms ISI but showing rapid recovery at 300 ms ISI, while the S2 signal was attenuated at a longer ISI of 500 ms [26]. In addition, S1 and S2 responses decreased rapidly from the first to fifth, and those for third, fourth, or fifth pulse were significantly attenuated compared with first pulse, and there were no significant differences among the responses to the third, fourth, and fifth pulse [26]. On the other hand, Huttunen reported that the first few P35m responses to 15 median nerve stimulus pulses were markedly reduced at an ISI of 100 ms and remain attenuated for the duration of the train, whereas P60m exhibited no significant attenuation, and N20m amplitude did not differ between the first and fifth stimulus but decreased significantly by the last (15th) stimulus [27].

4.3.3 Effect of Interfering Stimuli

SEF amplitude is modulated not only by stimulus intensity and frequency but by the intensity of interfering stimuli applied between test stimuli. Lim et al. (2012) compared the effects of two to four weak, moderate, and strong interfering stimuli (ISI from 0.5 to 1.5 s) delivered to the median nerve between test pulses (ISI from 3 to 4 s) and found that the P35m was reduced by interfering stimuli of moderate intensity and P60m by both moderate and strong interfering stimuli, while N20m amplitude was insensitive [28].

4.3.4 Gating Effect

Many reports have shown that SEFs induced by electrical stimulation are modified by voluntary movement, a phenomenon termed movement-related somatosensory gating. This gating appears particularly significant for cortical responses underlying the P35m. Kakigi et al. (1995) reported that N20m and P35m were reduced during active movement of the stimulated hand compared with the control (no movement) condition [29]. P35m is reduced in amplitude not only during movement but also during the preparatory period prior to voluntary movement, while N20m is not altered [30, 31]. Recently, Huttunen et al. (2012) reported a selective and robust decrease of P35m in response to median nerve stimulation during movement of the stimulated hand, while N20m and P60m were unchanged [27, 32]. In addition, they found that P35m was slightly enhanced by movement of the unstimulated hand [32].

In addition to movement, concurrent brushing of the ipsilateral palm or finger using foam rubber or coarse cloth reduced P35m generated by median or digital nerve stimulation [33, 34], while N20m and P60m responses were unchanged [34]. However, Kakigi et al. (1996) reported that N20m and P60m as well as P35m responses to median nerve stimulation were attenuated by ipsilateral tactile interfering stimuli [35].

4.3.5 Age-Related Changes

Several studies have reported a significant positive correlation between N20m amplitude and age, while P35m and P60m appeared age independent [36–38]. The common explanation for this increase in N20m amplitude is decreased cortical inhibition in the elderly. However, Stephen et al. found that in addition to a larger N20m with age, P35m also decreased in the elderly (>64 years) compared with younger adults (20–29) [39].

4.3.6 Effects of Clinical Drugs

SEFs are also modulated by medication. For example, oral administration of the GABAa agonist lorazepam, a short-acting benzodiazepine, slightly increased the source activity of N20m [40, 41] but reduced those of P35m and P60m [41]. Oral administration of the dopamine type 2 receptor (D2) antagonist haloperidol also decreased P60m slightly but had no effect on N20m and P35m [42]. The authors concluded that this phenomenon was because of blockade of tonic dopaminergic input to the cortex, which led to a reduction in GABAb-mediated current. In another study, P35m and P60m were significantly reduced after oral administration of the muscarinic antagonist scopolamine, while N20m amplitude was unchanged [43].

4.3.7 Effects of Stroke

There are several reports concerning SEF after acute or post-acute monohemispheric stroke [44–54]. Gallien et al. (2003) reported that no SEF response recorded in some patients immediately after stroke onset, and SEF was still not recorded 1 year after stroke from patients with poor recovery [47]. Wikstrom et al. (1999) reported that SEF deflections were markedly attenuated or absent in patients with pure sensory stroke or sensorimotor paresis and that the attenuation of SEF deflections was more clearly correlated with two-point discrimination than with joint position or vibration sense [44, 45]. In addition, they suggested that the recovery of discriminative touch after stroke was parallel to the increase in P35m or P60m [44, 45]. Rossini et al. (1998) examined the interhemispheric differences in latencies, source strengths, and source locations of early SEF components following median nerve stimulation in stroke patients [49, 51–56] and reported that lesions involving cortico-subcortical areas receiving sensory input from the hand resulted in substantial asymmetry of source locations and SEF waveforms between the unaffected and affected hemispheres [52]. They also indicated that the cortical hand area was significantly larger in the affected than that in the unaffected hemisphere (hand extension) and that the mean extension of the hand area on the affected side was 39 mm in the subcortical lesion group and 27 mm in the cortical lesion group, compared with only 16 mm in the healthy control group [52]. Moreover, the excessive interhemispheric asymmetry of the hand area strongly correlated with a higher level of clinical recovery [49, 51].

4.4 Nerve- and Stimulus Site-Specific Variations

In addition to median nerve stimulation, many investigators have reported the cortical location and/or time course of source activity in response to stimulation of the ulnar nerve [5, 14, 57], posterior tibial nerve [17, 58–62], peroneal nerve [61, 63], sural nerve [58, 61], and femoral nerve [61, 62], as well as in response to surface stimulation of the hands and fingers [18, 22, 64], tongue [65–71], lips [17, 72, 73], intraoral mucosa [74], ear [75–77], face [78–80], shoulder and posterior scalp [81], trunk [82], sacral surface [83], and dorsal penis [60]. Indeed, such studies in aggregate have revealed the somatosensory homunculus. Narici et al. (1991) mapped the somatosensory homunculus generated by median, femoral, tibial, and pudendal nerve stimulation [62], while Hari et al. (1993) defined the cortical representations of the leg, hand, fingers, lip, and tongue [64]. The mean distance between the source location of the SEF and the location of the somatosensory cortex obtained invasively has been reported to be approximately 10 mm in adults [84].

While electrical stimulation is a useful tool for site-specific stimulation of the body surface, electrical noise can contaminate early components of the SEF generated by face or neck stimulation. Thus, other stimulus modalities have been used to define the somatosensory homunculus, such as surface mechanical

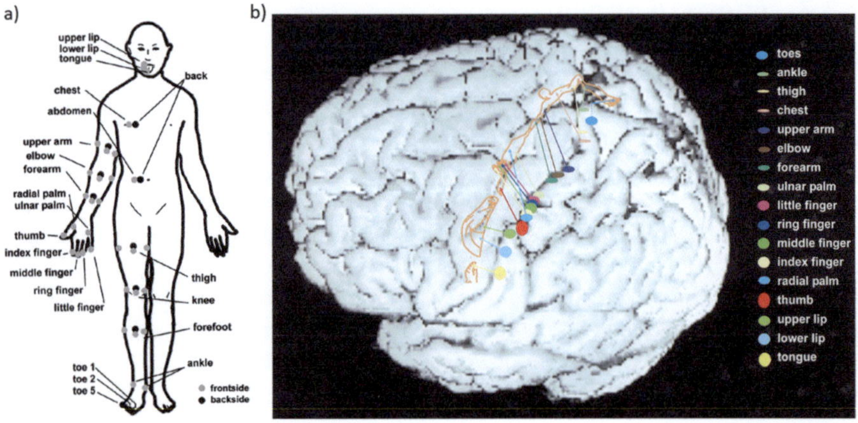

Fig. 4.3 Somatosensory homunculus as mapped by MEG. (**a**) Forty-three stimulated points of the right hemibody. The *gray circles* indicate stimulus points on the front side of the body, and *black circles* indicate stimulus points on the backside of the body. In the cases of arm and leg stimulation, lateral, medial, anterior, and posterior points were stimulated. (**b**) Detailed somatosensory receptive map represented by MEG. The three-dimensional brain image was reconstructed using the subject's MRI. Each receptive area located in the posterior bank of the central sulcus is projected onto the cortical surface. The size of each ellipse reflects the presumed size of the activated cortical area. Note that the receptive area for the toes is in the mesial side of the left hemisphere (Adapted from Nakamura et al. [86])

stimulation [85–88]. Indeed, Nakamura et al. defined the whole body somatosensory homunculus by SEF using mechanical stimulation [86] (Fig. 4.3).

The mean peak latencies of the first response following stimulation of various parts of the body are summarized in Table 4.1. The peak latency of the first response to median nerve and ulnar nerve stimulation at the wrist is almost similar and approximately 19–22 m. On the other hand, delay to first peak latency is 2–3 ms longer for the sural nerve (a pure sensory nerve) than for the posterior tibial nerve (a mixed nerve) despite a more proximal or equidistant site of stimulation [58, 61]. The longer latency for pure sensory nerve stimulation compared with that for mixed nerve stimulation is probably caused by an afferent volley in rapidly conducting muscle afferents. However, this difference may also be partly attributed to central factors [89].

4.5 Modality-Specific Characteristics

4.5.1 Mechanical Stimulation

SEF studies have used a variety of mechanical stimuli, including air puffs (pneumatic stimulation) (Fig. 4.4a) [78–80, 85, 86, 88, 90–97], brushes (Fig. 4.4b) [98–100], plastic pieces driven by airflow [101], mechanical pins driven by piezoelectric actuators (Fig. 4.4c) [22, 74, 102], and vibration buzzer (Piezo Undulative

Table 4.1 Peak latencies of the earliest SEF deflections following stimulation of various parts of the body

Nerve (N) or cutaneous stimulation	Stimulus parts	Mean peak latency (ms)
Median N.	Wrist	21.1 [14]
Ulnar N.	Wrist	21.6 [14]
Digital N.	2nd finger	25.1 [22]
Posterior portion of the shoulder	Shoulder	20.8 [81]
Mastoid	Mastoid	21.7 [81]
Cheek	Face	20.0 [81]
Lip (upper/lower)	Lip	21.9/21.3 [73]
Ear (helix/lobulus/tragus)	Ear	22.9/20.7/21.1 [75]
Femoral N.	Inguinal	26.6 [61]
Peroneal N.	Knee	30.0 [61]
Posterior tibial N.	Ankle	37.0–43 [17, 58, 60, 61]
Sural N.	Ankle	38.9–45 [58, 61]
Th6 dermatome	Trunk	25.3 [82]
Th8 dermatome	Trunk	26.2 [82]
Th10 dermatome	Trunk	26.3 [82]
Sacral surface	Trunk	30.2 [83]
Dorsal penile N.	Penis	63.8 [60]

Multifrequency Apparatus) [103]. Pneumatic stimulation is a useful tool to record the SEF in response to face or lip stimulation. However, the rise time for pneumatic stimulation is relatively long (above 10 ms), so the early phase of cortical activity cannot be measured as clearly compared with responses generated by electrical stimulation. In contrast, the rise time of mechanical pins driven by piezoelectric actuators is less than 1 ms, and the stimulus is precise and consistent. Therefore, this device is useful for investigating the early time course of cortical activity following life-like tactile sensation, tactile-off responses, and response to multiple stimuli distributed over a region for sensory paradigms like two-point discrimination [22, 102]. However, the earliest response (N20m) is smaller in amplitude and longer in latency following mechanical stimulation compared with electrical stimulation [22, 91], so we often cannot observe N20m or estimate the source location with acceptable goodness of fit following mechanical stimulation [22, 102]. This extra delay may be related to the time for skin indentation.

4.5.2 Laser Stimulation

Generally, acute pain is classified as first pain (sharp pain) associated with rapidly conducting myelinated Aδ fibers and second pain (burning pain) transmitted by slowly conducting unmyelinated C fibers. Yttrium aluminum garnet (YAG) and CO_2 lasers can stimulate both pathways together or selectively stimulate either Aδ- or C-nociceptive fibers [104–115]. Kakigi et al. (2003) reported the selective

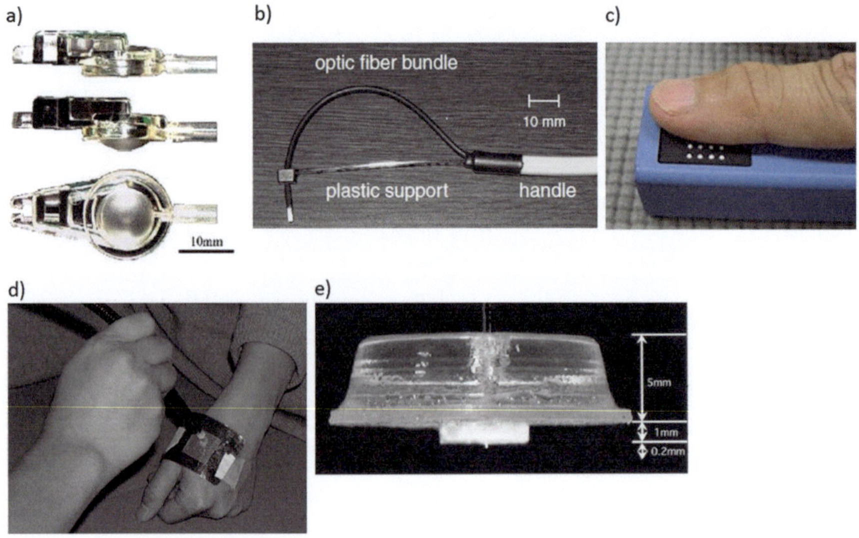

Fig. 4.4 Different types of stimulator. (**a**) Air-puff stimulator consisting of a small balloon (1 cm diameter) inflated by air pressure [78]. (**b**) Brush stimulator composed of a plastic handle, flexible support, and optic fiber bundle (diameter 1.5 mm) [99]. (**c**) Mechanical pins driven by piezoelectric actuators. The specifications of each pin are 1.3 mm in diameter and 0.8 mm in height for the protrusion. The distance between each pin is 2.4 mm [22]. (**d**) Laser stimulator. A *thin aluminum plate* (0.1 mm in depth, 40 mm in length, and 60 mm in width) was used as a spatial filter on the skin at the stimulation site. In a 25 × 25 mm square on this plate, *parallel lines* were drawn every 1 mm so that there were 26 × 26 intersections. A total of 676 (26 × 26) tiny holes were drilled at these intersections, each with a diameter of 0.4 mm, corresponding to an area of 0.125 mm^2 for each hole. The array of holes allowed the 2-mm laser beam to pass through 1–4 holes to reach the skin [111]. (**e**) Intraepidermal needle electrode. The needle electrode is a pushpin-type and consists of a holding plate, a stop device, and a needle tip 0.2 mm in length [120]

stimulation of C fibers of the skin in humans using a low-intensity CO_2 laser focused on a tiny area of the skin with a very thin aluminum plate perforated by tiny holes as a spatial filter (Fig. 4.4d) [111]. Laser stimulation delivered to a tiny skin area with low total energy is likely to activate predominantly C fibers [109, 111, 116]. Forss et al. (2005) compared cortical activation patterns and kinetics in response to specific Aδ- or C-fiber stimulation at the hand using a thulium-YAG laser and observed an earlier S2 response to Aδ-fiber stimulation, peaking at 165 ms versus 811 ms for C-fiber stimulation [114]. Activity of contralateral S1 is often not observed with laser stimulation [108, 110, 111, 114]. Kakigi et al. (2003), however, estimated both contralateral S1 and S2 activities using a two-dipole model during activation of C fibers by a CO_2 laser [111]. Nakata et al. (2008) investigated human pain processing using SEFs generated by YAG laser stimulation to the thigh to separate the activities of several cortical areas. Using a multidipole model, they detected the activities of contralateral S1, S2, and PPC as well as ipsilateral S2 and measured mean peak latencies of 152 ms in S1, 170 ms in contralateral S2, 183 ms in PPC, and 181 ms in ipsilateral S2 [115].

4.5.3 Intraepidermal Stimulation

A method utilizing a pushpin-like intraepidermal needle electrode to stimulate the epidermal area for preferentially activating Aδ fibers has been developed (Fig. 4.4e) [117–122]. Inui et al. (2003) recorded SEFs in response to both intraepidermal electrical stimulation (ES) and transcutaneous electrical stimulation (TS) [119]. While ES is painful and activates mainly noxious Aδ fibers, TS is nonpainful and activates primarily non-noxious Aβ fibers. TS activated two sources within S1, Brodmann areas 3b and 1. Activities from area 3b consisted of 20 ms and 30 ms responses, and those from area 1 consisted of three components peaking at 26, 36, and 49 ms following TS. In contrast, ES activated only one source within S1, area 1, with peaks at 88, 98, and 109 ms. The first detectable cortical activity induced by both ES and TS was in the contralateral S1, but the more delayed activities were in bilateral S2, insular cortex, cingulate cortex, and anterior medial temporal area and in ipsilateral S1 [120]. They concluded that the processing of noxious and non-noxious stimuli is similar with respect to source location except that there is no detectable activation within area 3b following (noxious) ES [119].

4.5.4 Intramuscular Motor Point Stimulation

Motor point stimulation is a useful tool to investigate muscle afferent feedback to the cortex [123, 124]. The most prominent and fastest magnetic field is observed 70–80 ms following motor point stimulation, referred to as the M70m [123, 124]. The source location of M70m was estimated to be medial and superior to that of N20m elicited by median nerve stimulation; N20m is accepted as a tangential source in area 3b, and that is probably located at area 4. The peak latency of M70m is too delayed compared with that of N20m when evoked by median nerve stimulation; therefore, M70m component must not reflect direct proprioceptive input to area 4; rather, this activity is possibly mediated by indirect input through area 2 and/or the cerebellum.

4.5.5 Cortical Responses to Passive Movement

Many investigators have reported movement-related cortical magnetic fields (MRCFs) following active movement [124–137], but relatively few have assessed response to passive movement reflecting proprioceptive feedback using MEG [135, 138–142]. Such studies have found three or four components in the contralateral hemisphere from 20 to 120 ms after passive movement onset [138–142]. The sources of these components were located in S1 [140] or in areas 3b and 4 [135, 138, 139, 141]. In addition, contralateral or bilateral S2 activity was observed from 80 to 180 ms [138, 140–142]. We recorded the time courses of cortical activities in response to active and passive movements using a whole-head MEG and estimated the corresponding source locations by a multidipole model [142]. Two SEF peaks were recorded after passive movement onset; the earliest component (PM1) peaked at 36.2 ms and the second component (PM2) peaked at 86.1 ms. The peak latency and source localization of PM1, estimated to be in area 4, were the same as those of movement evoked magnetic field 1 following active movement, while the sources of PM2 were estimated to be in area 4, supplementary motor area (SMA), and PPC of the contralateral hemisphere and in the bilateral secondary somatosensory cortex (S2). The peak latencies were obtained at 54–109 ms in SMA, 64–114 ms in PPC, and 84–184 ms in S2 (Fig. 4.5). These results agree with previous PET and fMRI

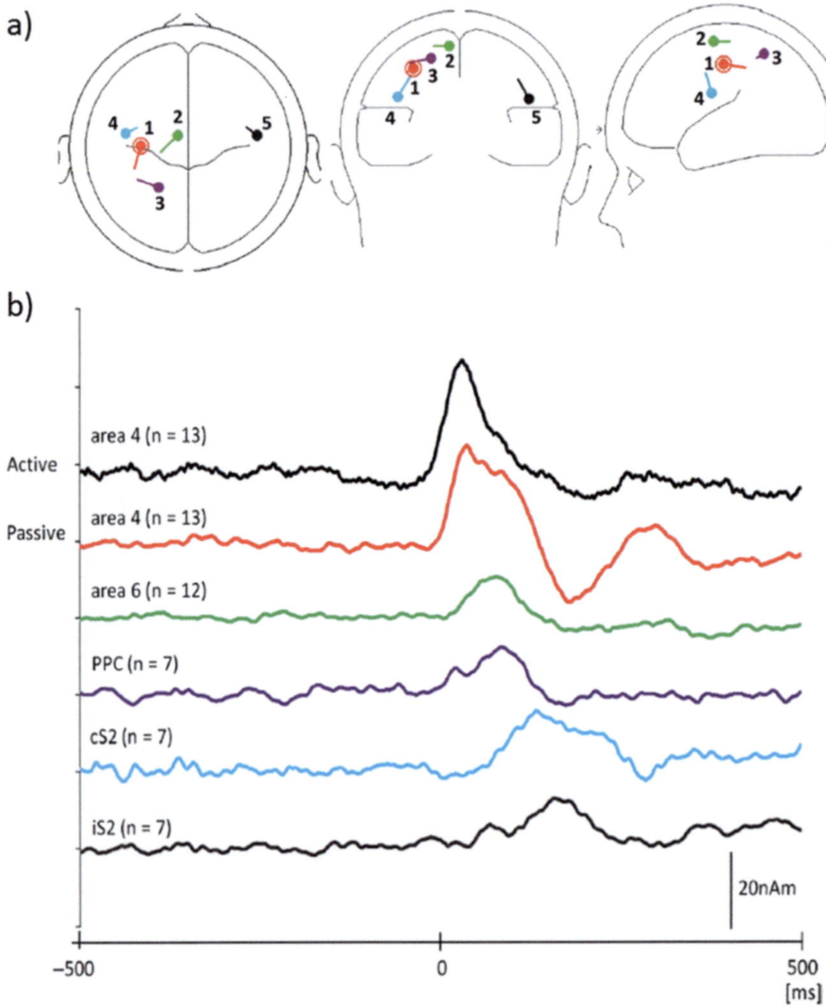

Fig. 4.5 Time course of source activities and the source locations induced by passive index finger movement [142]. (**a**) The locations of all sources following passive movement. 1 (*red dipole*), area 4; 2 (*green dipole*), SMA (area 6); 3 (*purple dipole*), PPC; 4 (*blue dipole*), contralateral secondary somatosensory cortex to the passive movement (cS2); and 5 (*dark gray* dipole), ipsilateral S2 to the passive movement (iS2). (**b**) Time course of the source activities following active (*black line*) and passive finger movements (*red*, *green*, *purple*, *blue*, and *dark gray lines*) (Adapted from Onishi H et al. [142])

studies, showing that passive movement activates the contralateral primary sensorimotor area, SMA, and PPC, as well as bilateral S2 [143–148]. The time courses of source activities, however, have not been clarified because PET and fMRI have lower temporal resolution than MEG. Alternatively, we can investigate the time courses of source activities in various regions of human cortex during passive finger movement using MEG with a multiple dipole analysis system.

References

1. Hari R, Forss N. Magnetoencephalography in the study of human somatosensory cortical processing. Philos Trans R Soc Lond B Biol Sci. 1999;354(1387):1145–54. doi:10.1098/rstb.1999.0470.
2. Kakigi R, Hoshiyama M, Shimojo M, Naka D, Yamasaki H, Watanabe S, Xiang J, Maeda K, Lam K, Itomi K, Nakamura A. The somatosensory evoked magnetic fields. Prog Neurobiol. 2000;61(5):495–523.
3. Hari R, Kaukoranta E. Neuromagnetic studies of somatosensory system: principles and examples. Prog Neurobiol. 1985;24(3):233–56.
4. Nagamine T, Makela J, Mima T, Mikuni N, Nishitani N, Satoh T, Ikeda A, Shibasaki H. Serial processing of the somesthetic information revealed by different effects of stimulus rate on the somatosensory-evoked potentials and magnetic fields. Brain Res. 1998;791(1–2):200–8.
5. Huttunen J, Hari R, Leinonen L. Cerebral magnetic responses to stimulation of ulnar and median nerves. Electroencephalogr Clin Neurophysiol. 1987;66(4):391–400.
6. Kakigi R. Somatosensory evoked magnetic fields following median nerve stimulation. Neurosci Res. 1994;20(2):165–74.
7. Forss N, Hari R, Salmelin R, Ahonen A, Hamalainen M, Kajola M, Knuutila J, Simola J. Activation of the human posterior parietal cortex by median nerve stimulation. Exp Brain Res. 1994;99(2):309–15.
8. Forss N, Merlet I, Vanni S, Hamalainen M, Mauguiere F, Hari R. Activation of human mesial cortex during somatosensory target detection task. Brain Res. 1996;734(1–2):229–35.
9. Mauguiere F, Merlet I, Forss N, Vanni S, Jousmaki V, Adeleine P, Hari R. Activation of a distributed somatosensory cortical network in the human brain. A dipole modelling study of magnetic fields evoked by median nerve stimulation. Part I: Location and activation timing of SEF sources. Electroencephalogr Clin Neurophysiol. 1997;104(4):281–9.
10. Huang MX, Aine C, Davis L, Butman J, Christner R, Weisend M, Stephen J, Meyer J, Silveri J, Herman M, Lee RR. Sources on the anterior and posterior banks of the central sulcus identified from magnetic somatosensory evoked responses using multistart spatio-temporal localization. Hum Brain Mapp. 2000;11(2):59–76.
11. Tiihonen J, Hari R, Hamalainen M. Early deflections of cerebral magnetic responses to median nerve stimulation. Electroencephalogr Clin Neurophysiol. 1989;74(4):290–6.
12. Wikstrom H, Huttunen J, Korvenoja A, Virtanen J, Salonen O, Aronen H, Ilmoniemi RJ. Effects of interstimulus interval on somatosensory evoked magnetic fields (SEFs): a hypothesis concerning SEF generation at the primary sensorimotor cortex. Electroencephalogr Clin Neurophysiol. 1996;100(6):479–87.
13. Kawamura T, Nakasato N, Seki K, Kanno A, Fujita S, Fujiwara S, Yoshimoto T. Neuromagnetic evidence of pre- and post-central cortical sources of somatosensory evoked responses. Electroencephalogr Clin Neurophysiol. 1996;100(1):44–50.
14. Huttunen J, Komssi S, Lauronen L. Spatial dynamics of population activities at S1 after median and ulnar nerve stimulation revisited: an MEG study. Neuroimage. 2006;32(3):1024–31. doi:10.1016/j.neuroimage.2006.04.196.

15. Sugawara K, Onishi H, Yamashiro K, Kojima S, Miyaguchi S, Kirimoto H, Tsubaki A, Tamaki H, Shirozu H, Kameyama S. The effect of anodal transcranial direct current stimulation over the primary motor or somatosensory cortices on somatosensory evoked magnetic fields. Clin Neurophysiol. 2014. doi:10.1016/j.clinph.2014.04.014.
16. Lin YY, Chen WT, Liao KK, Yeh TC, Wu ZA, Ho LT, Lee LS. Differential generators for N20m and P35m responses to median nerve stimulation. Neuroimage. 2005;25(4):1090–9. doi:10.1016/j.neuroimage.2004.12.047.
17. Hoshiyama M, Kakigi R, Koyama S, Watanabe S, Shimojo M. Activity in posterior parietal cortex following somatosensory stimulation in man: magnetoencephalographic study using spatio-temporal source analysis. Brain Topogr. 1997;10(1):23–30.
18. Inui K, Wang X, Tamura Y, Kaneoke Y, Kakigi R. Serial processing in the human somatosensory system. Cereb Cortex. 2004;14(8):851–7. doi:10.1093/cercor/bhh043.
19. Jousmaki V, Forss N. Effects of stimulus intensity on signals from human somatosensory cortices. Neuroreport. 1998;9(15):3427–31.
20. Torquati K, Pizzella V, Della Penna S, Franciotti R, Babiloni C, Rossini PM, Romani GL. Comparison between SI and SII responses as a function of stimulus intensity. Neuroreport. 2002;13(6):813–9.
21. Hoshiyama M, Kakigi R. Two evoked responses with different recovery functions in the primary somatosensory cortex in humans. Clin Neurophysiol. 2001;112(7):1334–42.
22. Onishi H, Sugawara K, Yamashiro K, Sato D, Suzuki M, Kirimoto H, Tamaki H, Murakami H, Kameyama S. Effect of the number of pins and inter-pin distance on somatosensory evoked magnetic fields following mechanical tactile stimulation. Brain Res. 2013;1535:78–88. doi:10.1016/j.brainres.2013.08.048.
23. Tsutada T, Tsuyuguchi N, Hattori H, Shimada H, Shimogawara M, Kuramoto T, Haruta Y, Matsuoka Y, Hakuba A. Determining the appropriate stimulus intensity for studying the dipole moment in somatosensory evoked fields: a preliminary study. Clin Neurophysiol. 1999;110(12):2127–30.
24. Lin YY, Shih YH, Chen JT, Hsieh JC, Yeh TC, Liao KK, Kao CD, Lin KP, Wu ZA, Ho LT. Differential effects of stimulus intensity on peripheral and neuromagnetic cortical responses to median nerve stimulation. Neuroimage. 2003;20(2):909–17. doi:10.1016/S1053-8119(03)00387-2.
25. Karhu J, Hari R, Paetau R, Kajola M, Mervaala E. Cortical reactivity in progressive myoclonus epilepsy. Electroencephalogr Clin Neurophysiol. 1994;90(2):93–102.
26. Hamada Y, Otsuka S, Okamoto T, Suzuki R. The profile of the recovery cycle in human primary and secondary somatosensory cortex: a magnetoencephalography study. Clin Neurophysiol. 2002;113(11):1787–93.
27. Huttunen J. In search of augmentation at human SI: somatosensory cortical responses to stimulus trains and their modulation by motor activity. Brain Res. 2010;1331:74–9. doi:10.1016/j.brainres.2010.03.058.
28. Lim M, Kim JS, Chung CK. Modulation of somatosensory evoked magnetic fields by intensity of interfering stimuli in human somatosensory cortex: an MEG study. Neuroimage. 2012;61(3):660–9. doi:10.1016/j.neuroimage.2012.04.003.
29. Kakigi R, Koyama S, Hoshiyama M, Watanabe S, Shimojo M, Kitamura Y. Gating of somatosensory evoked responses during active finger movements magnetoencephalographic studies. J Neurol Sci. 1995;128(2):195–204.
30. Wasaka T, Nakata H, Akatsuka K, Kida T, Inui K, Kakigi R. Differential modulation in human primary and secondary somatosensory cortices during the preparatory period of self-initiated finger movement. Eur J Neurosci. 2005;22(5):1239–47. doi:10.1111/j.1460-9568.2005.04289.x.
31. Kida T, Wasaka T, Inui K, Akatsuka K, Nakata H, Kakigi R. Centrifugal regulation of human cortical responses to a task-relevant somatosensory signal triggering voluntary movement. Neuroimage. 2006;32(3):1355–64. doi:10.1016/j.neuroimage.2006.05.015.

32. Huttunen J, Lauronen L. Intracortical modulation of somatosensory evoked fields during movement: evidence for selective suppression of postsynaptic inhibition. Brain Res. 2012;1459:43–51. doi:10.1016/j.brainres.2012.04.023.
33. Hashimoto I, Kimura T, Fukushima T, Iguchi Y, Saito Y, Terasaki O, Sakuma K. Reciprocal modulation of somatosensory evoked N20m primary response and high-frequency oscillations by interference stimulation. Clin Neurophysiol. 1999;110(8):1445–51.
34. Tanosaki M, Suzuki A, Takino R, Kimura T, Iguchi Y, Kurobe Y, Haruta Y, Hoshi Y, Hashimoto I. Neural mechanisms for generation of tactile interference effects on somatosensory evoked magnetic fields in humans. Clin Neurophysiol. 2002;113(5):672–80.
35. Kakigi R, Koyama S, Hoshiyama M, Kitamura Y, Shimojo M, Watanabe S, Nakamura A. Effects of tactile interference stimulation on somatosensory evoked magnetic fields. Neuroreport. 1996;7(2):405–8.
36. Huttunen J, Wikstrom H, Salonen O, Ilmoniemi RJ. Human somatosensory cortical activation strengths: comparison between males and females and age-related changes. Brain Res. 1999;818(2):196–203.
37. Hagiwara K, Ogata K, Okamoto T, Uehara T, Hironaga N, Shigeto H, Kira J, Tobimatsu S. Age-related changes across the primary and secondary somatosensory areas: an analysis of neuromagnetic oscillatory activities. Clin Neurophysiol. 2014;125(5):1021–9. doi:10.1016/j.clinph.2013.10.005.
38. Zappasodi F, Pasqualetti P, Tombini M, Ercolani M, Pizzella V, Rossini PM, Tecchio F. Hand cortical representation at rest and during activation: gender and age effects in the two hemispheres. Clin Neurophysiol. 2006;117(7):1518–28. doi:10.1016/j.clinph.2006.03.016.
39. Stephen JM, Ranken D, Best E, Adair J, Knoefel J, Kovacevic S, Padilla D, Hart B, Aine CJ. Aging changes and gender differences in response to median nerve stimulation measured with MEG. Clin Neurophysiol. 2006;117(1):131–43. doi:10.1016/j.clinph.2005.09.003.
40. Haueisen J, Heuer T, Nowak H, Liepert J, Weiller C, Okada Y, Curio G. The influence of lorazepam on somatosensory-evoked fast frequency (600 Hz) activity in MEG. Brain Res. 2000;874(1):10–4.
41. Huttunen J, Pekkonen E, Kivisaari R, Autti T, Kahkonen S. Modulation of somatosensory evoked fields from SI and SII by acute GABA A-agonism and paired-pulse stimulation. Neuroimage. 2008;40(2):427–34. doi:10.1016/j.neuroimage.2007.12.024.
42. Huttunen J, Kahkonen S, Kaakkola S, Ahveninen J, Pekkonen E. Effects of an acute D2-dopaminergic blockade on the somatosensory cortical responses in healthy humans: evidence from evoked magnetic fields. Neuroreport. 2003;14(12):1609–12. doi:10.1097/01.wnr.0000085689.46774.53.
43. Huttunen J, Jaaskelainen IP, Hirvonen J, Kaakkola S, Ilmoniemi RJ, Pekkonen E. Scopolamine reduces the P35m and P60m deflections of the human somatosensory evoked magnetic fields. Neuroreport. 2001;12(3):619–23.
44. Wikstrom H, Roine RO, Aronen HJ, Salonen O, Sinkkonen J, Ilmoniemi RJ, Huttunen J. Specific changes in somatosensory evoked magnetic fields during recovery from sensorimotor stroke. Ann Neurol. 2000;47(3):353–60.
45. Wikstrom H, Roine RO, Salonen O, Lund KB, Salli E, Ilmoniemi RJ, Aronen HJ, Huttunen J. Somatosensory evoked magnetic fields from the primary somatosensory cortex (SI) in acute stroke. Clin Neurophysiol. 1999;110(5):916–23.
46. Druschky K, Kaltenhauser M, Hummel C, Druschky A, Huk WJ, Neundorfer B, Stefan H. Post-apoplectic reorganization of cortical areas processing passive movement and tactile stimulation – a neuromagnetic case study. Neuroreport. 2002;13(18):2581–6. doi:10.1097/01.wnr.0000048922.00321.ee.
47. Gallien P, Aghulon C, Durufle A, Petrilli S, de Crouy AC, Carsin M, Toulouse P. Magnetoencephalography in stroke: a 1-year follow-up study. Eur J Neurol. 2003;10(4):373–82.
48. Oliviero A, Tecchio F, Zappasodi F, Pasqualetti P, Salustri C, Lupoi D, Ercolani M, Romani GL, Rossini PM. Brain sensorimotor hand area functionality in acute stroke: insights from

magnetoencephalography. Neuroimage. 2004;23(2):542–50. doi:10.1016/j.neuroimage.2004.06.040.
49. Tecchio F, Zappasodi F, Tombini M, Oliviero A, Pasqualetti P, Vernieri F, Ercolani M, Pizzella V, Rossini PM. Brain plasticity in recovery from stroke: an MEG assessment. Neuroimage. 2006;32(3):1326–34. doi:10.1016/j.neuroimage.2006.05.004.
50. Tecchio F, Pasqualetti P, Zappasodi F, Tombini M, Lupoi D, Vernieri F, Rossini PM. Outcome prediction in acute monohemispheric stroke via magnetoencephalography. J Neurol. 2007;254(3):296–305. doi:10.1007/s00415-006-0355-0.
51. Tecchio F, Zappasodi F, Tombini M, Caulo M, Vernieri F, Rossini PM. Interhemispheric asymmetry of primary hand representation and recovery after stroke: a MEG study. Neuroimage. 2007;36(4):1057–64. doi:10.1016/j.neuroimage.2007.02.058.
52. Rossini PM, Tecchio F, Pizzella V, Lupoi D, Cassetta E, Pasqualetti P. Interhemispheric differences of sensory hand areas after monohemispheric stroke: MEG/MRI integrative study. Neuroimage. 2001;14(2):474–85. doi:10.1006/nimg.2000.0686.
53. Tecchio F, Pasqualetti P, Pizzella V, Romani G, Rossini PM. Morphology of somatosensory evoked fields: inter-hemispheric similarity as a parameter for physiological and pathological neural connectivity. Neurosci Lett. 2000;287(3):203–6.
54. Rossini PM, Tecchio F, Pizzella V, Lupoi D, Cassetta E, Pasqualetti P, Romani GL, Orlacchio A. On the reorganization of sensory hand areas after mono-hemispheric lesion: a functional (MEG)/anatomical (MRI) integrative study. Brain Res. 1998;782(1–2):153–66.
55. Tecchio F, Rossini PM, Pizzella V, Cassetta E, Pasqualetti P, Romani GL. A neuromagnetic normative data set for hemispheric sensory hand cortical representations and their interhemispheric differences. Brain Res Brain Res Protoc. 1998;2(4):306–14.
56. Tecchio F, Rossini PM, Pizzella V, Cassetta E, Romani GL. Spatial properties and interhemispheric differences of the sensory hand cortical representation: a neuromagnetic study. Brain Res. 1997;767(1):100–8.
57. Vanni S, Rockstroh B, Hari R. Cortical sources of human short-latency somatosensory evoked fields to median and ulnar nerve stimuli. Brain Res. 1996;737(1–2):25–33.
58. Huttunen J, Kaukoranta E, Hari R. Cerebral magnetic responses to stimulation of tibial and sural nerves. J Neurol Sci. 1987;79(1–2):43–54.
59. Kakigi R, Koyama S, Hoshiyama M, Shimojo M, Kitamura Y, Watanabe S. Topography of somatosensory evoked magnetic fields following posterior tibial nerve stimulation. Electroencephalogr Clin Neurophysiol. 1995;95(2):127–34.
60. Nakagawa H, Namima T, Aizawa M, Uchi K, Kaiho Y, Yoshikawa K, Orikasa S, Nakasato N. Somatosensory evoked magnetic fields elicited by dorsal penile, posterior tibial and median nerve stimulation. Electroencephalogr Clin Neurophysiol. 1998;108(1):57–61.
61. Shimojo M, Kakigi R, Hoshiyama M, Koyama S, Kitamura Y, Watanabe S. Differentiation of receptive fields in the sensory cortex following stimulation of various nerves of the lower limb in humans: a magnetoencephalographic study. J Neurosurg. 1996;85(2):255–62. doi:10.3171/jns.1996.85.2.0255.
62. Narici L, Modena I, Opsomer RJ, Pizzella V, Romani GL, Torrioli G, Traversa R, Rossini PM. Neuromagnetic somatosensory homunculus: a non-invasive approach in humans. Neurosci Lett. 1991;121(1–2):51–4.
63. Kaukoranta E, Hari R, Hamalainen M, Huttunen J. Cerebral magnetic fields evoked by peroneal nerve stimulation. Somatosens Res. 1986;3(4):309–21.
64. Hari R, Karhu J, Hamalainen M, Knuutila J, Salonen O, Sams M, Vilkman V. Functional organization of the human first and second somatosensory cortices: a neuromagnetic study. Eur J Neurosci. 1993;5(6):724–34.
65. Karhu J, Hari R, Lu ST, Paetau R, Rif J. Cerebral magnetic fields to lingual stimulation. Electroencephalogr Clin Neurophysiol. 1991;80(6):459–68.
66. Maezawa H, Yoshida K, Matsuhashi M, Yokoyama Y, Mima T, Bessho K, Fujita S, Nagamine T, Fukuyama H. Evaluation of tongue sensory disturbance by somatosensory

evoked magnetic fields following tongue stimulation. Neurosci Res. 2011;71(3):244–50. doi:10.1016/j.neures.2011.07.1831.
67. Maezawa H, Yoshida K, Nagamine T, Matsubayashi J, Enatsu R, Bessho K, Fukuyama H. Somatosensory evoked magnetic fields following electric tongue stimulation using pin electrodes. Neurosci Res. 2008;62(2):131–9. doi:10.1016/j.neures.2008.07.004.
68. Sakamoto K, Nakata H, Yumoto M, Kakigi R. Somatosensory processing of the tongue in humans. Front Physiol. 2010;1:136. doi:10.3389/fphys.2010.00136.
69. Sakamoto K, Nakata H, Inui K, Perrucci MG, Del Gratta C, Kakigi R, Romani GL. A difference exists in somatosensory processing between the anterior and posterior parts of the tongue. Neurosci Res. 2010;66(2):173–9. doi:10.1016/j.neures.2009.10.013.
70. Sakamoto K, Nakata H, Kakigi R. Somatotopic representation of the tongue in human secondary somatosensory cortex. Clin Neurophysiol. 2008;119(9):2125–34. doi:10.1016/j.clinph.2008.05.003.
71. Sakamoto K, Nakata H, Kakigi R. Somatosensory-evoked magnetic fields following stimulation of the tongue in humans. Clin Neurophysiol. 2008;119(7):1664–73. doi:10.1016/j.clinph.2008.03.029.
72. Maezawa H, Matsuhashi M, Yoshida K, Mima T, Nagamine T, Fukuyama H. Evaluation of lip sensory disturbance using somatosensory evoked magnetic fields. Clin Neurophysiol. 2014;125(2):363–9. doi:10.1016/j.clinph.2013.07.017.
73. Hoshiyama M, Kakigi R, Koyama S, Kitamura Y, Shimojo M, Watanabe S. Somatosensory evoked magnetic fields following stimulation of the lip in humans. Electroencephalogr Clin Neurophysiol. 1996;100(2):96–104.
74. Tamura Y, Shibukawa Y, Shintani M, Kaneko Y, Ichinohe T. Oral structure representation in human somatosensory cortex. Neuroimage. 2008;43(1):128–35. doi:10.1016/j.neuroimage.2008.06.040.
75. Nihashi T, Kakigi R, Hoshiyama M, Miki K, Kajita Y, Yoshida J, Yatsuya H. Effect of tactile interference stimulation of the ear in human primary somatosensory cortex: a magnetoencephalographic study. Clin Neurophysiol. 2003;114(10):1866–78.
76. Nihashi T, Kakigi R, Okada T, Sadato N, Kashikura K, Kajita Y, Yoshida J. Functional magnetic resonance imaging evidence for a representation of the ear in human primary somatosensory cortex: comparison with magnetoencephalography study. Neuroimage. 2002;17(3):1217–26.
77. Nihashi T, Kakigi R, Kawakami O, Hoshiyama M, Itomi K, Nakanishi H, Kajita Y, Inao S, Yoshida J. Representation of the ear in human primary somatosensory cortex. Neuroimage. 2001;13(2):295–304. doi:10.1006/nimg.2000.0695.
78. Nguyen BT, Tran TD, Hoshiyama M, Inui K, Kakigi R. Face representation in the human primary somatosensory cortex. Neurosci Res. 2004;50(2):227–32. doi:10.1016/j.neures.2004.07.004.
79. Nguyen BT, Inui K, Hoshiyama M, Nakata H, Kakigi R. Face representation in the human secondary somatosensory cortex. Clin Neurophysiol. 2005;116(6):1247–53. doi:10.1016/j.clinph.2005.01.018.
80. Nevalainen P, Ramstad R, Isotalo E, Haapanen ML, Lauronen L. Trigeminal somatosensory evoked magnetic fields to tactile stimulation. Clin Neurophysiol. 2006;117(9):2007–15. doi:10.1016/j.clinph.2006.05.019.
81. Itomi K, Kakigi R, Hoshiyama M, Watanabe K. A unique area of the homonculus: the topography of the primary somatosensory cortex in humans following posterior scalp and shoulder stimulation. Brain Topogr. 2001;14(1):15–23.
82. Itomi K, Kakigi R, Maeda K, Hoshiyama M. Dermatome versus homunculus; detailed topography of the primary somatosensory cortex following trunk stimulation. Clin Neurophysiol. 2000;111(3):405–12.
83. Matsushita M, Nakasato N, Nakagawa H, Kanno A, Kaiho Y, Arai Y. Primary somatosensory evoked magnetic fields elicited by sacral surface electrical stimulation. Neurosci Lett. 2008;431(1):77–80. doi:10.1016/j.neulet.2007.11.025.

84. Schiffbauer H, Berger MS, Ferrari P, Freudenstein D, Rowley HA, Roberts TP. Preoperative magnetic source imaging for brain tumor surgery: a quantitative comparison with intraoperative sensory and motor mapping. Neurosurg Focus. 2003;15(1):E7.
85. Yang TT, Gallen CC, Schwartz BJ, Bloom FE. Noninvasive somatosensory homunculus mapping in humans by using a large-array biomagnetometer. Proc Natl Acad Sci U S A. 1993;90(7):3098–102.
86. Nakamura A, Yamada T, Goto A, Kato T, Ito K, Abe Y, Kachi T, Kakigi R. Somatosensory homunculus as drawn by MEG. Neuroimage. 1998;7(4 Pt 1):377–86. doi:10.1006/nimg.1998.0332.
87. Makela JP, Kirveskari E, Seppa M, Hamalainen M, Forss N, Avikainen S, Salonen O, Salenius S, Kovala T, Randell T, Jaaskelainen J, Hari R. Three-dimensional integration of brain anatomy and function to facilitate intraoperative navigation around the sensorimotor strip. Hum Brain Mapp. 2001;12(3):180–92.
88. Castillo EM, Papanicolaou AC. Cortical representation of dermatomes: MEG-derived maps after tactile stimulation. Neuroimage. 2005;25(3):727–33. doi:10.1016/j.neuroimage.2004.12.040.
89. Vogel P, Ruber P, Klein R. The latency difference of the tibial and sural nerve SEP: peripheral versus central factors. Electroencephalogr Clin Neurophysiol. 1986;65(4):269–75.
90. Huttunen J. Magnetic cortical responses evoked by tactile stimulation of the middle finger in man. Pflugers Arch. 1986;407(2):129–33.
91. Forss N, Salmelin R, Hari R. Comparison of somatosensory evoked fields to airpuff and electric stimuli. Electroencephalogr Clin Neurophysiol. 1994;92(6):510–7.
92. Hoshiyama M, Kakigi R, Koyama S, Kitamura Y, Shimoio M, Watanabe S. Somatosensory evoked magnetic fields after mechanical stimulation of the scalp in humans. Neurosci Lett. 1995;195(1):29–32.
93. Mertens M, Lutkenhoner B. Efficient neuromagnetic determination of landmarks in the somatosensory cortex. Clin Neurophysiol. 2000;111(8):1478–87.
94. Hoechstetter K, Rupp A, Meinck HM, Weckesser D, Bornfleth H, Stippich C, Berg P, Scherg M. Magnetic source imaging of tactile input shows task-independent attention effects in SII. Neuroreport. 2000;11(11):2461–5.
95. Hoechstetter K, Rupp A, Stancak A, Meinck HM, Stippich C, Berg P, Scherg M. Interaction of tactile input in the human primary and secondary somatosensory cortex – a magnetoencephalographic study. Neuroimage. 2001;14(3):759–67. doi:10.1006/nimg.2001.0855.
96. Simoes C, Mertens M, Forss N, Jousmaki V, Lutkenhoner B, Hari R. Functional overlap of finger representations in human SI and SII cortices. J Neurophysiol. 2001;86(4):1661–5.
97. Karageorgiou E, Koutlas IG, Alonso AA, Leuthold AC, Lewis SM, Georgopoulos AP. Cortical processing of tactile stimuli applied in quick succession across the fingertips: temporal evolution of dipole sources revealed by magnetoencephalography. Exp Brain Res. 2008;189(3):311–21. doi:10.1007/s00221-008-1425-6.
98. Lin YY, Kajola M. Neuromagnetic somatosensory responses to natural moving tactile stimulation. Can J Neurol Sci. 2003;30(1):31–5.
99. Jousmaki V, Nishitani N, Hari R. A brush stimulator for functional brain imaging. Clin Neurophysiol. 2007;118(12):2620–4. doi:10.1016/j.clinph.2007.08.024.
100. Hesse MD, Nishitani N, Fink GR, Jousmaki V, Hari R. Attenuation of somatosensory responses to self-produced tactile stimulation. Cereb Cortex. 2010;20(2):425–32. doi:10.1093/cercor/bhp110.
101. Hadoush H, Inoue K, Nakanishi K, Kurumadani H, Sunagawa T, Ochi M. Ipsilateral primary sensorimotor cortical response to mechanical tactile stimuli. Neuroreport. 2010;21(2):108–13. doi:10.1097/WNR.0b013e3283349a17.
102. Onishi H, Oyama M, Soma T, Kubo M, Kirimoto H, Murakami H, Kameyama S. Neuromagnetic activation of primary and secondary somatosensory cortex following tactile-on and tactile-off stimulation. Clin Neurophysiol. 2010;121(4):588–93. doi:10.1016/j.clinph.2009.12.022.

103. Mogilner A, Nomura M, Ribary U, Jagow R, Lado F, Rusinek H, Llinas R. Neuromagnetic studies of the lip area of primary somatosensory cortex in humans: evidence for an oscillotopic organization. Exp Brain Res. 1994;99(1):137–47.
104. Huttunen J, Kobal G, Kaukoranta E, Hari R. Cortical responses to painful CO2 stimulation of nasal mucosa; a magnetoencephalographic study in man. Electroencephalogr Clin Neurophysiol. 1986;64(4):347–9.
105. Kakigi R, Koyama S, Hoshiyama M, Kitamura Y, Shimojo M, Watanabe S. Pain-related magnetic fields following painful CO2 laser stimulation in man. Neurosci Lett. 1995;192 (1):45–8.
106. Kakigi R, Koyama S, Hoshiyama M, Kitamura Y, Shimojo M, Watanabe S. Pain-related brain responses following CO2 laser stimulation: magnetoencephalographic studies. Electroencephalogr Clin Neurophysiol Suppl. 1996;47:111–20.
107. Bragard D, Chen AC, Plaghki L. Direct isolation of ultra-late (C-fibre) evoked brain potentials by CO2 laser stimulation of tiny cutaneous surface areas in man. Neurosci Lett. 1996;209(2):81–4.
108. Watanabe S, Kakigi R, Koyama S, Hoshiyama M, Kaneoke Y. Pain processing traced by magnetoencephalography in the human brain. Brain Topogr. 1998;10(4):255–64.
109. Tran TD, Lam K, Hoshiyama M, Kakigi R. A new method for measuring the conduction velocities of Abeta-, Adelta- and C-fibers following electric and CO(2) laser stimulation in humans. Neurosci Lett. 2001;301(3):187–90.
110. Tran TD, Inui K, Hoshiyama M, Lam K, Qiu Y, Kakigi R. Cerebral activation by the signals ascending through unmyelinated C-fibers in humans: a magnetoencephalographic study. Neuroscience. 2002;113(2):375–86.
111. Kakigi R, Tran TD, Qiu Y, Wang X, Nguyen TB, Inui K, Watanabe S, Hoshiyama M. Cerebral responses following stimulation of unmyelinated C-fibers in humans: electro- and magneto-encephalographic study. Neurosci Res. 2003;45(3):255–75.
112. Qiu Y, Fu Q, Wang X, Tran TD, Inui K, Iwase S, Kakigi R. Microneurographic study of C fiber discharges induced by CO2 laser stimulation in humans. Neurosci Lett. 2003;353 (1):25–8.
113. Qiu Y, Inui K, Wang X, Nguyen BT, Tran TD, Kakigi R. Effects of distraction on magnetoencephalographic responses ascending through C-fibers in humans. Clin Neurophysiol. 2004;115(3):636–46. doi:10.1016/j.clinph.2003.10.017.
114. Forss N, Raij TT, Seppa M, Hari R. Common cortical network for first and second pain. Neuroimage. 2005;24(1):132–42. doi:10.1016/j.neuroimage.2004.09.032.
115. Nakata H, Tamura Y, Sakamoto K, Akatsuka K, Hirai M, Inui K, Hoshiyama M, Saitoh Y, Yamamoto T, Katayama Y, Kakigi R. Evoked magnetic fields following noxious laser stimulation of the thigh in humans. Neuroimage. 2008;42(2):858–68. doi:10.1016/j.neuroimage.2008.05.017.
116. Tran TD, Inui K, Hoshiyama M, Lam K, Kakigi R. Conduction velocity of the spinothalamic tract following CO2 laser stimulation of C-fibers in humans. Pain. 2002;95(1–2):125–31.
117. Inui K, Tran TD, Hoshiyama M, Kakigi R. Preferential stimulation of Adelta fibers by intra-epidermal needle electrode in humans. Pain. 2002;96(3):247–52.
118. Inui K, Tran TD, Qiu Y, Wang X, Hoshiyama M, Kakigi R. Pain-related magnetic fields evoked by intra-epidermal electrical stimulation in humans. Clin Neurophysiol. 2002;113 (2):298–304.
119. Inui K, Wang X, Qiu Y, Nguyen BT, Ojima S, Tamura Y, Nakata H, Wasaka T, Tran TD, Kakigi R. Pain processing within the primary somatosensory cortex in humans. Eur J Neurosci. 2003;18(10):2859–66.
120. Inui K, Tran TD, Qiu Y, Wang X, Hoshiyama M, Kakigi R. A comparative magnetoencephalographic study of cortical activations evoked by noxious and innocuous somatosensory stimulations. Neuroscience. 2003;120(1):235–48.
121. Wang X, Inui K, Qiu Y, Kakigi R. Cortical responses to noxious stimuli during sleep. Neuroscience. 2004;128(1):177–86. doi:10.1016/j.neuroscience.2004.06.036.

122. Kakigi R, Inui K, Tran DT, Qiu Y, Wang X, Watanabe S, Hoshiyama M. Human brain processing and central mechanisms of pain as observed by electro- and magneto-encephalography. J Chin Med Assoc. 2004;67(8):377–86.
123. Kimura T, Nishijo K, Hashimoto I. Somatosensory evoked potentials elicited by motor point stimulation. Electroencephalogr Clin Neurophysiol Suppl. 1999;49:73–6.
124. Onishi H, Oyama M, Soma T, Kirimoto H, Sugawara K, Murakami H, Kameyama S. Muscle-afferent projection to the sensorimotor cortex after voluntary movement and motor-point stimulation: an MEG study. Clin Neurophysiol. 2011;122(3):605–10. doi:10.1016/j.clinph.2010.07.027.
125. Cheyne D, Weinberg H. Neuromagnetic fields accompanying unilateral finger movements: pre-movement and movement-evoked fields. Exp Brain Res. 1989;78(3):604–12.
126. Cheyne D, Kristeva R, Deecke L. Homuncular organization of human motor cortex as indicated by neuromagnetic recordings. Neurosci Lett. 1991;122(1):17–20.
127. Kristeva R, Cheyne D, Deecke L. Neuromagnetic fields accompanying unilateral and bilateral voluntary movements: topography and analysis of cortical sources. Electroencephalogr Clin Neurophysiol. 1991;81(4):284–98.
128. Kristeva-Feige R, Walter H, Lutkenhoner B, Hampson S, Ross B, Knorr U, Steinmetz H, Cheyne D. A neuromagnetic study of the functional organization of the sensorimotor cortex. Eur J Neurosci. 1994;6(4):632–9.
129. Kristeva-Feige R, Rossi S, Pizzella V, Tecchio F, Romani GL, Erne S, Edrich J, Orlacchio A, Rossini PM. Neuromagnetic fields of the brain evoked by voluntary movement and electrical stimulation of the index finger. Brain Res. 1995;682(1–2):22–8.
130. Kristeva-Feige R, Rossi S, Pizzella V, Sabato A, Tecchio F, Feige B, Romani GL, Edrich J, Rossini PM. Changes in movement-related brain activity during transient deafferentation: a neuromagnetic study. Brain Res. 1996;714(1–2):201–8.
131. Kristeva-Feige R, Rossi S, Feige B, Mergner T, Lucking CH, Rossini PM. The bereitschaftspotential paradigm in investigating voluntary movement organization in humans using magnetoencephalography (MEG). Brain Res Brain Res Protoc. 1997;1(1):13–22.
132. Cheyne D, Endo H, Takeda T, Weinberg H. Sensory feedback contributes to early movement-evoked fields during voluntary finger movements in humans. Brain Res. 1997;771(2):196–202.
133. Nagamine T, Toro C, Balish M, Deuschl G, Wang B, Sato S, Shibasaki H, Hallett M. Cortical magnetic and electric fields associated with voluntary finger movements. Brain Topogr. 1994;6(3):175–83.
134. Hoshiyama M, Kakigi R, Berg P, Koyama S, Kitamura Y, Shimojo M, Watanabe S, Nakamura A. Identification of motor and sensory brain activities during unilateral finger movement: spatiotemporal source analysis of movement-associated magnetic fields. Exp Brain Res. 1997;115(1):6–14.
135. Woldag H, Waldmann G, Schubert M, Oertel U, Maess B, Friederici A, Hummelsheim H. Cortical neuromagnetic fields evoked by voluntary and passive hand movements in healthy adults. J Clin Neurophysiol. 2003;20(2):94–101.
136. Oishi M, Kameyama S, Fukuda M, Tsuchiya K, Kondo T. Cortical activation in area 3b related to finger movement: an MEG study. Neuroreport. 2004;15(1):57–62.
137. Onishi H, Soma T, Kameyama S, Oishi M, Fuijmoto A, Oyama M, Furusawa AA, Kurokawa Y. Cortical neuromagnetic activation accompanying two types of voluntary finger extension. Brain Res. 2006;1123(1):112–8. doi:10.1016/j.brainres.2006.09.033.
138. Xiang J, Hoshiyama M, Koyama S, Kaneoke Y, Suzuki H, Watanabe S, Naka D, Kakigi R. Somatosensory evoked magnetic fields following passive finger movement. Brain Res Cogn Brain Res. 1997;6(2):73–82.
139. Lange R, Nowak H, Haueisen J, Weiller C. Passive finger movement evoked fields in magnetoencephalography. Exp Brain Res. 2001;136(2):194–9.

140. Alary F, Simoes C, Jousmaki V, Forss N, Hari R. Cortical activation associated with passive movements of the human index finger: an MEG study. Neuroimage. 2002;15(3):691–6. doi:10.1006/nimg.2001.1010.
141. Druschky K, Kaltenhauser M, Hummel C, Druschky A, Huk WJ, Neundorfer B, Stefan H. Somatosensory evoked magnetic fields following passive movement compared with tactile stimulation of the index finger. Exp Brain Res. 2003;148(2):186–95. doi:10.1007/s00221-002-1293-4.
142. Onishi H, Sugawara K, Yamashiro K, Sato D, Suzuki M, Kirimoto H, Tamaki H, Murakami H, Kameyama S. Neuromagnetic activation following active and passive finger movements. Brain Behav. 2013;3(2):178–92. doi:10.1002/brb3.126.
143. Alary F, Doyon B, Loubinoux I, Carel C, Boulanouar K, Ranjeva JP, Celsis P, Chollet F. Event-related potentials elicited by passive movements in humans: characterization, source analysis, and comparison to fMRI. Neuroimage. 1998;8(4):377–90. doi:10.1006/nimg.1998.0377.
144. Albanese MC, Duerden EG, Bohotin V, Rainville P, Duncan GH. Differential effects of cognitive demand on human cortical activation associated with vibrotactile stimulation. J Neurophysiol. 2009;102(3):1623–31. doi:10.1152/jn.91295.2008.
145. Mima T, Terada K, Maekawa M, Nagamine T, Ikeda A, Shibasaki H. Somatosensory evoked potentials following proprioceptive stimulation of finger in man. Exp Brain Res. 1996;111(2):233–45.
146. Mima T, Sadato N, Yazawa S, Hanakawa T, Fukuyama H, Yonekura Y, Shibasaki H. Brain structures related to active and passive finger movements in man. Brain. 1999;122(Pt 10):1989–97.
147. Radovanovic S, Korotkov A, Ljubisavljevic M, Lyskov E, Thunberg J, Kataeva G, Danko S, Roudas M, Pakhomov S, Medvedev S, Johansson H. Comparison of brain activity during different types of proprioceptive inputs: a positron emission tomography study. Exp Brain Res. 2002;143(3):276–85. doi:10.1007/s00221-001-0994-4.
148. Weiller C, Juptner M, Fellows S, Rijntjes M, Leonhardt G, Kiebel S, Muller S, Diener HC, Thilmann AF. Brain representation of active and passive movements. Neuroimage. 1996;4(2):105–10. doi:10.1006/nimg.1996.0034.

Part IV

Auditory System

5 Basic Function

Masato Yumoto and Tatsuya Daikoku

Abstract

Magnetoencephalography (MEG) has been used to explore basic auditory function in humans. In this chapter, major streams of clinical MEG studies based on auditory basic function were briefly reviewed according to the stimulation-measurement paradigms and introduce our recent studies as one possibility to expand a range of auditory experimental paradigm. Repetition of an identical stimulus is the simplest form of stimulus sequence. A paired-click paradigm, in which inter-pair interval is the only statistical variable, has been most commonly used in clinical application as a probe for sensory gating impairment. Given that another stimulus with different physical property is infrequently inserted into a series of identical stimulus repetition, the sequence becomes well-known oddball paradigms, which can probe impairment of the sensory memory trace. The common target of these paradigms has been the neural correlates of impairment in neuropsychiatric disorders, mostly language related. More than two distinct stimuli can be organized into statistical structure similar to language and music. Tone sequences with higher-order structure can potentially probe impairment of higher brain function. Finally, recent experiments in the authors' laboratory are described as a seed for potential application.

Keywords

Paired-click paradigm • Oddball paradigm • Statistical learning • Markov stochastic model • Magnetoencephalography

---

M. Yumoto (✉) • T. Daikoku
Department of Clinical Laboratory, Graduate School of Medicine, The University of Tokyo, 7-3-1 Hongo, Bunkyo-ku, 113-8655 Tokyo, Japan
e-mail: yumoto-tky@umin.ac.jp

S. Tobimatsu, R. Kakigi (eds.), *Clinical Applications of Magnetoencephalography*,
DOI 10.1007/978-4-431-55729-6_5

5.1 Introduction

Magnetoencephalography (MEG) is a noninvasive technique for investigating neuronal activity in the living human brain and has been used to study human auditory function since 1990. MEG is suitable to study auditory function due to two major advantages. First, the MEG does not produce any sound during measurement. The neuromagnetometer system measuring MEG activity is a completely passive-mode machine that solely measures magnetic flux produced by the neuronal current in the brain without any radiation or production of physical particles or electromagnetic waves. It is a great advantage for auditory researchers that the MEG can record neural activity in complete silence, avoiding contamination of acoustic stimulation delivered to the subjects with machine noise. Second, the MEG can well resolve the signals produced by the auditory cortices located bilaterally in the temporal lobes. It is more difficult to resolve electric potentials produced by the auditory cortex in each hemisphere, considering the symmetrical scalp potential when either side of the auditory cortex is activated in isolation. Additionally, any auditory stimulation cannot solely activate a single side of the auditory cortex by the anatomical limitation. Auditory information from each ear does radiate both sides of the auditory cortex with contralateral weighting, without utilizing specialized experimental conditions such as frequency tagging [17]. Furthermore, the electric potential reflects activity in the subcortical regions as well as in the auditory cortex whereas MEG predominantly reflects cortical responses. Thus, the MEG is an indispensable tool to noninvasively measure and analyze neural signals from the auditory cortices with high temporal and spatial resolution.

A considerable amount of studies concerning human auditory system has so far been reported, addressing selective attention [39], auditory stream segregation [81], scene analysis [20], and information masking. In this chapter, we will briefly review clinical MEG studies concerning basic auditory function from the viewpoint of experimental paradigms and describe our recent trials to expand experimental paradigm specifically suitable to auditory research.

5.2 Repetition

Simple averaging of the evoked responses to repeated stimuli (e.g., a click, tone, or noise burst) has long been used in clinical studies. In adults, the most prominent deflection of the auditory evoked responses (AEFs) to tone bursts with enough duration presented with sufficient interstimulus intervals for response recovery is N1m (magnetic counterpart of N1 potential) peaking approximately 100 ms after stimulus onset [21, 33, 44, 51]. Although tone clicks or bursts with rapidly rising envelope can elicit earlier magnetic responses such as brain stem [15] and middle latency responses [40], clinical application has been limited [31].

The AEF components, as well as auditory evoked potentials (AEPs) [19], are known to change with maturation [25, 84]. In children until the age of 7 years, the most prominent AEP component is P1, from which the later components N1 and P2

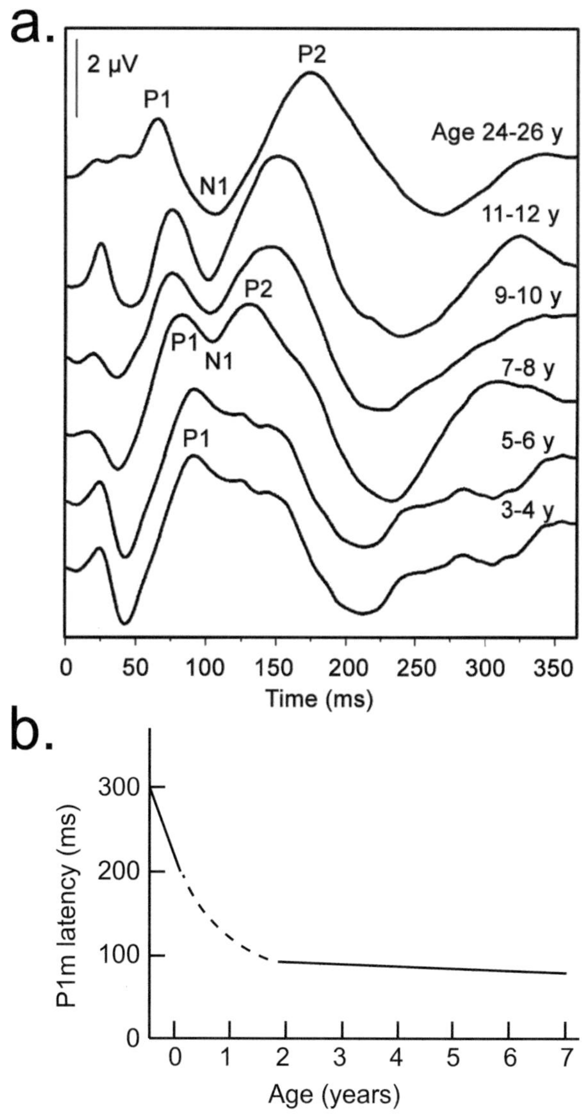

Fig. 5.1 (**a**) Maturation of auditory evoked responses (Revised from [19]). Evoked potentials were recorded at Cz referenced to the right mastoid from normal-hearing participants ($N = 8$–12 for each age group), while a 23-ms speech syllable was presented with an interstimulus interval of 2 s. Averaged responses were filtered with a band-pass 4–30 Hz filter and grand-averaged within each age group. (**b**) Developmental latency shortening of P1m revealed by the two previous studies (Revised from [25] and [84])

are branching (Fig. 5.1). The P1 (P1m) response shows a progressive and rapid decrease in latency along white matter maturation (i.e., myelination), which can be represented by increasing fractional anisotropy (FA) in diffusion tensor imaging. The white matter FA in the acoustic radiations of the auditory pathway, from the medial geniculate nucleus in the thalamus to the primary auditory cortex, is negatively correlated with P1m latency [60]. This latency shortening is delayed in children with autism [61, 62].

The source for N1m responses to pure tones in the left hemisphere is known to be localized posterior (~14 mm) to that in the right hemisphere [14], resulting at least partially from the larger planum temporale in the left hemisphere. A significant reduction of the right-sided N1m anteriority consistent with neuroanatomical deviation was first reported in patients with schizophrenia [63]. Recently, similar reduction of the anteriority was replicated in patients with autism and an association between language functioning and the degree of asymmetry was suspected [68].

Gamma-band oscillatory activity in MEG has been considered as a reliable marker for cognitive function. Oscillatory activity can be evoked or induced by a variety of cognitive tasks. In auditory modality, repetition of click trains has often been used as stimuli to record the auditory steady-state response (ASSR), which exhibits resonant frequencies in response to the click trains at approximately 40 Hz most frequently [50, 54]. Previous MEG studies reported reduced gamma-band ASSR in patients with chronic schizophrenia [76] and in patients with bipolar disorder [49].

5.3 Paired Repetition

Paired-click paradigm has been used for clinical research since 1980s. Sensory gating function, which is considered to represent a very early stage of attention, has been studied using the paired-click paradigm [2]. In this paradigm, identical clicks (or short sounds) are presented as a pair "S1-S2" typically separated with 0.5 s at intertrial intervals of several to c.a. 10 s [48]. In normal subjects, the responses to the S2 are significantly reduced compared with those to the S1, which has been interpreted as a representation of sensory gating function to suppress incoming irrelevant sensory inputs. The most reliable neurophysiological index of the response reduction is considered to be P50 (or the magnetic counterpart P50m), which is generated near the primary auditory cortex peaking approximately 50 ms after the stimulus onset. In patients with schizophrenia, this amplitude reduction is less pronounced compared with normal subjects, which has been believed to account for cognitive symptoms such as trouble focusing induced by sensory overload.

The sensory gating effect was also studied by using MEG [8, 9, 13, 24, 27, 78]. The MEG sensory gating deficit in schizophrenia was also replicated with left hemispheric dominance [74] and was presumed partially resulting from a complex alteration of information processing [55] and impaired gating process associated with alpha oscillation [43]. In the schizophrenia group, anterior hippocampal volume was smaller, and both the P50 and M50 gating ratios were larger (worse) than in controls [75]. Patients with schizophrenia showed significantly higher P50m gating ratios to human voices specifically in the left hemisphere.

Moreover, patients with higher left P50m gating ratios showed more severe auditory hallucinations, while patients with higher right P50m gating ratios showed more severe negative symptoms [23]. Acoustic startle prepulse inhibition (PPI) is known as another index of sensory (sensorimotor) gating. However, previous studies found little correlation between the two measures, suggesting independent aspects of brain inhibitory functions [7, 26, 28].

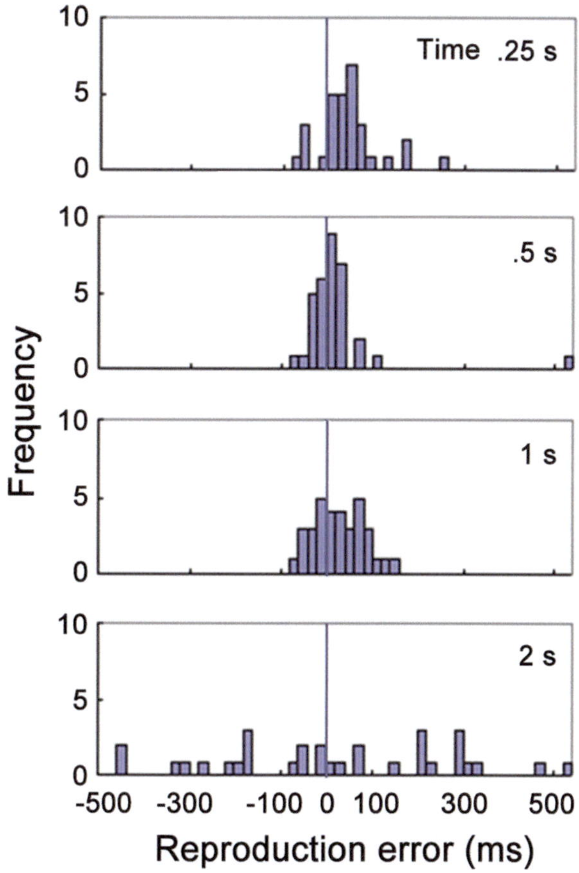

Fig. 5.2 A histogram of timing errors in a behavioral task of time-interval reproduction in a representative subject. Intervals of 0.25 and 0.5 s were well reproduced, whereas timing error increased drastically when the interval gets longer than 1 or 2 s. Behavioral performance of temporal reproduction correlated with neuromagnetic measures of P1m gating (unpublished data)

Several MEG studies were conducted to investigate the pathophysiology of stuttering [4, 6, 66, 67, 77]. Using this paired-click paradigm, a previous study reported that stutterers exhibited impaired left auditory sensory gating and expanded tonotopic organization in the right hemisphere, which was consistent with a significant increase in the gray matter volume of the right superior temporal gyrus revealed by voxel-based morphometry [37]. The degree of PPI, on the other hand, was not correlated with the effect of altered auditory feedback on stuttering [3].

In this paradigm, the inter-pair interval is the only variable parameter. The predictabilities of the occurrence timing of S1 and S2 are quite asymmetric. The forthcoming S2 appears 0.5 s after the most recent S1, whereas the S1 appears several to c.a. 10 s after the most recent S2. Previous studies on temporal reproduction and sensorimotor synchronization revealed hierarchically organized chronometric function in humans [42, 56]. The neural substrates for temporal prediction in audition in this time range may involve cortical and subcortical temporal processing network (Fig. 5.2) [69]. Function assessed by this paradigm may include not only attention-mediated gating but also temporal processing.

5.4 Oddball Paradigm

The oddball paradigm (first used by Squires et al. [72]) has been extensively used in a considerable range of MEG as well as EEG studies to probe human auditory system. A large part of the studies on human auditory function using oddball paradigms focused on the mismatch negativity (MMN) [46], which has been considered as an index of pre-attentive change detection processes occurred in the absence of directed attention. The conventional MMN is defined as a differential event-related component superimposed onto the response to oddballs. The MMN can hence be extracted by subtracting the responses to the standard stimuli from those to the rare stimuli in oddball paradigms. The MMN is one of the most promising neurophysiological candidates for biomarkers reflecting mental illness, such as schizophrenia (first reported by Shelley et al. [70]) and Alzheimer's disease (first reported by Pekkonen et al. [53]).

Previous studies showed that the MMNm (mismatch fields; the magnetic counterpart of the MMN) produced by speech sound rather than tonal stimuli exhibited more marked reduction in schizophrenia [35]. The power of the speech-sound MMNm in the left hemisphere was positively correlated with gray matter volume of the left planum temporale in patients with schizophrenia, implying that a significant reduction of the gray matter volume of the left planum temporale may underlie functional abnormalities of fundamental language-related processing in schizophrenia (Fig. 5.3) [82]. Latency for speech-sound MMNm in adults with autism was prolonged compared to the normal subjects [34]. The prolonged peak latency of speech-sound MMNm was replicated in children with autism [59], and the delay was most evident in those with concomitant language impairment, which may validate the speech-sound MMNm as a biomarker for pathology relevant to language ability.

MMN generation is regulated through the N-methyl-D-aspartate-type glutamate receptor [30], which was supported by a recent study showing that the MMNm was

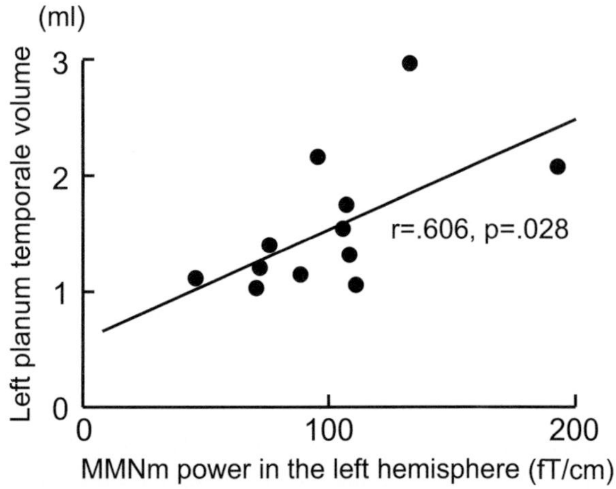

Fig. 5.3 Scatterplot depicting a correlation between the phonetic MMN power in the left hemisphere and the *gray* matter volume of the left planum temporale in patients with schizophrenia ($n = 13$). (Revised from [82])

modulated by genetic variations in metabotropic glutamate receptor 3 (GRM3) in healthy subjects [36]. It has been known that people with the variant of the GRM3 gene were at increased risk of developing bipolar disorder as well as schizophrenia [32]. The power of MMNm in the right hemisphere under the pure-tone condition was significantly delayed in patients with bipolar disorder [73]. MMNm elicited by pitch deviance could be a potential trait marker reflecting the global severity of bipolar disorder [71]. Recent advances in this research line can be found elsewhere [47].

In the research lines of basic studies, it was found that ensembles of physically different stimuli that share the same abstract feature could produce MMN (abstract-feature MMN) [64]. A notion of error signals in a framework of predictive processing may include MMN-like activities produced by cross-modal oddball paradigms such as visual-auditory [22, 79, 85, 87] and motor-auditory [86] links. In between the previous notion of stimulus repetition and oddball paradigms, roving standard paradigm has been receiving attention [10]. Using this paradigm, a recent study proposed two separate mechanisms involved in auditory memory trace formation [58].

5.5 Higher-Order Stochastic Sequence

Predictive coding frameworks of perception [16] tell us that most of the stimulus sequences used in AEF studies may have been too simple for our brain as a highly predictive organ. In other words, most of the event-related response studies may reflect the brain already adapted to (i.e., learned) the experimental paradigm (i.e., contrivances). Indeed, rule learning is achieved very rapidly even when complex rules are embedded in the sequence [5, 45].

The Markov chain [41] is one of the methods that can regulate statistical rules and systematically expand the scope of stochastic information processing in auditory system. The Markov chain model is a specific form of variable-order Markov models, which have been applied to a wide variety of research fields such as information theory, machine-learning, and human learning of artificial grammars [57]. The Markov property is described that the next state depends only on the recent state and not on the sequence of events that preceded it.

A conventional oddball paradigm in which standard (s) stimuli are repeatedly presented while infrequently replaced by the other rare (r) stimuli with a probability of p is described by the following two unconditional probability components: $\{\mathrm{P}(s) = 1 - \mathrm{p}, \mathrm{P}(r) = \mathrm{p}\}$. The oddball sequence is often constrained by pseudo-random replacement of rare stimuli so that the rare stimuli are not presented consecutively. The stimulus sequence in such a case is considered as a discrete-time Markov chain (DTMC) of one order and can be described by the following three conditional probability components: $\{\mathrm{P}(s|s) = (1 - 2\mathrm{p})/(1 - \mathrm{p}), \mathrm{P}(r|s) = \mathrm{p}/(1 - \mathrm{p}), \mathrm{P}(s|r) = 1\}$. The next stimulus "$s$ or r" is statistically defined by the most recent stimulus "s or r" (Fig. 5.4a). The DTMC, thus involving conventional

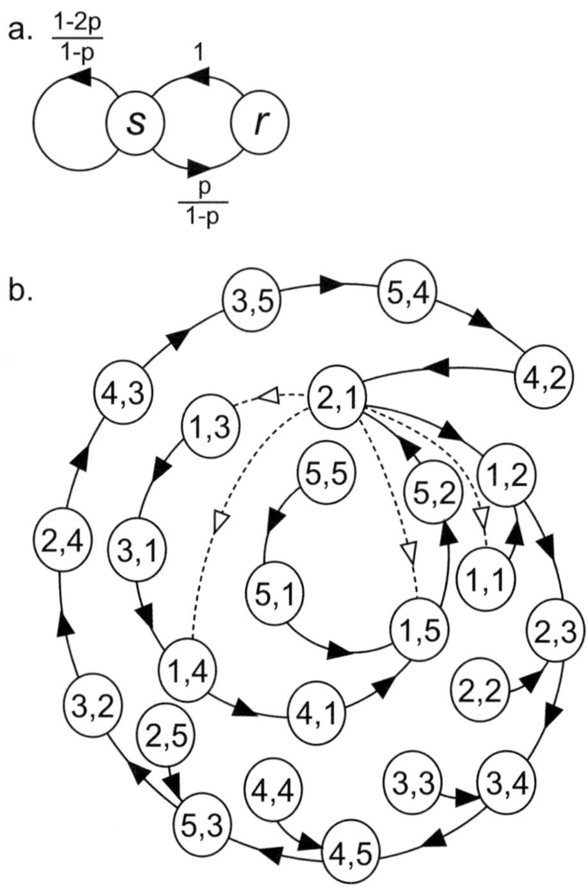

Fig. 5.4 (**a**) A state transition diagram of pseudo-random oddball sequences. From the law of total probability $P(r) = P(r|s)P(s) + P(r|r)P(r)$; $P(s) = P(s|s)P(s) + P(s|r)P(r)$, where $P(r|r) = 0$, $P(s|r) = 1$, $P(r) = p$, and $P(s) = 1 - p$, we have the transitional probabilities $P(s|s) = (1 - 2p)/(1 - p)$ and $P(r|s) = p/(1 - p)$ $(p < 0.5)$. (**b**) A state transition diagram of a second-order Markov chain used in our study. The *circled digits* indicate two adjacent tones: "3,5" indicates tone 5 and is followed by tone 3. The *arrows* on the *solid lines* indicate transitions with 80 % probability and those on the dashed lines indicate transitions with 5 % probability each from the state "2,1." All the other transition *arrows* with 5 % probability were ignored to avoid illegibility

pseudo-random oddball sequences, can systematically control the randomness and regularities embedded in stimulus sequence.

Although many behavioral studies investigated auditory sequence learning or statistical learning [65], surprisingly few neurophysiological studies have been conducted to date [1, 12, 52]. To the best of our knowledge, Furl et al. first conducted an MEG study using stimulus sequences based on a Markov stochastic model [18]. In this study, MEG responses were recorded while listeners were

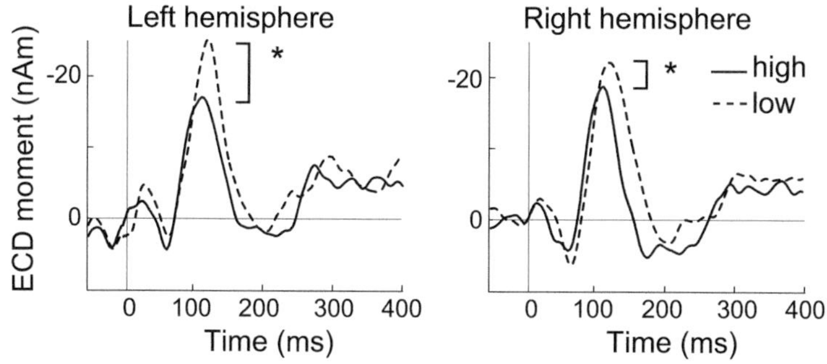

Fig. 5.5 Representative grand-averaged source-strength waveforms for the N1m responses ($N = 14$). The *solid lines* represent the responses to tones that appeared with higher transitional probability, and the *dashed lines* represent the responses to tones that appeared with lower transitional probability

paying attention to the auditory sequence. We carried out a study using Markov paradigm to clarify whether learning achievement was detected by a neurophysiological measure in ignoring condition, taking clinical susceptibility of ignoring condition into consideration. A difficulty of the sequence learning, which is parallel to the time that is needed to learn the sequence, depends on variabilities of stimuli, transition patterns, and the order of the Markov chains.

After pilot studies, we set the parameters so that the learning effect could be captured within the measurement period of 20–30 min. Figure 5.4b indicates an example of stochastic sequences used in our recent study [11].

During exposure sequence progression, the responses to the tones that appeared with higher transitional probability decayed, whereas those to the tones that appeared with lower probability retained their amplitude. Temporal profiles of the source waveforms seem astonishingly similar to those depicted in previous MMN/MMNm studies (Fig. 5.5). The amplitude decay, which may reflect learning achievement, was more rapid in attending condition compared to ignoring condition (Fig. 5.6). Learning effect detected in between the first and second thirds of the explicit sequence was preserved in the last third of the sequence, in which spectral shifts occurred without changing relative pitch intervals (i.e., transposition), suggesting that the participants could recognize the transposed sequence as the same melody already learned in the preceding exposure sequence. In ignoring condition, the MEG data indicated that the participants could finally learn the melody.

The temporal profiles of the difference waveforms obtained from the responses to tones that appeared with higher and lower transitional probabilities were quite similar to the MMNm. The difference observed in this study cannot be explained by short-term effects based on the assumption that there are distinct change-specific neurons in the auditory cortex that elicit the MMN. On the other hand, the adaptation hypothesis assuming that preceding stimuli adapt feature-specific neurons was proposed to interpret the MMN [29].

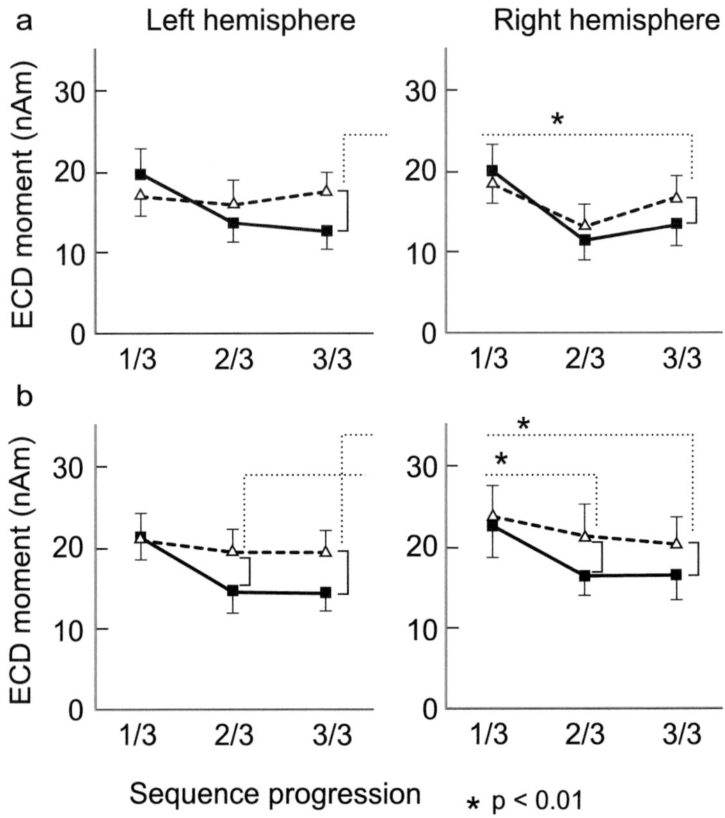

Fig. 5.6 Time courses of N1m amplitude changes along sequence progression in the ignoring (**a**) and attending conditions (**b**) ($N = 14$). The *solid lines* link N1m amplitudes for tones that appeared with higher transitional probability, and the *dashed lines* link N1m amplitudes for tones that appeared with lower transitional probability. The *error bars* indicate the standard error of the mean

The present results suggest that the adaptation to the statistical structure embedded in tone sequences may extend longer in timescale than sensory memory, which has been considered to work as a comparator between immediate past and forthcoming stimulus in previous studies using conventional oddball paradigm. A possible comparator that works during the statistical learning in this time range may involve the hippocampus [38].

Recently, Yaron et al. found that neurons in the rat auditory cortex were sensitive to the detailed structure of sound sequences over timescales of minutes, by controlling periodicity of deviant occurrence in oddball sequences [83]. On the other hand, Wilson et al. found a difference between humans and macaque monkeys despite a considerable level of cross-species correspondence [80]. These findings suggest that, although the mammals share auditory statistical learning mechanism expanding a timescale of minutes, there may be some difference between humans and monkeys when a statistical structure of the auditory sequence becomes as complex as human languages.

One of the main functions of our brain is to convert ensembles of individual occurrences in our environment into integrated knowledge by accumulating incidence and extracting hidden rules in order to better cope with future environmental changes. Introduction of higher-order stochastic rules into neurophysiological measure may give us a new scope for clinical pathological evaluation.

5.6 Conclusion

Clinical studies related to basic auditory function were briefly reviewed and our recent trials were described as a possible way to further investigate human auditory function. Any auditory sequence has statistics in the temporal and spectral axes. In discrete tone sequences, simple repetition $\{P(s) = 1\}$ and random sequence $\{P(Si) = 1/i\}$ are both ends of a broad statistical spectrum of the auditory events occurred in our living environment. Auditory sequences such as language and music have a hierarchical statistical structure, which can be interpreted as a context inspired by the human brain. Conversely, the human brain also processes sensory inputs in context-seeking fashion, which leads to integration of knowledge necessary to cope with forthcoming events at least cost. In any experimental approach to the human brain, such learning feature of the brain cannot be neglected. It must be noted that various levels of learning (e.g., gaiting, adaptation, reasoning) that involve distinct coupling of multiple brain regions, depending on the experimental paradigms or tasks, may underlie acquired neurophysiological data.

References

1. Abla D, Katahira K, Okanoya K. On-line assessment of statistical learning by event-related potentials. J Cogn Neurosci. 2008;20(6):952–64. doi:10.1162/jocn.2008.20058.
2. Adler LE, Pachtman E, Franks RD, Pecevich M, Waldo MC, Freedman R. Neurophysiological evidence for a defect in neuronal mechanisms involved in sensory gating in schizophrenia. Biol Psychiatry. 1982;17(6):639–54.
3. Alm PA. Stuttering and sensory gating: a study of acoustic startle prepulse inhibition. Brain Lang. 2006;97(3):317–21. doi:10.1016/j.bandl.2005.12.001.
4. Beal DS, Quraan MA, Cheyne DO, Taylor MJ, Gracco VL, De Nil LF. Speech-induced suppression of evoked auditory fields in children who stutter. NeuroImage. 2011;54 (4):2994–3003. doi:10.1016/j.neuroimage.2010.11.026.
5. Bendixen A, Schroger E. Memory trace formation for abstract auditory features and its consequences in different attentional contexts. Biol Psychol. 2008;78(3):231–41. doi:10.1016/j.biopsycho.2008.03.005.
6. Biermann-Ruben K, Salmelin R, Schnitzler A. Right rolandic activation during speech perception in stutterers: a MEG study. NeuroImage. 2005;25(3):793–801. doi:10.1016/j.neuroimage.2004.11.024.
7. Braff DL, Light GA, Swerdlow NR. Prepulse inhibition and P50 suppression are both deficient but not correlated in schizophrenia patients. Biol Psychiatry. 2007;61(10):1204–7. doi:10.1016/j.biopsych.2006.08.015.

8. Clementz BA, Blumenfeld LD, Cobb S. The gamma band response may account for poor P50 suppression in schizophrenia. Neuroreport. 1997;8(18):3889–93.
9. Clementz BA, Dzau JR, Blumenfeld LD, Matthews S, Kissler J. Ear of stimulation determines schizophrenia-normal brain activity differences in an auditory paired-stimuli paradigm. Eur J Neurosci. 2003;18(10):2853–8.
10. Cowan N, Winkler I, Teder W, Naatanen R. Memory prerequisites of mismatch negativity in the auditory event-related potential (ERP). J Exp Psychol Learn Mem Cogn. 1993;19 (4):909–21.
11. Daikoku T, Yatomi Y, Yumoto M. Implicit and explicit statistical learning of tone sequences across spectral shifts. Neuropsychologia. 2014;63:194–204. doi:10.1016/j.neuropsychologia. 2014.08.028.
12. Daikoku T, Yatomi Y, Yumoto M. Statistical learning of music- and language-like sequences and tolerance for spectral shifts. Neurobiol Learn Mem. 2015;118:8–19. doi:10.1016/j.nlm. 2014.11.001.
13. Edgar JC, Huang MX, Weisend MP, Sherwood A, Miller GA, Adler LE, Canive JM. Interpreting abnormality: an EEG and MEG study of P50 and the auditory paired-stimulus paradigm. Biol Psychol. 2003;65(1):1–20.
14. Elberling C, Bak C, Kofoed B, Lebech J, Saermark K. Auditory magnetic fields from the human cerebral cortex: location and strength of an equivalent current dipole. Acta Neurol Scand. 1982;65(6):553–69.
15. Erne SN, Scheer HJ, Hoke M, Pantew C, Lutkenhoner B. Brainstem auditory evoked magnetic fields in response to stimulation with brief tone pulses. Int J Neurosci. 1987;37(3–4):115–25.
16. Friston K. A theory of cortical responses. Philos Trans R Soc Lond B Biol Sci. 2005;360 (1456):815–36. doi:10.1098/rstb.2005.1622.
17. Fujiki N, Jousmaki V, Hari R. Neuromagnetic responses to frequency-tagged sounds: a new method to follow inputs from each ear to the human auditory cortex during binaural hearing. J Neurosci. 2002;22(3):RC205.
18. Furl N, Kumar S, Alter K, Durrant S, Shawe-Taylor J, Griffiths TD. Neural prediction of higher-order auditory sequence statistics. NeuroImage. 2011;54(3):2267–77. doi:10.1016/j. neuroimage.2010.10.038.
19. Gilley PM, Sharma A, Dorman M, Martin K. Developmental changes in refractoriness of the cortical auditory evoked potential. Clin Neurophysiol: Off J Int Fed Clin Neurophysiol. 2005;116(3):648–57. doi:10.1016/j.clinph.2004.09.009.
20. Gutschalk A, Dykstra AR. Functional imaging of auditory scene analysis. Hear Res. 2014;307:98–110. doi:10.1016/j.heares.2013.08.003.
21. Hari R, Pelizzone M, Makela JP, Hallstrom J, Leinonen L, Lounasmaa OV. Neuromagnetic responses of the human auditory cortex to on- and offsets of noise bursts. Audiology. 1987;26 (1):31–43.
22. Herholz SC, Lappe C, Knief A, Pantev C. Neural basis of music imagery and the effect of musical expertise. Eur J Neurosci. 2008;28(11):2352–60. doi:10.1111/j.1460-9568.2008. 06515.x.
23. Hirano Y, Hirano S, Maekawa T, Obayashi C, Oribe N, Monji A, Kasai K, Kanba S, Onitsuka T. Auditory gating deficit to human voices in schizophrenia: a MEG study. Schizophr Res. 2010;117(1):61–7. doi:10.1016/j.schres.2009.09.003.
24. Hirano Y, Onitsuka T, Kuroki T, Matsuki Y, Hirano S, Maekawa T, Kanba S. Auditory sensory gating to the human voice: a preliminary MEG study. Psychiatry Res. 2008;163(3):260–9. doi:10.1016/j.pscychresns.2007.07.002.
25. Holst M, Eswaran H, Lowery C, Murphy P, Norton J, Preissl H. Development of auditory evoked fields in human fetuses and newborns: a longitudinal MEG study. Clin Neurophysiol: Off J Int Fed Clin Neurophysiol. 2005;116(8):1949–55. doi:10.1016/j.clinph.2005.04.008.
26. Hong LE, Summerfelt A, Wonodi I, Adami H, Buchanan RW, Thaker GK. Independent domains of inhibitory gating in schizophrenia and the effect of stimulus interval. Am J Psychiatry. 2007;164(1):61–5. doi:10.1176/appi.ajp.164.1.61.

27. Huang MX, Edgar JC, Thoma RJ, Hanlon FM, Moses SN, Lee RR, Paulson KM, Weisend MP, Irwin JG, Bustillo JR, Adler LE, Miller GA, Canive JM. Predicting EEG responses using MEG sources in superior temporal gyrus reveals source asynchrony in patients with schizophrenia. Clin Neurophysiol: Off J Int Fed Clin Neurophysiol. 2003;114(5):835–50.
28. Inui K, Tsuruhara A, Nakagawa K, Nishihara M, Kodaira M, Motomura E, Kakigi R. Prepulse inhibition of change-related P50m no correlation with P50m gating. Springerplus. 2013;2:588. doi:10.1186/2193-1801-2-588.
29. Jaaskelainen IP, Ahveninen J, Bonmassar G, Dale AM, Ilmoniemi RJ, Levanen S, Lin FH, May P, Melcher J, Stufflebeam S, Tiitinen H, Belliveau JW. Human posterior auditory cortex gates novel sounds to consciousness. Proc Natl Acad Sci U S A. 2004;101(17):6809–14. doi:10.1073/pnas.0303760101.
30. Javitt DC, Steinschneider M, Schroeder CE, Arezzo JC. Role of cortical N-methyl-D-aspartate receptors in auditory sensory memory and mismatch negativity generation: implications for schizophrenia. Proc Natl Acad Sci U S A. 1996;93(21):11962–7.
31. Kaga K, Kurauchi T, Yumoto M, Uno A. Middle-latency auditory-evoked magnetic fields in patients with auditory cortex lesions. Acta Otolaryngol. 2004;124(4):376–80.
32. Kandaswamy R, McQuillin A, Sharp SI, Fiorentino A, Anjorin A, Blizard RA, Curtis D, Gurling HM. Genetic association, mutation screening, and functional analysis of a Kozak sequence variant in the metabotropic glutamate receptor 3 gene in bipolar disorder. JAMA Psychiatry. 2013;70(6):591–8. doi:10.1001/jamapsychiatry.2013.38.
33. Kasai K, Asada T, Yumoto M, Takeya J, Matsuda H. Evidence for functional abnormality in the right auditory cortex during musical hallucinations. Lancet. 1999;354(9191):1703–4. doi:10.1016/S0140-6736(99)05213-7.
34. Kasai K, Hashimoto O, Kawakubo Y, Yumoto M, Kamio S, Itoh K, Koshida I, Iwanami A, Nakagome K, Fukuda M, Yamasue H, Yamada H, Abe O, Aoki S, Kato N. Delayed automatic detection of change in speech sounds in adults with autism: a magnetoencephalographic study. Clin Neurophysiol: Off J Int Fed Clin Neurophysiol. 2005;116(7):1655–64. doi:10.1016/j.clinph.2005.03.007.
35. Kasai K, Yamada H, Kamio S, Nakagome K, Iwanami A, Fukuda M, Yumoto M, Itoh K, Koshida I, Abe O, Kato N. Neuromagnetic correlates of impaired automatic categorical perception of speech sounds in schizophrenia. Schizophr Res. 2003;59(2–3):159–72.
36. Kawakubo Y, Suga M, Tochigi M, Yumoto M, Itoh K, Sasaki T, Kano Y, Kasai K. Effects of metabotropic glutamate receptor 3 genotype on phonetic mismatch negativity. PLoS One. 2011;6(10):e24929. doi:10.1371/journal.pone.0024929.
37. Kikuchi Y, Ogata K, Umesaki T, Yoshiura T, Kenjo M, Hirano Y, Okamoto T, Komune S, Tobimatsu S. Spatiotemporal signatures of an abnormal auditory system in stuttering. NeuroImage. 2011;55(3):891–9. doi:10.1016/j.neuroimage.2010.12.083.
38. Kumaran D, Maguire EA. Match mismatch processes underlie human hippocampal responses to associative novelty. J Neurosci. 2007;27(32):8517–24. doi:10.1523/JNEUROSCI.1677-07.2007.
39. Lee AK, Larson E, Maddox RK, Shinn-Cunningham BG. Using neuroimaging to understand the cortical mechanisms of auditory selective attention. Hear Res. 2014;307:111–20. doi:10.1016/j.heares.2013.06.010.
40. Lutkenhoner B, Krumbholz K, Lammertmann C, Seither-Preisler A, Steinstrater O, Patterson RD. Localization of primary auditory cortex in humans by magnetoencephalography. NeuroImage. 2003;18(1):58–66.
41. Markov AA. Extension of the limit theorems of probability theory to a sum of variables connected in a chain, Markov Chains, vol. 1. Wiley; 1971.
42. Mates J, Muller U, Radil T, Poppel E. Temporal integration in sensorimotor synchronization. J Cogn Neurosci. 1994;6(4):332–40. doi:10.1162/jocn.1994.6.4.332.
43. Mathiak K, Ackermann H, Rapp A, Mathiak KA, Shergill S, Riecker A, Kircher TT. Neuromagnetic oscillations and hemodynamic correlates of P50 suppression in schizophrenia. Psychiatry Res. 2011;194(1):95–104. doi:10.1016/j.pscychresns.2011.01.001.

44. Muhlnickel W, Elbert T, Taub E, Flor H. Reorganization of auditory cortex in tinnitus. Proc Natl Acad Sci U S A. 1998;95(17):10340–3.
45. Naatanen R, Astikainen P, Ruusuvirta T, Huotilainen M. Automatic auditory intelligence: an expression of the sensory-cognitive core of cognitive processes. Brain Res Rev. 2010;64 (1):123–36. doi:10.1016/j.brainresrev.2010.03.001.
46. Naatanen R, Gaillard AW, Mantysalo S. Early selective-attention effect on evoked potential reinterpreted. Acta Psychol (Amst). 1978;42(4):313–29.
47. Naatanen R, Shiga T, Asano S, Yabe H. Mismatch negativity (MMN) deficiency: a breakthrough biomarker in predicting psychosis onset. Int J Psychophysiol. 2015;95(3):338–44. doi:10.1016/j.ijpsycho.2014.12.012.
48. Nagamoto HT, Adler LE, Waldo MC, Freedman R. Sensory gating in schizophrenics and normal controls: effects of changing stimulation interval. Biol Psychiatry. 1989;25(5):549–61.
49. Oda Y, Onitsuka T, Tsuchimoto R, Hirano S, Oribe N, Ueno T, Hirano Y, Nakamura I, Miura T, Kanba S. Gamma band neural synchronization deficits for auditory steady state responses in bipolar disorder patients. PLoS One. 2012;7(7):e39955. doi:10.1371/journal.pone.0039955.
50. Otsuka A, Yumoto M, Kuriki S, Hotehama T, Nakagawa S. Frequency characteristics of neuromagnetic auditory steady-state responses to sinusoidally amplitude-modulated sweep tones. Clin Neurophysiol: official journal of the International Federation of Clinical Neurophysiology. 2015. doi:10.1016/j.clinph.2015.05.002 (in press).
51. Pantev C, Hoke M, Lutkenhoner B, Lehnertz K, Kumpf W. Tinnitus remission objectified by neuromagnetic measurements. Hear Res. 1989;40(3):261–4.
52. Paraskevopoulos E, Kuchenbuch A, Herholz SC, Pantev C. Statistical learning effects in musicians and non-musicians: an MEG study. Neuropsychologia. 2012;50(2):341–9. doi:10.1016/j.neuropsychologia.2011.12.007.
53. Pekkonen E, Jousmaki V, Kononen M, Reinikainen K, Partanen J. Auditory sensory memory impairment in Alzheimer's disease: an event-related potential study. Neuroreport. 1994;5 (18):2537–40.
54. Picton TW, John MS, Dimitrijevic A, Purcell D. Human auditory steady-state responses. Int J Audiol. 2003;42(4):177–219.
55. Popov T, Jordanov T, Weisz N, Elbert T, Rockstroh B, Miller GA. Evoked and induced oscillatory activity contributes to abnormal auditory sensory gating in schizophrenia. NeuroImage. 2011;56(1):307–14. doi:10.1016/j.neuroimage.2011.02.016.
56. Poppel E. Pre-semantically defined temporal windows for cognitive processing. Philos Trans R Soc Lond B Biol Sci. 2009;364(1525):1887–96. doi:10.1098/rstb.2009.0015.
57. Reber AS. Implicit learning of artificial grammars. J Verbal Learn Verbal Behav. 1967;6:855–63.
58. Recasens M, Leung S, Grimm S, Nowak R, Escera C. Repetition suppression and repetition enhancement underlie auditory memory-trace formation in the human brain: an MEG study. NeuroImage. 2015;108:75–86. doi:10.1016/j.neuroimage.2014.12.031.
59. Roberts TP, Cannon KM, Tavabi K, Blaskey L, Khan SY, Monroe JF, Qasmieh S, Levy SE, Edgar JC. Auditory magnetic mismatch field latency: a biomarker for language impairment in autism. Biol Psychiatry. 2011;70(3):263–9. doi:10.1016/j.biopsych.2011.01.015.
60. Roberts TP, Khan SY, Blaskey L, Dell J, Levy SE, Zarnow DM, Edgar JC. Developmental correlation of diffusion anisotropy with auditory-evoked response. Neuroreport. 2009;20 (18):1586–91. doi:10.1097/WNR.0b013e3283306854.
61. Roberts TP, Lanza MR, Dell J, Qasmieh S, Hines K, Blaskey L, Zarnow DM, Levy SE, Edgar JC, Berman JI. Maturational differences in thalamocortical white matter microstructure and auditory evoked response latencies in autism spectrum disorders. Brain Res. 2013;1537:79–85. doi:10.1016/j.brainres.2013.09.011.
62. Roberts TP, Schmidt GL, Egeth M, Blaskey L, Rey MM, Edgar JC, Levy SE. Electrophysiological signatures: magnetoencephalographic studies of the neural correlates

of language impairment in autism spectrum disorders. Int J Psychophysiol. 2008;68 (2):149–60. doi:10.1016/j.ijpsycho.2008.01.012.
63. Rojas DC, Teale P, Sheeder J, Simon J, Reite M. Sex-specific expression of Heschl's gyrus functional and structural abnormalities in paranoid schizophrenia. Am J Psychiatry. 1997;154 (12):1655–62.
64. Saarinen J, Paavilainen P, Schoger E, Tervaniemi M, Naatanen R. Representation of abstract attributes of auditory stimuli in the human brain. Neuroreport. 1992;3(12):1149–51.
65. Saffran JR, Aslin RN, Newport EL. Statistical learning by 8-month-old infants. Science. 1996;274(5294):1926–8.
66. Salmelin R, Schnitzler A, Schmitz F, Freund HJ. Single word reading in developmental stutterers and fluent speakers. Brain. 2000;123(Pt 6):1184–202.
67. Salmelin R, Schnitzler A, Schmitz F, Jancke L, Witte OW, Freund HJ. Functional organization of the auditory cortex is different in stutterers and fluent speakers. Neuroreport. 1998;9 (10):2225–9.
68. Schmidt GL, Rey MM, Oram Cardy JE, Roberts TP. Absence of M100 source asymmetry in autism associated with language functioning. Neuroreport. 2009;20(11):1037–41. doi:10.1097/WNR.0b013e32832e0ca7.
69. Schwartze M, Tavano A, Schroger E, Kotz SA. Temporal aspects of prediction in audition: cortical and subcortical neural mechanisms. Int J Psychophysiol. 2012;83(2):200–7. doi:10.1016/j.ijpsycho.2011.11.003.
70. Shelley AM, Ward PB, Catts SV, Michie PT, Andrews S, McConaghy N. Mismatch negativity: an index of a preattentive processing deficit in schizophrenia. Biol Psychiatry. 1991;30 (10):1059–62.
71. Shimano S, Onitsuka T, Oribe N, Maekawa T, Tsuchimoto R, Hirano S, Ueno T, Hirano Y, Miura T, Kanba S. Preattentive dysfunction in patients with bipolar disorder as revealed by the pitch-mismatch negativity: a magnetoencephalography (MEG) study. Bipolar Disord. 2014;16 (6):592–9. doi:10.1111/bdi.12208.
72. Squires NK, Squires KC, Hillyard SA. Two varieties of long-latency positive waves evoked by unpredictable auditory stimuli in man. Electroencephalogr Clin Neurophysiol. 1975;38 (4):387–401.
73. Takei Y, Kumano S, Maki Y, Hattori S, Kawakubo Y, Kasai K, Fukuda M, Mikuni M. Preattentive dysfunction in bipolar disorder: a MEG study using auditory mismatch negativity. Prog Neuropsychopharmacol Biol Psychiatry. 2010;34(6):903–12. doi:10.1016/j.pnpbp.2010.04.014.
74. Thoma RJ, Hanlon FM, Moses SN, Edgar JC, Huang M, Weisend MP, Irwin J, Sherwood A, Paulson K, Bustillo J, Adler LE, Miller GA, Canive JM. Lateralization of auditory sensory gating and neuropsychological dysfunction in schizophrenia. Am J Psychiatry. 2003;160 (9):1595–605.
75. Thoma RJ, Hanlon FM, Petropoulos H, Miller GA, Moses SN, Smith A, Parks L, Lundy SL, Sanchez NM, Jones A, Huang M, Weisend MP, Canive JM. Schizophrenia diagnosis and anterior hippocampal volume make separate contributions to sensory gating. Psychophysiology. 2008;45(6):926–35. doi:10.1111/j.1469-8986.2008.00692.x.
76. Tsuchimoto R, Kanba S, Hirano S, Oribe N, Ueno T, Hirano Y, Nakamura I, Oda Y, Miura T, Onitsuka T. Reduced high and low frequency gamma synchronization in patients with chronic schizophrenia. Schizophr Res. 2011;133(1–3):99–105. doi:10.1016/j.schres.2011.07.020.
77. Walla P, Mayer D, Deecke L, Thurner S. The lack of focused anticipation of verbal information in stutterers: a magnetoencephalographic study. NeuroImage. 2004;22(3):1321–7. doi:10.1016/j.neuroimage.2004.03.029.
78. Weiland BJ, Boutros NN, Moran JM, Tepley N, Bowyer SM. Evidence for a frontal cortex role in both auditory and somatosensory habituation: a MEG study. NeuroImage. 2008;42 (2):827–35. doi:10.1016/j.neuroimage.2008.05.042.

79. Widmann A, Kujala T, Tervaniemi M, Kujala A, Schroger E. From symbols to sounds: visual symbolic information activates sound representations. Psychophysiology. 2004;41(5):709–15. doi:10.1111/j.1469-8986.2004.00208.x.
80. Wilson B, Smith K, Petkov CI. Mixed-complexity artificial grammar learning in humans and macaque monkeys: evaluating learning strategies. Eur J Neurosci. 2015;41(5):568–78. doi:10.1111/ejn.12834.
81. Winkler I, Denham S, Mill R, Bohm TM, Bendixen A. Multistability in auditory stream segregation: a predictive coding view. Philos Trans R Soc Lond B Biol Sci. 2012;367 (1591):1001–12. doi:10.1098/rstb.2011.0359.
82. Yamasue H, Yamada H, Yumoto M, Kamio S, Kudo N, Uetsuki M, Abe O, Fukuda R, Aoki S, Ohtomo K, Iwanami A, Kato N, Kasai K. Abnormal association between reduced magnetic mismatch field to speech sounds and smaller left planum temporale volume in schizophrenia. NeuroImage. 2004;22(2):720–7. doi:10.1016/j.neuroimage.2004.01.042.
83. Yaron A, Hershenhoren I, Nelken I. Sensitivity to complex statistical regularities in rat auditory cortex. Neuron. 2012;76(3):603–15. doi:10.1016/j.neuron.2012.08.025.
84. Yoshimura Y, Kikuchi M, Ueno S, Shitamichi K, Remijn GB, Hiraishi H, Hasegawa C, Furutani N, Oi M, Munesue T, Tsubokawa T, Higashida H, Minabe Y. A longitudinal study of auditory evoked field and language development in young children. NeuroImage. 2014;101:440–7. doi:10.1016/j.neuroimage.2014.07.034.
85. Yumoto M, Matsuda M, Itoh K, Uno A, Karino S, Saitoh O, Kaneko Y, Yatomi Y, Kaga K. Auditory imagery mismatch negativity elicited in musicians. Neuroreport. 2005;16 (11):1175–8.
86. Yumoto M, Ogata E, Mizuochi T, Karino S, Kuan C, Itoh K, Yatomi Y. A motor-auditory cross-modal oddball paradigm revealed top-down modulation of auditory perception in a predictive context. Int Cogress Ser. 2007;1300:93–6.
87. Yumoto M, Uno A, Itoh K, Karino S, Saitoh O, Kaneko Y, Yatomi Y, Kaga K. Audiovisual phonological mismatch produces early negativity in auditory cortex. Neuroreport. 2005;16 (8):803–6.

Clinical Applications

6

Hidehiko Okamoto

Abstract

Recent neuroimaging and neurophysiological data suggest that maladaptive cortical reorganization in the human auditory cortex can lead to the emergence of hearing disorders such as subjective tinnitus. Reversing maladaptive cortical reorganization by suitable behavioral training can be a useful treatment strategy to alleviate detrimental auditory symptoms. Here, we report our recent studies on reorganized brain activity measured in hearing-impaired people using magnetoencephalography (MEG): (1) tailor-made notched music therapy (TMNMT) against tinnitus, (2) broadened population-level frequency tuning by inappropriate portable music player usage, and (3) constraint-induced sound therapy (CIST) against sudden sensorineural hearing loss. The results obtained indicated that rehabilitation approaches were effective in the treatment of hearing disorders induced by maladaptive cortical reorganization. MEG itself does not produce any sound that interferes with the brain responses elicited by test sounds. Therefore, it provides important insights into the auditory neural mechanisms occurring in the brains of healthy and hearing-impaired people. MEG is a useful and noninvasive technique that contributes to the establishment of new clinical applications in the auditory system.

Keywords

Lateral inhibition • Sudden sensorineural hearing loss • Music • Rehabilitation • Tinnitus

H. Okamoto (✉)
Department of Integrative Physiology, National Institute for Physiological Sciences, 38 Nishigo-Naka, Myodaiji, Okazaki 444-8585, Japan

Department of Physiological Sciences, SOKENDAI (The Graduate University for Advanced Studies), Hayama, Japan
e-mail: hokamoto@nips.ac.jp

S. Tobimatsu, R. Kakigi (eds.), *Clinical Applications of Magnetoencephalography*,
DOI 10.1007/978-4-431-55729-6_6

6.1 Introduction

The cortex in adult humans is not hard-wired; it can be reorganized by behaviors and experiences. Cortical reorganization in the human auditory cortex can be beneficial. For example, the skills of professional musicians have been attributed to reorganized cortical maps (e.g., auditory cortex expansion in musicians with absolute pitch [1]). Although cortical reorganization is generally considered to be beneficial, it may also be disadvantageous [2]. Cortical reorganization has been shown to play an important role in the emergence of detrimental symptoms such as tinnitus [3] and phantom limb pain [4]. When cortical reorganization induces symptoms that worsen the quality of life, suitable behavioral treatment approaches are used to reverse this maladaptive reorganization [2]. Recent neurophysiological studies reported that the reversal of maladaptive cortical reorganization by behavioral training may alleviate subjective detrimental symptoms, even in the adult brain [5, 6]. For example, in the somatosensory domain, phantom limb pain, which is caused by maladaptive changes in the somatotopic map after amputation [4], can be relieved by using functional prostheses [7]. In the case of motor function, focal hand dystonia, which causes involuntary muscular contractions accompanied by maladaptive reorganization in the motor cortex [8], is effectively treated by constraint-induced movement therapy [9]. These findings have provided a basis for the adaptation of similar treatment strategies to the auditory system. In this chapter, we have discussed clinical applications based on cortical plasticity in the auditory system. We also introduced our recent studies on brain activity measured in hearing-impaired people using magnetoencephalography (MEG): (1) tailor-made notched music therapy against tinnitus [10–12], (2) broadened population-level frequency tuning by inappropriate portable music player usage [13], and (3) constraint-induced sound therapy against sudden sensorineural hearing loss [14].

6.2 Clinical Applications to the Auditory System

6.2.1 Tailor-Made Notched Music Therapy (TMNMT)

Tinnitus is a widespread disease, especially in industrially advanced nations [15, 16]. It is mostly a subjective symptom that can be defined as the conscious and unpleasant perception of a sound that lacks an objective physical source [17]. Although subjective tinnitus itself is not a fatal disease, it markedly reduces the quality of life of many people [18, 19]. However, the neural mechanisms underlying the generation of tinnitus have not yet been elucidated in detail, and causal treatments for tinnitus are very limited. Human neuroimaging studies recently reported that subjective tinnitus appeared to be caused by maladaptive reorganization in the auditory cortex [3, 20, 21]. Auditory cortex areas deprived of afferent neural inputs from the periphery were not silenced following neural damage to the auditory pathway, but they were found to be rewired with excitatory neural inputs from neighboring neurons [4, 22–24]. The lateral inhibition of these

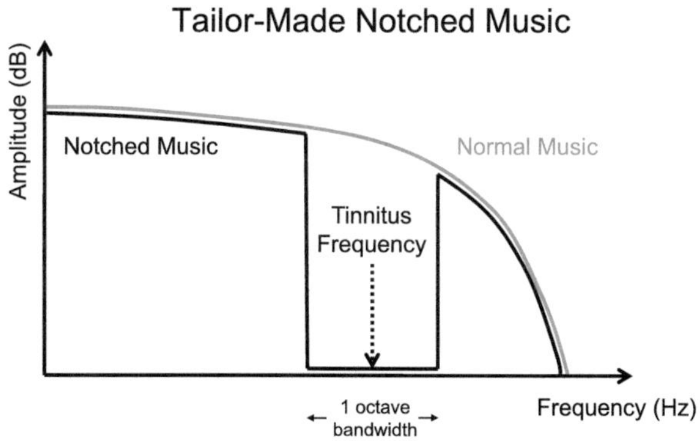

Fig. 6.1 Schematic model of the frequency spectrum of tailor-made notched music (*black line*). Music energy around individual tinnitus frequencies was eliminated from the normal music spectrum (*gray line*)

auditory cortex areas from neighboring cortical neurons was shown to be reduced [25, 26]. This loss of lateral inhibition in the auditory system may lead to the emergence of tinnitus [27]. We herein described a longitudinal, double-blinded treatment strategy, which consisted of regular listening to tailor-made notched music (TMNMT: see Fig. 6.1).

6.2.1.1 TMNMT Protocols

Participants

We recruited chronic tinnitus patients via different routes including local newspaper advertisements and flyers in local otorhinolaryngology clinics. All participants fulfilled the following criteria: (i) chronic tinnitus, (ii) lateralized tinnitus, (iii) tonal tinnitus, (iv) tinnitus frequency $\leq$8 kHz, (v) no severe hearing impairment, and (vi) no neurological or psychiatric complications. They were fully informed about the execution and goals of the study and gave written informed consent in accordance with procedures approved by the Ethics Commission of the Medical Faculty, University of Muenster. Participants were double-blindly assigned to one of two groups: (i) "target" notched music treatment or (ii) "placebo" notched music treatment. Eight tinnitus patients in both the "target" and "placebo" groups completed the 12-month training.

Measurement of Subjective Tinnitus Characteristics

It was essential to estimate the frequency of tinnitus for the TMNMT procedure. Tinnitus pitch matching was performed at least four times in order to ascertain whether the frequency of tinnitus was stable. Patients who exhibited a stable tinnitus frequency could participate in the study. Subjective tinnitus loudness was accessed via a visual analogue scale once per week. A baseline of tinnitus loudness was determined according to a period of at least 4 weeks prior to the study.

Auditory-Evoked Field Measurements

Auditory-evoked magnetic fields were recorded using a whole-head MEG system. MEG measurements were performed (1) prior to the study and then (2) 6 months and (3) 12 months after starting the study. During MEG measurements, two different frequency sound stimuli were delivered randomly to either the left or right ear. One of these stimuli corresponded to the patient's individual tinnitus frequency; the other stimulus had a frequency of 500 Hz as the control stimulus. Stimuli had a duration of 1 s with an initial 0.3-s continuous pure tone followed by a 0.7-s fully 40-Hz amplitude-modulated tone in order to simultaneously record both N1m (originating from the non-primary auditory cortex [28–30]) and the auditory steady-state response (ASSR) (generated in the primary auditory cortex [28, 29, 31]). The control stimulus was presented at a sensation level of 45 dB; the tinnitus frequency stimulus was matched in loudness to the control stimulus prior to the baseline MEG measurement.

We calculated source locations and orientations based on auditory-evoked fields and then calculated the ratios between the source strengths elicited by the tinnitus frequency stimulus and control stimulus. We normalized the source strength ratios measured 6 and 12 months after starting the study to the source strength ratios at the baseline.

Music Manipulation Procedure

Participants in the "target" and "placebo" groups provided their favorite music CDs. We filtered the music for each patient individually according to one of two filtering protocols: (i) "target" or (ii) "placebo." In the "target" filtering manipulation, the frequency component within one octave width centered at the individual tinnitus frequency was eliminated from the original music energy spectrum (Fig. 6.1). In the "placebo" filtering manipulation, a moving filter of one octave width, which did not influence the frequency component around the individual tinnitus frequency, was applied. Participants were asked to listen to their manipulated music for 2 h every day via headphones over the course of 1 year.

6.2.1.2 Behavioral and Neurophysiological Outcomes of TMNMT

No significant differences were observed in age, tinnitus duration, tinnitus frequency, or subjective tinnitus loudness between the "target" and "placebo" groups. Moreover, auditory-evoked response ratios at study baseline were similar between groups. In the "target" group, the subjective tinnitus loudness was significantly lower after 12 months of treatment than that at baseline. In contrast, no significant differences were observed from baseline for tinnitus loudness in the "placebo" group. A significant interaction was found for tinnitus loudness between groups ("target" vs. "placebo") and the time point of measurement (baseline vs. after 12 months of treatment). Both the ASSR and N1m source strength ratios in the "target" group were significantly reduced after 12 months. In contrast, no significant difference from baseline was observed in the ASSR or N1m ratio in the "placebo" group. These results indicated that the "target" notched music effectively reduced subjective tinnitus loudness and neural activity corresponding to the tinnitus frequency within the primary and non-primary auditory cortex.

6.2.2 Broadened Population-Level Frequency Tuning by Inappropriate Portable Music Player Usage

Loud sounds are known to impair hearing ability and cause various conditions such as hearing loss, hyperacusis, and tinnitus [20, 32, 33]. Occupational noise exposure mainly attracted attention in industrialized countries [34]; however, portable music players (PMP) have become one of the main sources of loud sounds and are regularly used by large numbers of people [35]. Adolescents often use PMPs in noisy environments such as on the train or bus at very high volumes to overcome the surrounding noise and fulfill their appetite for music [36]. Hearing impairments have been examined in most previous studies on intense PMP usage by measuring behavioral and subjective outcomes. However, inappropriate PMP usage may cause subtle damage that conventional measurements, which require the attention of the participant, may overlook. Human neuroimaging is a useful technique that objectively measures neural activity during distracted listening. Previous studies [37–41] succeeded in estimating population-level frequency tuning during distracted listening by measuring the auditory-evoked responses elicited by tonal test stimuli that were embedded within band-eliminated noises (c.f. Fig. 6.2). Auditory-evoked responses were found to be influenced less by simultaneously presented band-eliminated noises when population-level frequency tuning was sharp [37, 39, 40]. By using MEG, we objectively measured the auditory-evoked responses of two groups of young subjects (PMP "exposed" group and "control" group) to examine the effects of inappropriate PMP usage on population-level frequency tuning in the human auditory cortex.

6.2.2.1 Participants

In this study, we recruited two types of people: (1) "exposed" people who had listened to intense music via PMP regularly for at least the last 2 years and (2) "control" people who had not used PMP regularly. The two groups were matched in age and gender. The study was performed in accordance with the

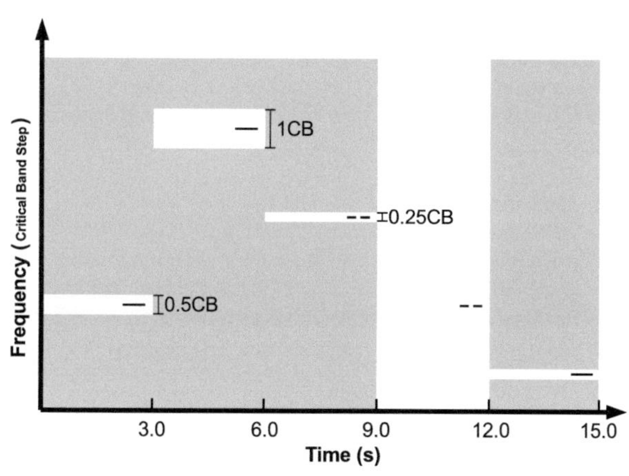

Fig. 6.2 Experimental design used to investigate population-level frequency tuning in the human auditory cortex. The *black line* and *gray* area represent a test pure tone and simultaneously presented band-eliminated noise, respectively. The eliminated bandwidth of a noise (*white areas*) was either 1/4, 1/2, or 1 critical band (CB) (Modified from Okamoto et al. [13])

Declaration of Helsinki, and all participants gave written informed consent for their participation.

6.2.2.2 MEG Measurement of Population-Level Frequency Tuning

We measured the auditory-evoked fields elicited by a pure tone presented either in silence or in band-eliminated noise (Fig. 6.2). Pure tones were presented in a pseudo-randomized manner in order to avoid expectation effects on the auditory-evoked fields [38]. The simultaneously presented noise had an eliminated band with widths of either 1/4, 1/2, or 1 critical band [42] (Fig. 6.2). Behavioral and MEG data were analyzed after pooling across all pure tones. Half of the pure tones had a silent gap in the middle. The 1,000-Hz pure tone was presented at a sensation level of 38 dB. The other pure tones were adjusted to have an identical power to the 1,000-Hz pure tone. The power of these noises was 12 dB higher than that of the pure tones.

MEG measurements were performed under two attentional conditions (distracted vs. focused listening). In the first MEG session, participants were asked to memorize the contents of a silent movie. In the second MEG session, participants had to press a response button when they noticed a silent gap within the pure tones. The auditory-evoked fields elicited by the pure tones were selectively averaged with respect to each band-eliminated noise condition (BEN_1/4CB, BEN_1/2CB, BEN_1CB, and no BEN) and not with respect to the pure tone frequencies. The neural sources of the N1m response [43] were estimated based on the no-noise condition, and the source strength waveforms were calculated for each band-eliminated noise condition and each attentional condition. The source strength waveforms of the exposed and control groups were similar when the participants performed the auditory task; however, they were different during distracted listening (Fig. 6.3).

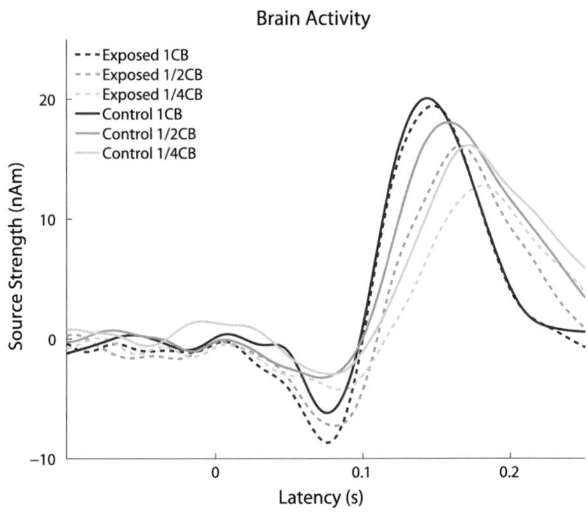

Fig. 6.3 Grand averaged source strength waveforms during distracted listening. *Solid* and *dashed* lines represent the control group and exposed group, which intensely and regularly used a portable music player (PMP), respectively. The *gray scale* codes the band-eliminated noise (BEN) condition (*black* = 1 critical band (CB), *dark gray* = 1/2 CB, *light gray* = 1/4 CB) (Modified from Okamoto et al. [13])

The normalized N1m source strengths in the distracted and focused listening conditions were separately analyzed via a repeated-measure ANOVA using the two factors BEN-TYPE (BEN_1/4CB, BEN_1/2CB, BEN_1CB) and GROUP ("exposed" vs. "control"). The repeated-measure ANOVA revealed a significant interaction between GROUP and BEN-TYPE ($p < 0.03$) in the distracted listening condition, but not between GROUP and BEN-TYPE in the focused listening condition. These results indicated that there was no significant difference in the neurophysiological outcomes between groups in the focused listening condition, whereas population-level frequency tuning was significantly broadened in the "exposed" group during distracted listening [37, 39, 40].

6.2.2.3 Behavioral Outcomes

Hearing thresholds for pure tone audiometry and the hearing-in-noise test [44] results did not differ significantly between the "exposed" and "control" groups. The reaction time and error rate during the MEG measurement in the focused listening condition were not significantly different between groups. Therefore, we did not detect any significant behavioral differences between the "exposed" and "control" groups when the participants performed auditory tasks.

6.2.3 Constraint-Induced Sound Therapy (CIST) Against Sudden Sensorineural Hearing Loss (SSHL)

Sudden sensorineural hearing loss (SSHL) is a hearing-impaired condition that is characterized by acute idiopathic sensorineural hearing loss [45]. Epidemiological studies previously estimated that the rates of SSHL treatment accepters were approximately 5–30 cases per 100,000 people per year [46–49]. However, the mechanism underlying the emergence of SSHL remains unknown, and the standard treatment for SSHL is still being debated [47, 50–53]. Neuroimaging studies recently demonstrated neural activity changes in the human auditory cortex after the onset of SSHL [54–56]. The extent of these cortical activity changes in the human brain inversely correlated with hearing recovery [57]. Therefore, cortical reorganization induced by SSHL may be maladaptive for the recovery of hearing in the affected ear. In order to reduce this maladaptive change, the potential consequences of "constraint-induced therapy," which urges patients to employ impaired body parts and at the same time disrupt them to use healthy counterparts, need to be considered [58–60]. Neuroimaging studies demonstrated that "constraint-induced movement therapy [59]" reduced maladaptive cortical reorganization in cortical areas [5, 61–63]. Therefore, we applied "constraint-induced sound therapy (CIST) [14]" to SSHL patients by plugging the canal of the unaffected ear and urging patients to listen to music with the impaired ear (Fig. 6.4). Treatment outcomes were assessed by the pure tone audiogram and brain activity measured by MEG.

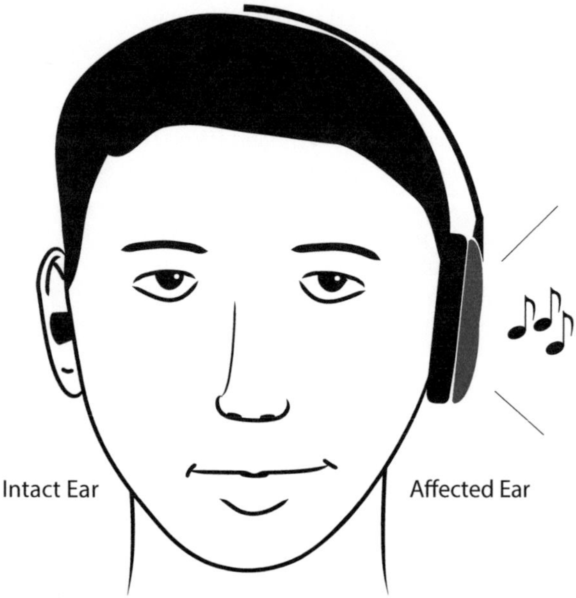

Fig. 6.4 Schematic illustration of constraint-induced sound therapy (CIST). The canal of the intact ear was plugged. Music was only presented to the affected ear (Modified from Okamoto et al. [14]: Drawing courtesy of Lothar Lagemann)

6.2.3.1 CIST Procedure

Participants

The CIST group consisted of 22 SSHL patients who received CIST in addition to corticosteroid therapy, while the control group consisted of 31 SSHL patents who received corticosteroid therapy only. All patients were fully informed about the execution and goals of the study and gave written informed consent in accordance with the Declaration of Helsinki. No significant differences were observed in the age and gender of the participants. The intact ear canal of patients in the CIST group was plugged for the whole day during their hospitalization. The CIST group also had to listen to classical music for 6 h per day with the affected ear by using a closed-type headphone. They were instructed to adjust the sound level and equalizer settings by themselves in order to make the music sound as natural as possible. This adjustment increased sound power around the affected frequencies. No adverse events associated with CIST were observed.

Pure Tone Audiogram

The hearing threshold levels of both the affected and intact ears were measured at the first entrance examination, when participants had been discharged from hospital (second measurement), and when they visited us as outpatients 1–6 months after being discharged (third measurement). The mean threshold differences between the affected and intact ears were similar between the groups before the treatment. However, the CIST group showed significantly better hearing recovery at the second and third hearing threshold examinations.

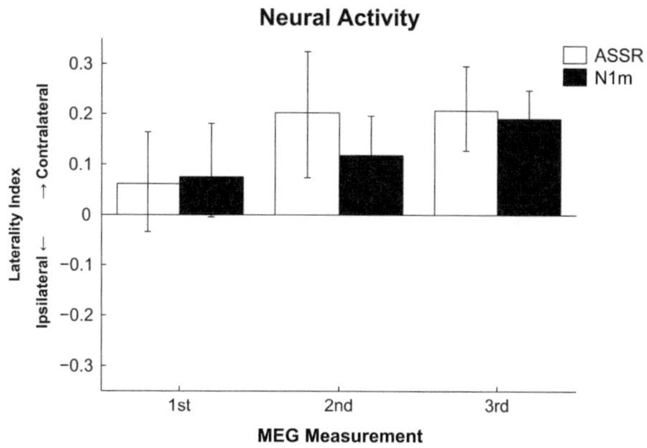

Fig. 6.5 Laterality indices (LIs) of the auditory steady state (ASSR: *white bars*) and N1m (*black bars*) responses obtained in sudden sensorineural hearing loss (SSHL) patients who received constraint-induced sound therapy at the first (*left*: entering hospital), second (*center*: discharged from hospital), and third (*right*: 3 months after being discharged) examinations. The error bars denote the 95 % confidence intervals (Adopted from Okamoto et al. 2014 [54])

Brain Activity Measurements

We measured auditory-evoked fields in six patients in the CIST group in three MEG sessions (before the treatment, at the end of hospitalization, and 3 months after being discharged). We measured brain responses elicited by a monaurally presented test stimulus. The test stimulus was a 40-Hz fully amplitude-modulated tone with the carrier frequency of the patient's worst hearing frequency in the affected ear. The auditory-evoked fields showed clear ASSR and N1m responses. In order to investigate the degree of cortical reorganization, we calculated the laterality indices of neural activity in the ipsilateral and contralateral hemispheres. The laterality indices of the ASSR and N1m responses were calculated as follows: ((source strengths elicited in the contralateral hemisphere) − (source strengths elicited in the ipsilateral hemisphere))/((source strengths elicited in the contralateral hemisphere) + (source strengths elicited in the ipsilateral hemisphere)). Figure 6.5 shows the laterality indices of ASSR and N1m responses in the CIST group. The laterality indices of both ASSR and N1m significantly increased over time. In the third examination, they reached approximately 0.2, which was similar to those of normal hearing people [64, 65].

6.3 General Remarks

6.3.1 Maladaptive Auditory Cortex Reorganization

Neurophysiological studies identified auditory cortex reorganization in hearing-impaired people [54, 66, 67]. This reorganization is considered to reflect the remodeling of central auditory neural connections triggered by the deprivation of their original sensory inputs from the cochlea [5, 68]. Due to the deprivation of bottom-up input, auditory neurons become excited by other neural inputs. Therefore, tonotopic maps in the auditory cortex appear to become disarranged with tinnitus, and these changes are basically responsible for the perception of tinnitus [69, 70]. In the case of SSHL, auditory cortex neurons that lose their bottom-up neural inputs from the affected cochlea may be rewired with new neural connections originating from the intact cochlea.

We utilized previous findings on use-dependent cortical plasticity to design a procedure suited to reduce the detrimental symptoms of tinnitus and improve the quality of life of patients. Behavioral training, the aim of which is to reverse maladaptive cortical reorganization, was effective for tinnitus and SSHL patients. The "target" notched music applied to tinnitus patients initiated reorganization of the auditory cortex area corresponding to the tinnitus frequency. Target notched music did not stimulate the tinnitus frequency area within the tonotopic map but activated surrounding neurons. Thus, neurons corresponding to the tinnitus frequency were suppressed via lateral inhibitory neural inputs originating from the surroundings [71]. SSHL patients who received CIST wore an earplug in the intact ear, while the affected ear was extensively stimulated with music, which might prevent maladaptive cortical reorganization related to SSHL.

Neural networks are composed not only of excitatory neurons but also of inhibitory neurons [39, 72–75]. Inhibitory neural networks contribute to sharp frequency tuning in the auditory system, leading to better auditory performance. These inhibitory networks were previously found to be sensitive to noise trauma [76, 77]. Therefore, damage limited to the inhibitory neural networks may broaden frequency tuning without elevating the hearing threshold. The broadened auditory frequency tuning in intense PMP users during distracted listening suggests that regular exposure to loud music via PMP led to plastic changes in tonotopic maps in the auditory cortex.

6.3.2 Impact of MEG Outcomes

The cortical source strength measured by MEG reflects the number of synchronously activated neurons. We measured neural activity related to auditory signals without asking participants to perform auditory tasks. The MEG data obtained provided us with objective indices on what happens in our brain due to behavioral training (e.g., TMNMT and CIST) or the inappropriate usage of PMP. The MEG results in the TMNMT study strongly suggested that the activity of auditory cortex neurons, which correspond to the tonotopic areas generating the tinnitus sensation, cumulatively decreased after target notched music listening. The auditory-evoked fields of the intense PMP users indicated that population-level frequency tuning was significantly broader than that of nonusers during distracted listening. However, no significant difference was observed between the PMP "exposed" and "control" groups in the behavioral measures that required auditory attention. Moreover, the auditory-evoked responses during focused listening were similar between the "exposed" and "control" groups. These results suggested that inappropriate PMP usage induced broadened population-level frequency tuning, which was easily compensated for by auditory-focused attention [37, 39, 40] and, as a consequence, be overlooked by standard behavioral measures. The MEG results obtained in SSHL patients demonstrated that neural activity elicited by monaural sounds became contralateral hemispheric dominant following CIST, similar to normal hearing subjects [57]. This result indicated that CIST assisted in neural reconnection between the affected cochlea and contralateral auditory cortex. The rewired neural connection led to the reversal of maladaptive cortical changes caused by the onset of SSHL.

6.3.3 Conclusion

Our treatment strategy based on cortical reorganization was derived from recent neurophysiological findings. We focused on reversing the maladaptive reorganization of specific auditory areas related to pathological symptoms by means of intensive and enjoyable rehabilitation approaches that were cost-effective and had minimal side effects. MEG is a useful and noninvasive technique for investigating cortical reorganization in the human brain. MEG itself does not produce any sound that interferes with the brain responses elicited by test sounds. Therefore, it provides important insights into the auditory neural mechanisms occurring in the brains of healthy and hearing-impaired people, leading to the establishment of new treatment strategies for hearing disorders.

References

1. Pantev C, et al. Increased auditory cortical representation in musicians. Nature. 1998; 392(6678):811–4.
2. Elbert T, Heim S. A light and a dark side. Nature. 2001;411(6834):139.
3. Muhlnickel W, et al. Reorganization of auditory cortex in tinnitus. Proc Natl Acad Sci U S A. 1998;95(17):10340–3.
4. Flor H, et al. Phantom-limb pain as a perceptual correlate of cortical reorganization following arm amputation. Nature. 1995;375(6531):482–4.
5. Elbert T, Rockstroh B. Reorganization of human cerebral cortex: the range of changes following use and injury. Neuroscientist. 2004;10(2):129–41.
6. Taub E, Uswatte G, Elbert T. New treatments in neurorehabilitation founded on basic research. Nat Rev Neurosci. 2002;3(3):228–36.
7. Weiss T, et al. Decrease in phantom limb pain associated with prosthesis-induced increased use of an amputation stump in humans. Neurosci Lett. 1999;272(2):131–4.
8. Elbert T, et al. Alteration of digital representations in somatosensory cortex in focal hand dystonia. Neuroreport. 1998;9(16):3571–5.
9. Candia V, et al. Constraint-induced movement therapy for focal hand dystonia in musicians. Lancet. 1999;353(9146):42.
10. Okamoto H, et al. Listening to tailor-made notched music reduces tinnitus loudness and tinnitus-related auditory cortex activity. Proc Natl Acad Sci U S A. 2010;107(3):1207–10.

11. Stracke H, Okamoto H, Pantev C. Customized notched music training reduces tinnitus loudness. Commun Integr Biol. 2010;3(3):274–7.
12. Teismann H, Okamoto H, Pantev C. Short and intense tailor-made notched music training against tinnitus: the tinnitus frequency matters. PLoS One. 2011;6(9):e24685.
13. Okamoto H, et al. Broadened population-level frequency tuning in human auditory cortex of portable music player users. PLoS One. 2011;6(3):e17022.
14. Okamoto H, et al. Constraint-induced sound therapy for sudden sensorineural hearing loss – behavioral and neurophysiological outcomes. Sci Rep. 2014;4:3927.
15. Heller A. Classification and epidemiology of tinnitus. Otolaryngol Clin N Am. 2003; 36(2):239.
16. Lockwood AH, Salvi RJ, Burkard RF. Tinnitus. N Engl J Med. 2002;347(12):904–10.
17. McFadden D, National Research Council (U.S.). Working Group 89., *Tinnitus:facts, theories, and treatments*. Washington, DC: National Academy Press; 1982. xi, 150 p.
18. Hinton DE, et al. Tinnitus among Cambodian refugees: relationship to PTSD severity. J Trauma Stress. 2006;19(4):541–6.
19. Fagelson MA. The association between tinnitus and posttraumatic stress disorder. Am J Audiol. 2007;16(2):107–17.
20. Eggermont J, Roberts L. The neuroscience of tinnitus. Trends Neurosci. 2004;27(11):676–82.
21. Dietrich V, et al. Cortical reorganization in patients with high frequency cochlear hearing loss. Hear Res. 2001;158(1–2):95–101.
22. Buonomano DV, Merzenich MM. Cortical plasticity: from synapses to maps. Annu Rev Neurosci. 1998;21:149–86.
23. Pons TP, et al. Massive cortical reorganization after sensory deafferentation in adult macaques. Science. 1991;252(5014):1857–60.
24. Kaas JH, et al. Reorganization of retinotopic cortical maps in adult mammals after lesions of the retina. Science. 1990;248(4952):229–31.
25. Rajan R. Receptor organ damage causes loss of cortical surround inhibition without topographic map plasticity. Nat Neurosci. 1998;1(2):138–43.
26. Jacobs KM, Donoghue JP. Reshaping the cortical motor map by unmasking latent intracortical connections. Science. 1991;251(4996):944–7.
27. Diesch E, et al. Interaction among the components of multiple auditory steady-state responses: enhancement in tinnitus patients, inhibition in controls. Neuroscience. 2010;167(2):540–53.
28. Eggermont JJ, Ponton CW. The neurophysiology of auditory perception: from single units to evoked potentials. Audiol Neurootol. 2002;7(2):71–99.
29. Engelien A, et al. A combined functional in vivo measure for primary and secondary auditory cortices. Hear Res. 2000;148(1–2):153–60.
30. Lutkenhoner B, Steinstrater O. High-precision neuromagnetic study of the functional organization of the human auditory cortex. Audiol Neurootol. 1998;3(2–3):191–213.
31. Pantev C, et al. Tonotopic organization of the sources of human auditory steady-state responses. Hear Res. 1996;101(1–2):62–74.
32. Rosler G. Progression of hearing-loss caused by occupational noise. Scand Audiol. 1994; 23(1):13–37.
33. Clark WW, Bohne BA. Effects of noise on hearing. JAMA. 1999;281(17):1658–9.
34. Morata TC, et al. Effects of occupational exposure to organic-solvents and noise on hearing. Scand J Work Environ Health. 1993;19(4):245–54.
35. Danhauer JL, et al. Survey of college students on iPod use and hearing health. J Am Acad Audiol. 2009;20(1):5–27.
36. Fligor BJ, Cox LC. Output levels of commercially available portable compact disc players and the potential risk to hearing. Ear Hear. 2004;25(6):513–27.
37. Kauramäki J, Jääskeläinen IP, Sams M. Selective attention increases both gain and feature selectivity of the human auditory cortex. PLoS One. 2007;2(9):e909.
38. Okamoto H, et al. Bottom-up driven involuntary auditory evoked field change: constant sound sequencing amplifies but does not sharpen neural activity. J Neurophysiol. 2010;103(1):244–9.

39. Okamoto H, et al. Attention improves population-level frequency tuning in human auditory cortex. J Neurosci. 2007;27(39):10383–90.
40. Okamoto H, et al. Frequency-specific modulation of population-level frequency tuning in human auditory cortex. BMC Neurosci. 2009;10(1):1.
41. Sams M, Salmelin R. Evidence of sharp frequency tuning in the human auditory cortex. Hear Res. 1994;75(1–2):67–74.
42. Zwicker E, Fastl H. Psychoacoustics : facts and models. 2nd updated ed. Springer series in information sciences 22. Berlin/New York: Springer; 1999. xii, 416 p.
43. Näätänen R, Picton T. The N1 wave of the human electric and magnetic response to sound: a review and an analysis of the component structure. Psychophysiology. 1987;24(4):375–425.
44. Nilsson M, Soli SD, Sullivan JA. Development of the hearing in noise test for the measurement of speech reception thresholds in quiet and in noise. J Acoust Soc Am. 1994;95(2):1085–99.
45. National Institute of Health. Sudden deafness. Vol. 00-4757. Bethesda: NIH Publication; 2000.
46. Byl Jr FM. Sudden hearing loss: eight years' experience and suggested prognostic table. Laryngoscope. 1984;94(5 Pt 1):647–61.
47. Nosrati-Zarenoe R, Arlinger S, Hultcrantz E. Idiopathic sudden sensorineural hearing loss: results drawn from the Swedish national database. Acta Otolaryngol. 2007;127(11):1168–75.
48. Teranishi M, et al. Thirty-year trends in sudden deafness from four nationwide epidemiological surveys in Japan. Acta Otolaryngol. 2007;127(12):1259–65.
49. Wu CS, Lin HC, Chao PZ. Sudden sensorineural hearing loss: evidence from Taiwan. Audiol Neuro Otol. 2006;11(3):151–6.
50. Cinamon U, Bendet E, Kronenberg J. Steroids, carbogen or placebo for sudden hearing loss: a prospective double-blind study. Eur Arch Otorhinolaryngol. 2001;258(9):477–80.
51. Conlin AE, Parnes LS. Treatment of sudden sensorineural hearing loss II. A meta-analysis. Arch Otolaryngol Head Neck Surg. 2007;133(6):582–6.
52. Conlin AE, Parnes LS. Treatment of sudden sensorineural hearing loss I. A systematic review. Arch Otolaryngol Head Neck Surg. 2007;133(6):573–81.
53. Nosrati-Zarenoe R, Hultcrantz E. Corticosteroid treatment of idiopathic sudden sensorineural hearing loss: randomized triple-blind placebo-controlled trial. Otol Neurotol. 2012;33(4): 523–31.
54. Li LPH, et al. Healthy-side dominance of cortical neuromagnetic responses in sudden hearing loss. Ann Neurol. 2003;53(6):810–5.
55. Morita T, et al. A recovery from enhancement of activation in auditory cortex of patients with idiopathic sudden sensorineural hearing loss. Neurosci Res. 2007;58(1):6–11.
56. Suzuki M, et al. Cortical representation of hearing restoration in patients with sudden deafness. Neuroreport. 2002;13(14):1829–32.
57. Li LPH, et al. Neuromagnetic index of hemispheric asymmetry prognosticating the outcome of sudden hearing loss. PLoS One. 2012;7(4):e35055.
58. Blanton S, Wolf SL. An application of upper-extremity constraint-induced movement therapy in a patient with subacute stroke. Phys Ther. 1999;79(9):847–53.
59. Taub E, Uswatte G, Pidikiti R. Constraint-induced movement therapy: a new family of techniques with broad application to physical rehabilitation – a clinical review. J Rehabil Res Dev. 1999; 36(3):237–51.
60. Wolf SL, et al. Effect of constraint-induced movement therapy on upper extremity function 3 to 9 months after stroke – The EXCITE randomized clinical trial. JAMA-J Am Med Assoc. 2006;296(17):2095–104.
61. Elbert T, et al. New developments in stroke rehabilitation based on behavioral and neuroscientific principles: constraint-induced therapy. Nervenarzt. 2003;74(4):334–42.
62. Liepert J, et al. Motor cortex plasticity during constraint-induced movement therapy in stroke patients. Neurosci Lett. 1998;250(1):5–8.
63. Nelles G. Cortical reorganization – effects of intensive therapy – results from prospective functional imaging studies. Restor Neurol Neurosci. 2004;22(3–5):239–44.

64. Pantev C, et al. Comparison between simultaneously recorded auditory-evoked magnetic fields and potentials elicited by ipsilateral, contralateral and binaural tone burst stimulation. Audiology. 1986;25(1):54–61.
65. Tiihonen J, et al. Interaural interaction in the human auditory-cortex. Audiology. 1989;28(1): 37–48.
66. Bilecen D, et al. Cortical reorganization after acute unilateral hearing loss traced by fMRI. Neurology. 2000;54(3):765–7.
67. Burton H, et al. Activation lateralization in human core, belt, and parabelt auditory fields with unilateral deafness compared to normal hearing. Brain Res. 2012;1454:33–47.
68. Rajan R, Irvine DRF. Neuronal responses across cortical field A1 in plasticity induced by peripheral auditory organ damage. Audiol Neuro Otol. 1998;3(2–3):123–44.
69. Eggermont JJ. Pathophysiology of tinnitus. Prog Brain Res. 2007;166:19–35.
70. Eggermont JJ. The role of sound in adult and developmental auditory cortical plasticity. Ear Hear. 2008;29(6):819–29.
71. Norena A, Micheyl C, Chery-Croze S. An auditory negative after-image as a human model of tinnitus. Hear Res. 2000;149(1–2):24–32.
72. Pantev C, et al. Lateral inhibition and habituation of the human auditory cortex. Eur J Neurosci. 2004;19(8):2337–44.
73. Von Békésy G. Sensory inhibition. Herbert Sidney Langfeld memorial lectures 1965. Princeton: Princeton University Press; 1967. x, 265 p.
74. Oswald AMM, Schiff ML, Reyes AD. Synaptic mechanisms underlying auditory processing. Curr Opin Neurobiol. 2006;16(4):371–6.
75. Wehr M, Zador AM. Balanced inhibition underlies tuning and sharpens spike timing in auditory cortex. Nature. 2003;426(6965):442–6.
76. Calford MB, Rajan R, Irvine DRF. Rapid changes in the frequency tuning of neurons in cat auditory-cortex resulting from pure-tone-induced temporary threshold shift. Neuroscience. 1993;55(4):953–64.
77. Norena AJ, Tomita M, Eggermont JJ. Neural changes in cat auditory cortex after a transient pure-tone trauma. J Neurophysiol. 2003;90(4):2387–401.

Part V
Visual System

7 Basic Function

Kensaku Miki and Ryusuke Kakigi

Abstract

The human visual system consists of two main separate pathways, the ventral (what) and dorsal (where) pathways, which is similar to that in monkeys. The ventral pathway is involved in the processing of shapes and objects as well as color perception, whereas the dorsal pathway is responsible for processing location and motion as well as three-dimensional shape perception. Magnetoencephalography (MEG) achieves very high temporal and spatial resolution and, thus, is useful for examining the temporal and spatial characteristics of the human visual system. MEG studies previously demonstrated the following: (1) P100m was the most important component of the primary visual cortex activity and was affected by the properties of stimuli; (2) in the ventral pathway, the inferior temporal area was crucial for the perception of colors, shapes, faces, and characters; and (3) in the dorsal pathway, the occipitotemporal area, the human MT/V5 area homologue, was identified as being the most important for the perception of motion.

The ventral and dorsal pathways are both involved in face perception. The fusiform gyrus in the ventral pathway was previously shown to be activated by the static face. Furthermore, the activity of the right fusiform gyrus was affected more by the inversion of features whereas that of the left fusiform gyrus was affected more by a disruption in the spatial relationship between facial contours and features. On the other hand, the occipitotemporal area in the dorsal pathway

K. Miki (✉) • R. Kakigi
Department of Integrative Physiology, National Institute for Physiological Sciences, 38 Nishigonaka Myodaiji, Okazaki 444-8585, Aichi, Japan

Department of Physiological Sciences, School of Life Science, SOKENDAI (The Graduate University for Advanced Studies), Hayama 240-0193, Kanagawa, Japan
e-mail: kensaku@nips.ac.jp

S. Tobimatsu, R. Kakigi (eds.), *Clinical Applications of Magnetoencephalography*, DOI 10.1007/978-4-431-55729-6_7

was found to be activated by facial movements and influenced by whether these movements appeared in the contours and/or features of the face.

Keywords
MEG • Primary visual cortex • Ventral pathway • Dorsal pathway • Fusiform • MT/V5 • P100

7.1 The Visual System

The neurophysiological properties of the visual system have been investigated in detail in animals such as monkeys and cats. Many cortical areas have been identified including the IT (inferior temporal) area, STS (superior temporal sulcus) area, and MT (middle temporal)/V5 area. Previous studies using a noninvasive neuroimaging method revealed that the visual system in humans was similar to that in monkeys [1, 2].

The human visual system consists of two main separate pathways [3, 4]. The first pathway begins from cones in the retina, which process colors and receive information through the primary visual (striate) cortex via the parvocellular layers of the lateral geniculate nucleus (LGN). This pathway is called the "ventral (what) pathway" and processes shapes and objects and plays a role in color perception. On the other hand, the second pathway begins from rods in the retina, which process lightness and receive information through the primary visual area via the magnocellular layers of LGN. This pathway is called "the dorsal (where) pathway" and processes location and motion and also plays a role in three-dimensional shape perception [5].

Magnetoencephalography (MEG) achieves very high temporal and spatial resolution. Previous MEG studies revealed the temporal and spatial characteristics of the human visual system. In MEG studies of the primary visual area, evoked magnetic components were identified by checkerboard pattern reversal stimuli [6–8]. In a previous MEG study [6], N75m, P100m, and N145m were clearly observed around the midoccipital position, while a very small component, P50m, was occasionally detected prior to N75m. The equivalent current dipole (ECD) of the P50m, N75m, and P100m component was estimated in the striate cortex, while the ECD of N145m was estimated in the extrastriate cortex.

P100m is the most important biomarker for primary visual cortex activity among the components evoked by pattern reversal stimuli. In an MEG study on responses to pattern reversal left half-field central (0–2°, 0–5°) and left half-field peripheral stimulation (2–15°, 5–15°) [7], the P100m dipole to the central field was localized more posterior than that to the peripheral stimulation. These findings were attributed to the retinotopic organization of the visual cortex. P100m is also affected by the properties of stimuli, for example, the check size. Nakamura et al. [8] used half-field stimuli with or without central occlusion with check sizes of 15′, 30′, 60′, 90′, and 180′ of the visual arc and recorded the P100m component. The latencies for the smaller checks were significantly longer than those for the larger checks, while the amplitudes for the smaller checks were significantly smaller than those for the larger checks when using a stimulation with central occlusion, but not that without central occlusion. However, the equivalent current dipoles were located around the calcarine fissure and did not differ significantly in location with the check size.

Previous MEG studies demonstrated that the perception of characters and shapes was related to the ventral pathway, such as the IT area [9–12]. Koyama et al. [9] recorded MEG in Japanese subjects in order to investigate the recognition of Japanese characters (kanji and kana) and letters of the alphabet. The first evoked component was found to peak after approximately 180 ms under all conditions, and the estimated ECDs were located in the IT area. These findings revealed that the IT area was essential for character discrimination.

Since the activity of the primary visual areas to luminance changes is strong and long-lasting, it is frequently difficult to clearly identify IT area activity, even when using MEG with high spatial resolution. To overcome this issue, we newly developed a random dots blinking (RDB) method that uses temporal changes in the patterns of a large number of small dots to present stimuli without a change in luminance during the presentation of an object (e.g., a circle, a letter, or a schematic face) and, consequently, reduces activities in the primary and other higher visual areas and detects activity in the IT area only [10, 13–17]. Okusa et al. [10] investigated the visual recognition of characters using the RDB method. One clear component, with a peak latency of approximately 300 ms, was identified and the discrimination accuracy rate increased as the character display duration became longer. The source of this component was estimated in the IT area. These findings showed that the RDB method was useful for investigating IT area activity in relation to visual perception.

Apparent motion is the perception of the realistic smooth motion of an object that first flashes at one place and then at another. Kaneoke et al. [18] reported the presence of a localized cortical area, corresponding to the human homologue of MT/V5, which was exclusively sensitive to apparent motion stimuli, which was identical to that for smooth motion. Kawakami et al. [19] measured magnetic responses to apparent motion stimuli and compared the findings obtained with a subjective rating of the quality of perceived motion with various stimulus timings. The strength (ECD moment) of the response varied with the stimulus timing, with the maximum value being at the inter-stimulus interval 0 ms, and was related to the subjective rating of the quality of motion. Another MEG study reported that the response to downward motion in the upper visual field was significantly larger than that to upward motion, but not in the lower visual field [20]. These findings indicated that human MT/V5 had directional preference for downward motion versus upward motion in the upper visual field.

7.2 Face Perception

7.2.1 Static Face Perception

The face contains a lot of information that is relevant to our daily lives, such as information about age, sex, and familiarity, and plays an important role in social communication. Recent studies investigated static face perception using neuroimaging methods, including electroencephalography (EEG), MEG, and functional magnetic resonance imaging (fMRI). For example, EEG demonstrated that a negative component was evoked at approximately 170 ms during object perception, and this component was termed N170 [21, 22]. N170 was shown to be larger during the viewing of faces than during the observation of other objects, such as cars or chairs [23].

The temporal and spatial processing of face perception in 12 normal subjects was previously investigated by MEG [11, 12]. Five different kinds of stimuli, (1) *face with opened eyes*, (2) *face with closed eyes*, (3) *eyes*, (4) *scrambled face*, and (5) *hand*, were used. Subjects were asked to count the *hand* stimuli and declare the number after each session to avoid attention effects caused by the P300 component on responses to *face* and *eyes* stimuli.

Two components, 1M and 2M, were identified in the right and left hemispheres, and the mean peak latencies of the 1M and 2M components were approximately 132 and 179 ms, respectively, (Fig. 7.1). The 1M component to all kinds of this stimulus was recorded in all subjects. This component appeared to be simple phase reversed. The 2M component was clearly identified from the right hemisphere in ten out of the 12 subjects and appeared to overlap with the 1M component in five subjects. The 2M component was clearly observed in response to the *face with opened eyes* or *closed eyes* in all ten subjects, but to the *eyes* in only three subjects with a smaller amplitude. The 2M component was not observed in response to the *scrambled face* and *hand*. On the other hand, the 2M component was clearly identified in five and two out of the ten subjects in response to the *face* (*with opened eyes* and *with closed eyes*) and *eyes* stimuli from the left hemisphere, respectively. The 2M latency to the *eyes* stimuli was significantly longer than those to the *face with opened eyes* and *face with closed eyes*. No significant difference was observed in the 2M latency between the *face with opened eyes* and *face with closed eyes*. The source analysis revealed that the 1M component was generated in the primary visual cortex in the bilateral hemispheres, while the 2M component was generated in the inferior temporal cortex around the fusiform gyrus (Fig. 7.2). These findings suggested that (1) the fusiform gyrus played an important role in the holistic and/or configural process of human static face perception and (2) the right hemisphere was more dominant.

A phenomenon exists that is unique to humans and nonhuman primates. Psychological studies previously reported that face recognition was more difficult when inverted faces were presented rather than upright faces and named this phenomenon the face inversion effect. These findings suggested that face inversion disrupted the holistic and/or configural processing of facial information [24, 25].

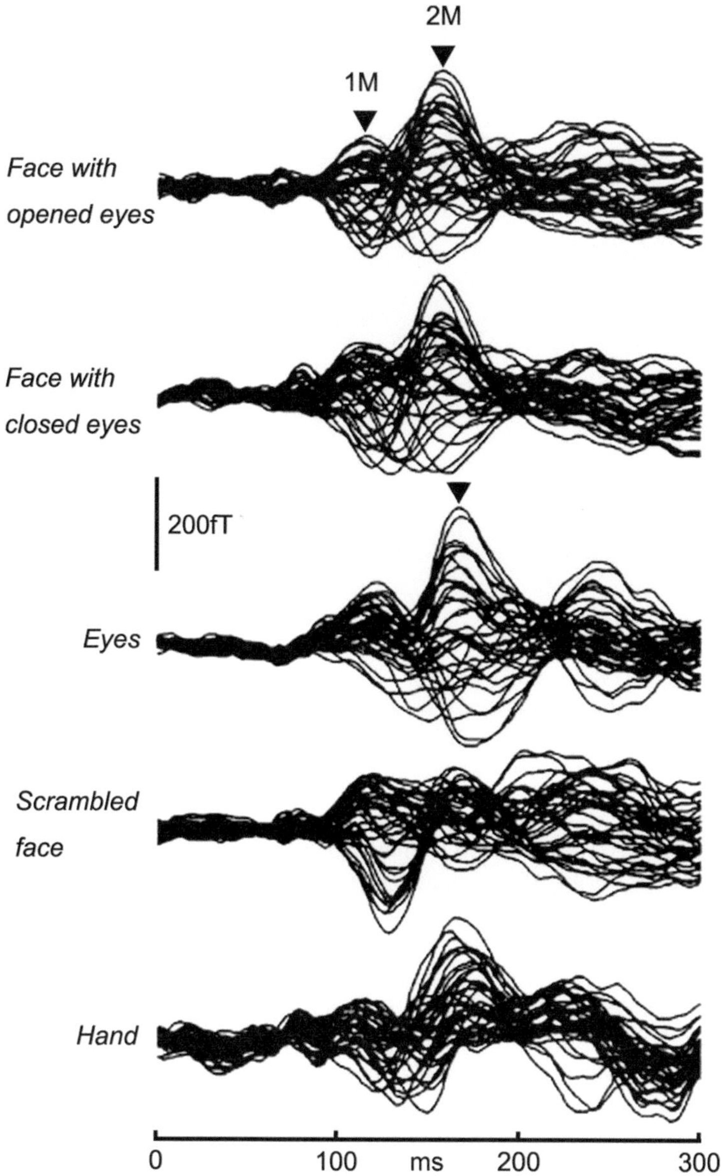

Fig. 7.1 MEG waveforms of a representative subject in response to five categories of stimuli: *face with opened eyes*, *face with closed eyes*, *eyes*, *scrambled face*, and *hand*. MEG waveforms recorded from 37 channels from the right hemisphere were superimposed. Two components, 1M and 2M, were identified. The 1M component was recorded for all stimuli, whereas the 2M component was not recorded for *scrambled face* or *hand*. The 2M latency to *eyes* stimuli was significantly longer than those to *face with opened eyes* and *face with closed eyes*. No significant difference was observed in the 2M latency between *face with opened eyes* and *face with closed eyes* (Adapted from [11])

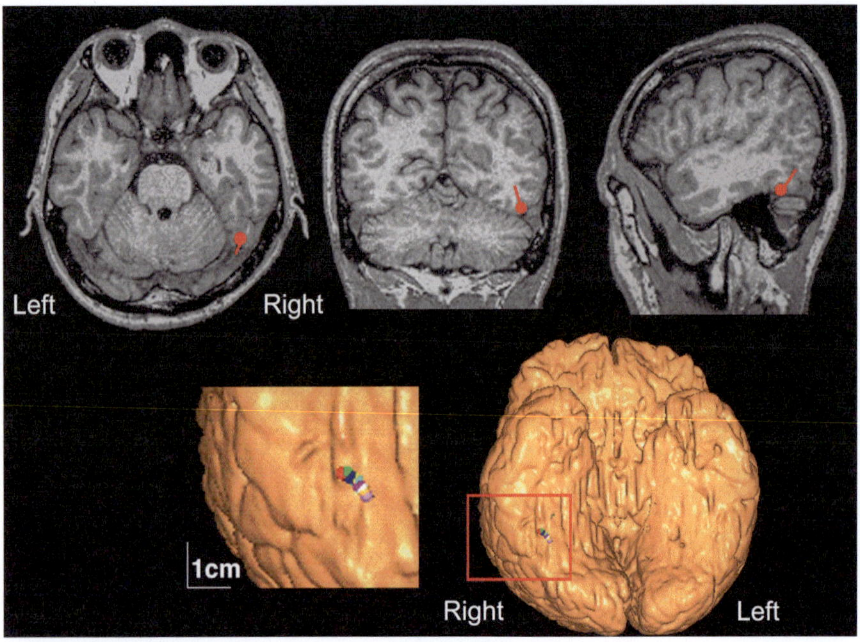

Fig. 7.2 The ECD of the 2M component to *face with opened eyes* recorded from the right hemisphere overlapped on two-dimensional and three-dimensional MRI in a representative subject (Adapted from [12])

In EEG studies, N170 was shown to be affected by inversion. In previous studies [21, 26–28], N170 was longer in latency and larger in amplitude for inverted faces than for upright faces. MEG studies have also examined face inversion effects. Watanabe et al. [29] reported that the latency of the component evoked by inverted faces was longer in the right hemisphere and shorter in the left hemisphere than that evoked by upright faces. In addition, the latency of N170 was longer for scrambled features (e.g., eyes, nose, and mouth), in which the spatial relationship between the facial contours (e.g, the hair, chin, and ears) and features was disrupted [22, 27], than for upright faces. In our MEG study [30], we investigated (a) how brain activity related to static face perception was modulated by inverting the contours and/or features of the human face and (b) what information within the face was important for processing static face perception. We used MEG and compared the cortical activities evoked by viewing an upright face, inverted face, and face in which the spatial relationship between the contours and features was disrupted. We focused on the activity in the fusiform gyrus for static face perception. We studied ten right-handed adults with normal or corrected visual acuity. We used the following three conditions (Fig. 7.3):

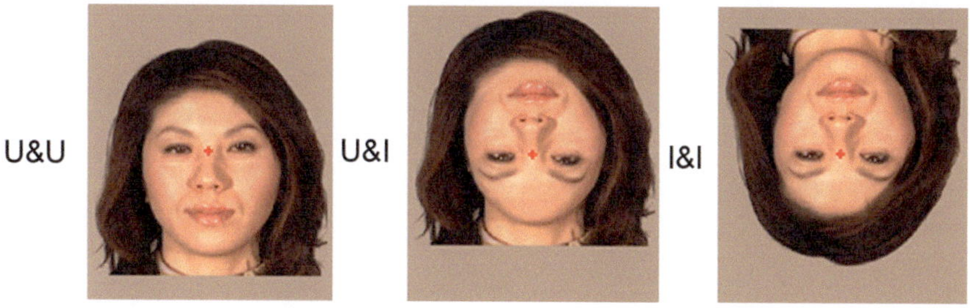

Fig. 7.3 Examples of stimulus conditions. (1) U&U, upright contour (hair and chin) and upright features (eyes, nose, and mouth); (2) U&I, upright contour and inverted features; and (3) I&I, inverted contour and inverted features (Adapted from [30])

1. U&U (upright contours and upright features): Contours (hair and chin) and features (eyes, nose, and mouth) upright.
2. U&I (upright contours and inverted features): Contours remained upright; however, the features were mirrored and inverted relative to the U&U condition, while the spatial relationship among the eyes, nose, and mouth was not changed.
3. I&I (inverted contours and inverted features): A mirrored and inverted form of the image used in the U&U condition.

Figure 7.4 shows the waveforms recorded from 204 gradiometers of a representative subject following the U&U (upright contour and upright features) condition and the waveforms in all conditions at representative sensors, in which the largest component was identified for the U&U condition in each occipital or temporal area of the same subject. The sources were estimated to lie in the fusiform area from MEG responses under all conditions. The latency of the activity of the fusiform area was significantly longer for U&I (upright contour and inverted features; $p < 0.05$) and I&I (inverted contour and inverted features; $p < 0.05$) than for U&U in the right hemisphere and significantly longer for U&I than for U&U ($p < 0.01$) and I&I ($p < 0.05$) in the left hemisphere (Table 7.1). No significant differences were observed in the dipole moment (strength) among the three conditions. These results demonstrated the following: in static face perception, the activity of the right fusiform area was affected more by the inversion of features, while that of the left fusiform area was affected more by a disruption in the spatial relationship between facial contours and features.

7.2.2 Facial Movement Perception

Facial movements are also useful for social communication in humans. For example, the direction of the eye gaze is used to assess the social attention of others, and it becomes markedly easier to understand speech when the mouth movements of the speaker can be observed. In a previous MEG study [31], a specific region for the perception of eye movements was detected within the occipitotemporal area

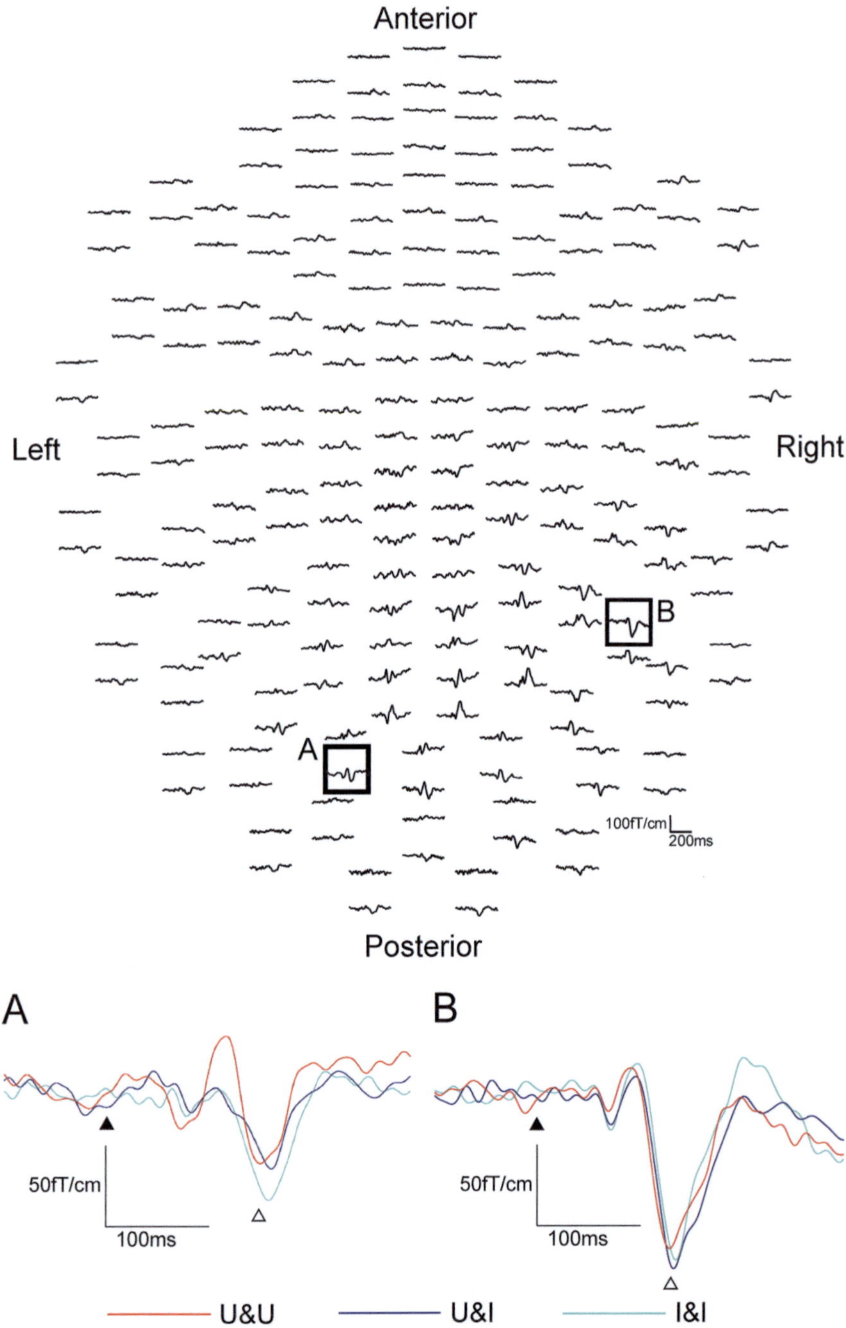

Fig. 7.4 The upper image shows the waveforms recorded from 204 gradiometers of a representative subject following the U&U (upright contour and upright features) condition. The head was viewed from the top. In each response at 102 pairs of gradiometers, the upper response showed the field along the latitude of the gradiometers, while the lower one showed that along the longitude of

Table 7.1 Dipole moments (nAm) and peak latencies (ms) for U&U, and U&I, and I&I. Means and standard deviations (SDs) for U&U, U&I, and I&I in right and left hemispheres (Adapted from [30])

	Left ($n = 10$)		Right ($n = 10$)	
	(nAm)	(ms)	(nAm)	(ms)
U&U	21.9 ± 13.5	143.2 ± 19.7	28.6 ± 21.1	133.6 ± 18.5
U&I	20.3 ± 11.8	162.2 ± 21.0[a, b]	35.5 ± 18.5	148.4 ± 22.4[c]
I&I	21.9 ± 14.5	148.4 ± 13.6	34.4 ± 18.5	151.4 ± 24.1[c]

[a]$p < 0.01$ significantly different from U&U
[b]$p < 0.05$ significantly different from I&I
[c]$p < 0.05$ significantly different from U&U

corresponding to the human MT/V5, and this was different from motion in general. We examined the temporal characteristics of the brain activity elicited by viewing mouth movements (opening and closing) and compared them to those of eye aversion movements and motion in general [32]. Seventeen right-handed adults with normal or corrected visual acuity participated in this study. We used apparent motion, in which the first stimulus (S1) was replaced by a second stimulus (S2) with no inter-stimulus interval as follows:

1. M-OP: The mouth is opening.
2. M-CL: The mouth is closing.
3. EYES: The eyes are averted.
4. RADIAL: A radial stimulus moving inward.

A large clear component, 1M, was elicited by all conditions (M-OP, M-CL, EYES, and RADIAL) within 200 ms of the stimulus onset (Fig. 7.5). Concerning the peak latency of 1M, the means and standard deviations were 159.8 ± 17.3, 161.9 ± 15.0, 161.2 ± 18.9, and 140.1 ± 18.0 ms for M-OP, M-CL, EYES, and RADIAL in the right hemisphere, respectively, and 162.4 ± 11.6, 160.9 ± 9.8, 164.6 ± 14.2, and 138.4 ± 9.0 ms for M-OP, M-CL, EYES, and RADIAL in the left, respectively. The latency for RADIAL was significantly shorter than that for the facial motion conditions (M-OP, M-CL, and EYES) ($p < 0.05$). No significant differences were observed in 1M latency between M-OP, M-CL, or EYES. In the

Fig. 7.4 (continued) the gradiometers. The lower image shows enhanced waveforms at sensors A and B in the upper image, which showed the largest component for the U&U condition in each occipital or temporal area, recorded in red for U&U, blue for U&I (upright contour and inverted features), and light blue for I&I (inverted contour and inverted features). (**a**) Waveforms at sensor A on the left hemisphere in the upper image. (**b**) Waveforms at sensor B on the right in the upper image. Black arrows show the stimulus onset and white arrows the response chosen for further analyses. Responses after the stimulus onset were clearly longer in latency for U&I and I&I than for U&U in the right hemisphere and were longer in latency for U&I than for U&U and I&I in the left hemisphere for this subject (Adapted from [30])

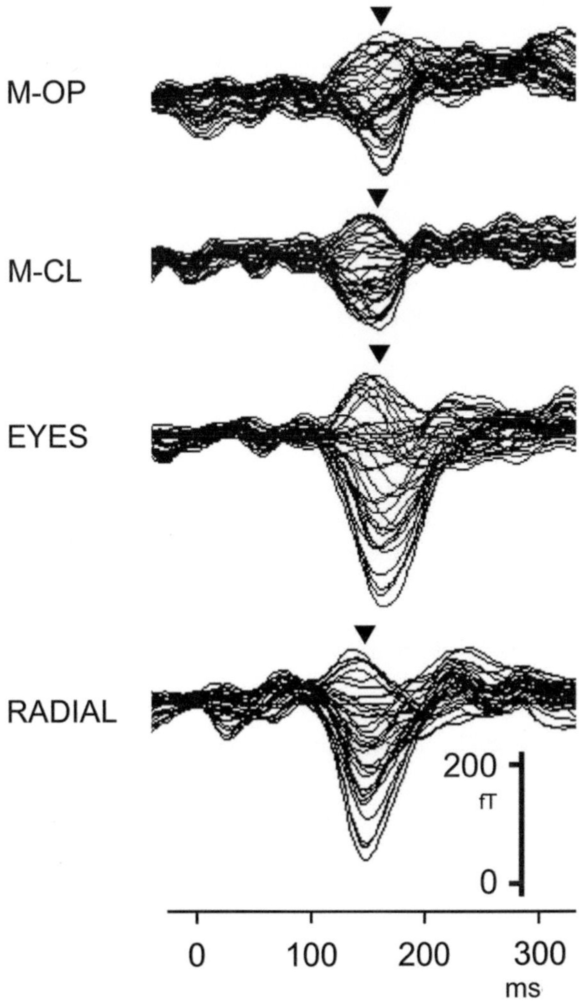

Fig. 7.5 Right hemisphere MEG activity shown in a 37-channel superimposed display for all conditions. In a representative subject, 1 M peak latencies were 154.8, 156.7, 162.5, and 148.1 ms for M-OP, M-CL, EYES, and RADIAL, respectively. Associated maximum root mean square (RMS) values were 62.6, 66.7, 122.0, and 119.4 fT(Adapted from [32])

source analysis, the sources were estimated to lie in the occipitotemporal area, the human MT/V5 area homologue, from 1M under all conditions (Fig. 7.6).

The means and standard deviations of the dipole moment of the estimated dipoles from 1M were 7.9 ± 1.9, 7.8 ± 3.2, 10.0 ± 6.8, and 13.8 ± 4.9 nAm for M-OP, M-CL, EYES, and RADIAL in the right hemisphere, respectively, and 7.4 ± 2.8, 6.7 ± 3.0, 9.3 ± 4.3, and 13.6 ± 1.8 nAm for M-OP, M-CL, EYES, and RADIAL in the left hemisphere, respectively. No significant differences were observed in the dipole moment (strength) for M-OP, M-CL, and EYES between either hemispheres. However, M-OP and M-CL were significantly smaller than RADIAL ($p < 0.05$) in the right hemisphere, while M-OP, M-CL, and EYES were significantly smaller than RADIAL ($p < 0.05$) in the left hemisphere.

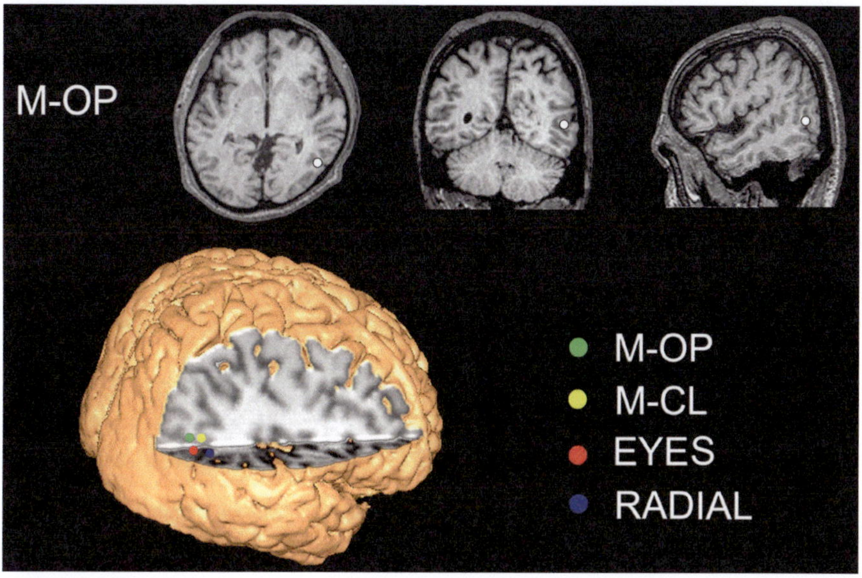

Fig. 7.6 Right hemisphere locations for ECDs under the M-OP condition overlaid on axial, coronal, and sagittal MRI slices and the volume-rendered brain under M-OP, M-CL, EYES, and RADIAL conditions of a representative subject (Adapted from [32])

The results of this study indicated that the occipitotemporal (human MT/V5) area was active in the perception of both mouth and eye movements. Furthermore, viewing mouth and eye movements did not elicit significantly different activities in the occipitotemporal (human MT/V5) area, which suggested that the perception of the movement of facial parts was processed in the same manner, and this is different from motion in general.

However, the main factor(s) causing differences in the recognition of facial versus general movement has yet to be elucidated in detail. Many studies investigated the effects of facial contours and features using a static face. A previous study reported that it took longer to recognize an eyes-only stimulus or only facial features (eyes, nose, and mouth) than a full-face stimulus with contours [11], and the contours of the face were important in static face recognition. However, to the best of our knowledge, the effects of facial contours and features on facial movement recognition have not yet been investigated. Therefore, we examined the effects of facial contours and features on early occipitotemporal activity evoked by facial movement [33]. In this study, we used a schematic face because a simple schematic drawing with a circle for a contour, two dots for eyes, and a straight line for lips was recognized as a face even though each individual component of the drawing by itself was not. Previous studies using a schematic face showed that N170 was evoked by schematic faces as well as photographs of real faces [27, 28].

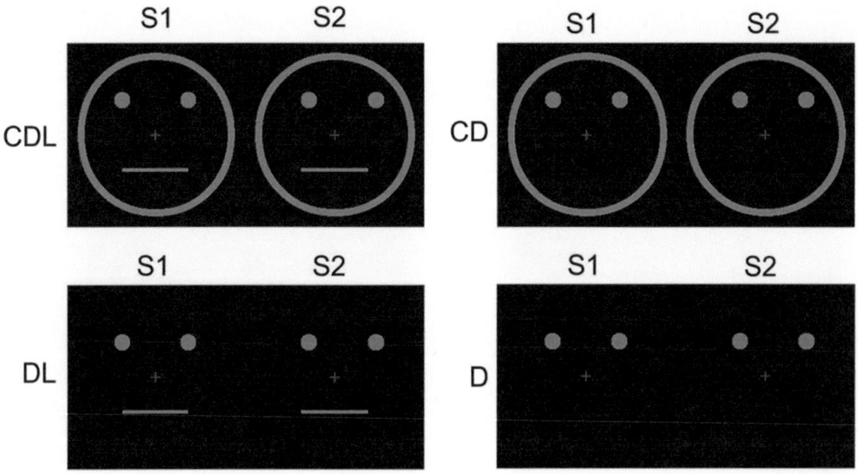

Fig. 7.7 Examples of stimulus conditions. (1) CDL: schematic face consisting of a contour, two dots, and a horizontal line, (2) CD: the contour and two dots, (3) DL: two dots and a horizontal line, and (4) D: two dots only (Adapted from [33])

Thirteen right-handed adults with normal or corrected visual acuity participated in this study. We used apparent motion where the first stimulus, S1, was replaced by a second stimulus, S2, with no inter-stimulus interval and presented the following four conditions (Fig. 7.7). All the subjects described the smooth movement of the eyes or dots with this paradigm:

1. CDL: A schematic face consisting of a facial contour, two dots, and a horizontal line
2. CD: The contour and two dots
3. DL: Two dots and a horizontal line
4. D: Two dots only

The subjects described the simple movement of dots for D, whereas eye movement for CDL, CD, and DL though movement modalities was the same under all conditions. In source modeling, we used a single equivalent current dipole (ECD) model [34] within 145–220 ms of the stimulus onset.

Clear MEG responses were elicited under all conditions (CDL, CD, DL, and D) at the sensors in the bilateral occipitotemporal area (Fig. 7.8). The means and standard deviations of the peak latency of the estimated dipole were 179.3 ± 26.3, 183.0 ± 16.9, 180.9 ± 20.8, and 180.3 ± 23.7 ms for CDL, CD, DL, and D in the right hemisphere, respectively, and 180.2 ± 14.9, 180.5 ± 24.8, 174.0 ± 24.9, and 177.7 ± 20.3 ms for CDL, CD, DL, and D in the left hemisphere, respectively. No significant differences were observed between any of these conditions.

The means and standard deviations of the dipole moment were 14.4 ± 6.2, 11.2 ± 7.9, 9.6 ± 6.5, and 10.3 ± 5.3 nAm for CDL, CD, DL, and D in the right

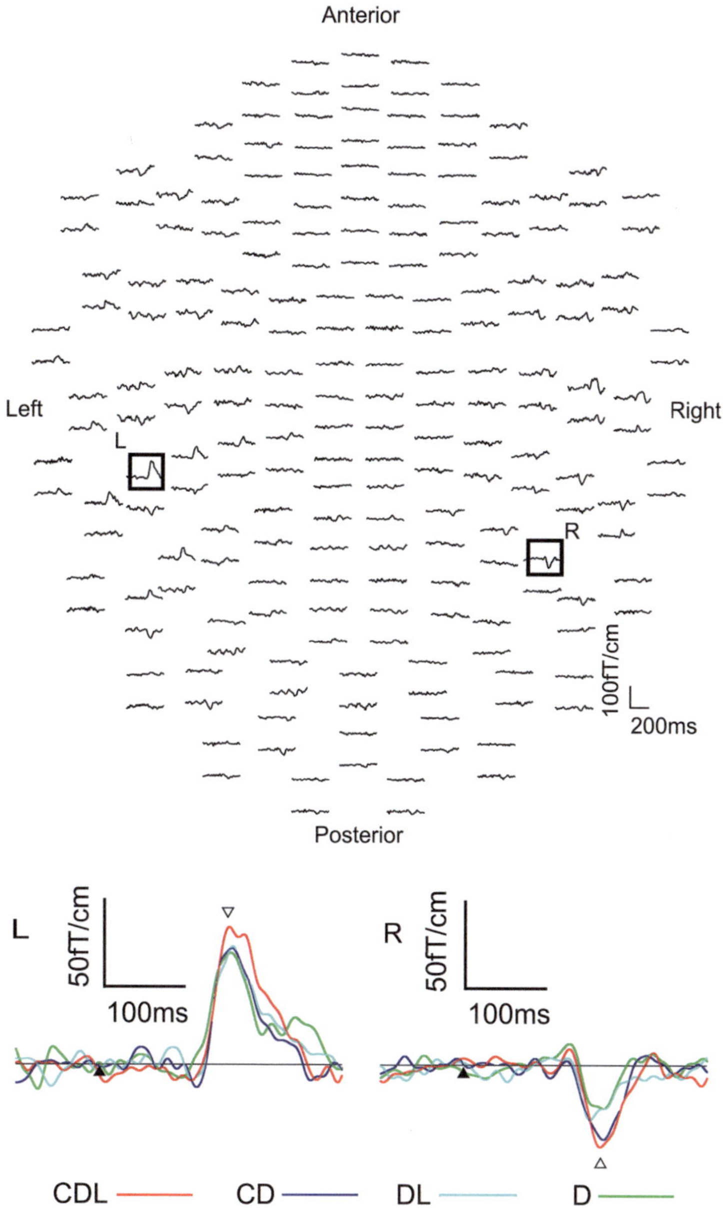

Fig. 7.8 Waveforms recorded from 204 gradiometers in a representative subject following S2 onset (eyes movement) under CDL conditions. The head was viewed from the top. The lower image shows waveforms at sensors L and R in the upper image, which showed a clear component in each hemisphere was recorded in red for CDL, blue for CD, light blue for DL, and green for D. L: representative waveforms at sensor L on the left hemisphere of the upper image. R: representative waveforms at sensor R on the right of the upper image. Black arrows are the S2 onset. The main response after the S2 onset, indicated by white arrows, to CDL was larger in amplitude than the others. The polarity of each sensor may have been due to the current of the dipole (Adapted from [33])

hemisphere, respectively, and 12.7 ± 6.7, 11.1 ± 6.1, 9.6 ± 5.4, and 8.9 ± 5.5 nAm for CDL, CD, DL, and D in the left hemisphere, respectively. The moment was significantly larger for CDL than for CD ($p < 0.05$), DL ($p < 0.01$), and D ($p < 0.01$) in the right hemisphere and for CDL than for DL and D ($p < 0.01$) in the left hemisphere.

Our results in this study demonstrated specific information processing for eye movements, which was different from motion in general, and activity in the occipitotemporal (human MT/V5) area related to this processing was influenced by whether movements appeared with the contours and/or features of the face.

References

1. Van Essen DC, Lewis JW, Drury HA, Hadjikhani N, Tootell RB, Bakircioglu M, Miller MI. Mapping visual cortex in monkeys and humans using surface-based atlases. Vision Res. 2001;41:1359–78.
2. Zeki S, Watson JDG, Lueck CJ, Friston KJ, Kennard C, Frackowiak RSJ. A direct demonstration of functional specialization in human visual cortex. J Neurosci. 1991;11:641–9.
3. Livingstone M, Hubel D. Segregation of form, color, movement, and depth: anatomy, physiology, and perception. Science. 1988;240:740–9.
4. Tootell RBH, Dale AM, Sereno MI, Malach R. New images from human visual cortex. Trends Neurosci. 1996;19:481–9.
5. Nakamura M, Kaneoke Y, Watanabe S, Kakigi R. Visual information process in Williams syndrome: intact motion detection accompanied by typical visuospatial dysfunctions. Eur J Neurosci. 2002;16:1810–8.
6. Nakamura A, Kakigi R, Hoshiyama M, Koyama S, Kitamura Y, Shimojo M. Visual evoked cortical magnetic fields to pattern reversal stimulation. Cogn Brain Res. 1997;6:9–22.
7. Brecelj J, Kakigi R, Koyama S, Hoshiyama M. Visual evoked magnetic responses to central and peripheral stimulation: simultaneous VEP recordings. Brain Topogr. 1998;10:227–37.
8. Nakamura M, Kakigi R, Okusa T, Hoshiyama M, Watanabe K. Effects of check size on pattern reversal visual evoked magnetic field and potential. Brain Res. 2000;872:77–86.
9. Koyama S, Kakigi R, Hoshiyama M, Kitamura Y. Reading of Japanese Kanji (morphograms) and Kana (syllabograms): a magnetoencephalographic study. Neuropsychologia. 1998;36:83–98.
10. Okusa T, Kaneoke Y, Koyama S, Kakigi R. Random dots blinking: a new approach to elucidate the activities of the extrastriate cortex in humans. NeuroReport. 1998;9:3961–5.
11. Watanabe S, Kakigi R, Koyama S, Kirino E. It takes longer to recognize the eyes than the whole face in humans. NeuroReport. 1999;10:2193–8.
12. Watanabe S, Kakigi R, Koyama S, Kirino E. Human face perception traced by magneto- and electro-encephalography. Brain Res Cogn Brain Res. 1999;8:125–42.
13. Okusa T, Kakigi R, Osaka N. Cortical activity related to cue-invariant shape perception in humans. Neuroscience. 2000;98:615–24.
14. Noguchi Y, Inui K, Kakigi R. Temporal dynamics of neural adaptation effect in the human visual ventral stream. J Neurosci. 2004;24:6283–90.
15. Noguchi Y, Kakigi R. Neural mechanisms of visual backward masking revealed by high temporal resolution imaging of human brain. Neuroimage. 2005;27:178–87.
16. Noguchi Y, Kakigi R. Time representations can be made from nontemporal information in the brain: an MEG study. Cereb Cortex. 2006;16:1797–808.
17. Noguchi Y, Tanabe HC, Sadato N, Hoshiyama M, Kakigi R. Voluntary attention changes the speed of perceptual neural processing. Eur J Neurosci. 2007;25:3163–72.

18. Kaneoke Y, Bundou M, Koyama S, Suzuki H, Kakigi R. Human cortical area responding to stimuli in apparent motion. NeuroReport. 1997;8:677–82.
19. Kawakami O, Kaneoke Y, Kakigi R. Perception of apparent motion is related to the neural activity in the human extrastriate cortex as measured by magnetoencephalography. Neurosci Lett. 2000;285:135–8.
20. Naito T, Kaneoke Y, Osaka N, Kakigi R. Asymmetry of the human visual field in response to apparent motion. Brain Res. 2000;865:221–6.
21. Bentin S, Allison T, Puce A, Perez E, McCarthy G. Electrophysiological studies of face perception in humans. J Cogn Neurosci. 1996;8:551–65.
22. George N, Evans J, Fiori N, Davidoff J, Renault B. Brain events related to normal and moderately scrambled faces. Brain Res Cogn Brain Res. 1996;4:65–76.
23. Rossion B, Jacques C. Does physical interstimulus variance account for early electrophysiological face sensitive responses in the human brain? Ten lessons on the N170. Neuroimage. 2008;39:1959–79.
24. Mondloch CJ, Le Grand R, Maurer D. Configural face processing develops more slowly than featural face processing. Perception. 2002;31:553–66.
25. Tanaka JW, Farah MJ. Parts and wholes in face recognition. Q J Exp Psychol A. 1993;46:225–45.
26. Itier RJ, Taylor MJ. N170 or N1? Spatiotemporal differences between object and face processing using ERPs. Cereb Cortex. 2004;14:132–42.
27. Latinus M, Taylor MJ. Face processing stage: impact of difficulty and the separation of effect. Brain Res. 2006;1123:179–87.
28. Sagiv N, Bentin S. Structural encoding of human and schematic faces: holistic and part-based processes. J Cogn Neurosci. 2001;13:937–51.
29. Watanabe S, Kakigi R, Puce A. The spatiotemporal dynamics of the face inversion effect: a magneto- and electro-encephalographic study. Neuroscience. 2003;116:879–95.
30. Miki K, Takeshima Y, Watanabe S, Honda Y, Kakigi R. Effects of inverting contour and features on processing for static and dynamic face perception: an MEG study. Brain Res. 2011;1383:230–41.
31. Watanabe S, Kakigi R, Puce A. Occipitotemporal activity elicited by viewing eye movements: a magnetoencephalographic study. Neuroimage. 2001;13:351–63.
32. Miki K, Watanabe S, Kakigi R, Puce A. Magnetoencephalographic study of occipitotemporal activity elicited by viewing mouth movements. Clin Neurophysiol. 2004;115:1559–74.
33. Miki K, Watanabe S, Honda Y, Nakamura M, Kakigi R. Effects of face configuration and features on early occipitotemporal activity when viewing eye movement. Neuroimage. 2007;35:1624–35.
34. Hämäläinen M, Hari R, Ilmoniemi RJ, Knuutila J, Lounasmaa OV. Magnetoencephalography—theory, instrumentation, and applications to non-invasive studies of the working brain. Rev Mod Phys. 1993;65:413–97.

Clinical Applications

8

Takao Yamasaki, Saeko Inamizu, Yoshinobu Goto, and Shozo Tobimatsu

Abstract

Various types of visual disturbance are often observed in several neurological and psychiatric disorders. In humans, visual information is processed simultaneously via multiple parallel pathways, each of which performs multiple steps of hierarchical processing. Visual evoked and event-related magnetic fields (VEFs/ERFs) can noninvasively provide rich spatiotemporal information of visual processing in the visual pathways and visual cortex. By manipulating the visual stimulus parameters of VEFs/ERFs, we can separately explore the function of each pathway at different hierarchical levels based on the distinct physiological characteristics of each visual pathway. In this review, we first summarize current concepts of parallel visual pathways in humans and the relationship between the impairment of visual pathways and visual processing disorders. Next, we describe the clinical applications of VEFs/ERFs for several neurological and psychiatric conditions involving specific types of visual dysfunction such as visual field deficits and impairments of face, word, and motion perception. From this review, we would like to stress that VEFs/ERFs are a valuable neuroimaging

T. Yamasaki, M.D., Ph.D. (✉)
Department of Clinical Neurophysiology, Neurological Institute, Graduate School of Medical Sciences, Kyushu University, 3-1-1 Maidashi, Higashi-ku, Fukuoka 812-8582, Japan

Department of Neurology, Minkodo Minohara Hospital, 3553 Kanaide, Sasaguri-machi, Kasuya-gun, Fukuoka 811-2402, Japan
e-mail: yamasa@neurophy.med.kyushu-u.ac.jp

S. Inamizu • S. Tobimatsu
Department of Clinical Neurophysiology, Neurological Institute, Graduate School of Medical Sciences, Kyushu University, 3-1-1 Maidashi, Higashi-ku, Fukuoka 812-8582, Japan

Y. Goto
Department of Occupational Therapy, School of Health Sciences at Fukuoka, International University of Health and Welfare, 137-1 Enokizu, Okawa, Fukuoka 831-8501, Japan

S. Tobimatsu, R. Kakigi (eds.), *Clinical Applications of Magnetoencephalography*,
DOI 10.1007/978-4-431-55729-6_8

modality for the evaluation of visual dysfunction in various neurological and psychiatric disorders.

Keywords
Magnetoencephalography (MEG) • Visual evoked magnetic fields (VEFs) • Event-related magnetic fields (ERFs) • Parallel visual pathways • Clinical application

8.1 Introduction

The human visual system separates different types of information into anatomically segregated parallel streams of processing [1–3]. In primates, the tracts from the retina to the primary visual cortex (V1) are clearly split into two pathways: the magnocellular (M) and parvocellular (P) pathways. These pathways differ markedly along several anatomical and physiological dimensions. The higher levels of visual cortex also contain two streams of processing, each of which includes multiple visual areas and mediates different visual behaviors. The dorsal stream includes areas in the parietal cortex and is important for vision related to motion or spatial relationships. Conversely, the ventral stream includes visual areas in the temporal lobe and is more involved in the analysis of form and color [1–3]. Therefore, specific types of visual symptoms are caused when a specific part of the parallel visual pathways is not working normally.

Various types of visual disturbance are often observed in several neurological and psychiatric disorders [3], so that the detection of specific types of visual impairment can become a clue to the identification of the causative brain lesion or clinical diagnosis. It is acknowledged that neurophysiological techniques such as electroencephalography (EEG) and magnetoencephalography (MEG) offer reproducible and quantitative data on brain function [2]. MEG is a noninvasive technique that measures the magnetic fields generated by the neuronal activity of the brain, and the primary source of the magnetic signals is considered to be the same as that of the electrical signals recorded by EEG. However, unlike electrical potentials, magnetic fields penetrate the skull and scalp undistorted. In comparison with EEG, the biggest advantage of MEG is its high-spatial resolution (millimeter), while both EEG and MEG provide excellent temporal resolution (millisecond) [4, 5]. Accordingly, MEG measurements to record visual evoked magnetic fields (VEFs) and event-related magnetic fields (ERFs) with optimal visual stimuli are ideally suited to examine the rapidly changing patterns of brain activity that underlie visual dysfunction in neurological and psychiatric disorders.

In this review, we first summarize current concepts of the human visual system and its dysfunction. Next, we describe the clinical applications of VEFs and ERFs for several neurological and psychiatric conditions involving specific types of visual dysfunction. The purpose of this review is to stress the usefulness of VEFs and ERFs for evaluation of visual function in various neurological and psychological disorders.

8.2 Relationship Between Parallel Visual Pathways and Visual Processing Disorders

8.2.1 Lower-Level Visual Pathway

The visual pathway consists of the retina, the optic nerve, the fibers of an optic nerve traveling through the optic chiasm to the lateral geniculate nucleus (LGN), and optic radiations terminating in V1 of the occipital lobe [6] (Fig. 8.1a). Each optic nerve contains fibers from only one retinal ganglion cell. The optic chiasm contains fibers from the nasal parts of the retinas of both eyes. These fibers cross to the opposite side of the brain at the optic chiasm. The fibers from the temporal part of each eye do not cross at the optic chiasm, pass through the LGN on the same side of the brain, and continue back to the occipital lobe (Fig. 8.1a). Therefore, damage along these pathways causes a variety of visual field deficits [7]. For example,

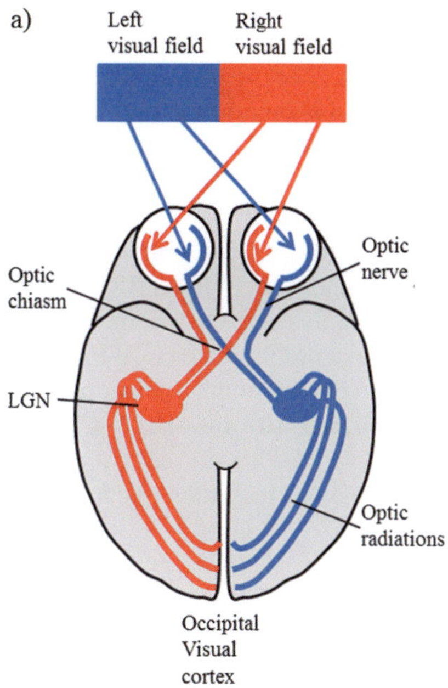

Fig. 8.1 The parallel visual pathways in humans. The lower-level visual pathways (**a**) The higher-level visual pathways (**b**) Abbreviations: *d-d stream* dorso-dorsal stream, *v-d stream* ventrodorsal stream, *LGN* lateral geniculate nucleus, *V1, 2, 3, 4, and 6* primary, secondary, tertiary, quaternary, and senary visual cortices, *V3A* V3 accessory, *V5/MT* quinary visual cortex/middle temporal area, *MST* medial superior temporal area, *IPL* inferior parietal lobule, *SPL* superior parietal lobule, *IT* inferior temporal cortex

lesions affecting the optic nerve from the retina to the optic chiasm cause loss of vision of the unilateral eye. Compressive lesions of the optic chiasm cause loss of vision in the temporal halves of the bilateral visual fields (bitemporal hemianopsia). Lesions affecting the optic tract, the LGN, or the optic radiation to V1 cause loss of vision in the temporal half of the contralateral visual field (homonymous hemianopsia) [7].

The visual projection from retina to cortex consists of two major parallel divisions: the P- and M-pathways [1–3] (Fig. 8.1b). The P-pathway is characterized by its high-spatial resolution, color sensitivity, low-contrast sensitivity, and low-temporal resolution. Thus, this pathway is important for color and form perception. In contrast, the M-pathway is characterized by high-temporal resolution, high-contrast sensitivity, color insensitivity, and low-spatial resolution and plays an important role in motion detection [2, 3]. Accordingly, lesions of the P-pathway impair color vision and high-spatial frequency form vision, whereas lesions of the M-pathway impair high-temporal frequency flicker and motion perception [8].

8.2.2 Higher-Level Visual Pathway

Two major cortical visual pathways have been proposed: the ventral (or temporal) and dorsal (or parietal) streams [2, 3] (Fig. 8.1b). The ventral stream is also commonly referred to as the "what" system because it is involved in the identification of objects, while the dorsal stream is called the "where" system because of its involvement in processing spatial location. The lower-level P- and M-pathways approximately correspond to the two systems, with the P-pathway projecting primarily to the ventral stream and the M-pathway providing the primary input to the dorsal stream. After V1, the ventral (P) pathway projects to the fusiform gyrus (V4) and the inferior temporal cortex, where it contains category-specific regions, including the fusiform face area (FFA [9]) and visual word form area (VWFA [10]). Conversely, the dorsal (M) stream projects to the V5/middle temporal area (MT) and the ventrodorsal (v-d) and dorso-dorsal (d-d) streams. The v-d stream includes the inferior parietal lobule (IPL), while the d-d stream consists of the superior parietal lobule (SPL). Both dorsal streams play an important role in motion detection [3].

A number of clinical syndromes including visual object agnosia, prosopagnosia, and pure alexia are related to lesions of the fusiform gyrus within the ventral stream [3]. Patients with prosopagnosia cannot recognize familiar faces or learn new faces [3]. Most patients with prosopagnosia suffer from bilateral damage to the inferior occipitotemporal region including FFA [11, 12], but there are also numerous reports of prosopagnosia after unilateral right hemisphere lesions [13]. Patients with pure alexia (reading disabilities) have severely impaired recognition of single-word forms but are still able to read letter by letter [12, 14]. These deficits are attributed to damage to a specific area located in the left mid-fusiform gyrus or disconnection between V1 and the VWFA [12]. Conversely, several syndromes are linked to lesions in the dorsal stream [3]. One of these syndromes is akinetopsia, which refers to the loss of perception of visual motion with other visual functions, such as perception of form and color, remaining intact [15]. Akinetopsia is thought to be caused by bilateral damage of area V5/MT [16], while unilateral damage to area V5/MT may lead to more subtle deficits in motion processing or hemiakinetopsia [17].

8.3 Clinical Applications of VEFs/ERFs

As mentioned above, visual information is processed simultaneously via multiple parallel pathways or channels, and a functional specialization in the visual system exists so that different attributes of the visual scene are processed in an anatomically separate part of the visual cortex [2]. Therefore, modulating the attributes of the visual stimuli can enable us to explore the human visual function from the retina to higher cortical areas.

8.3.1 VEFs for Lower-Level Visual Pathway

For testing the lower-level visual pathway, methods using VEFs are appropriate. VEFs elicited by monocular or binocular stimulation have various advantages in comparison with visual evoked potentials (VEPs) elicited in a similar manner [4]. Unlike VEPs, the responses of VEFs in the bilateral occipital lobes are clearly and easily distinguishable from each other using visual stimuli for the full, the left-half or the right-half visual field [18]. Any unilateral abnormality in the visual cortex or interhemispherical difference between the bilateral cortices can be easily and accurately detected. In particular, the estimation accuracy of the signal source is higher for partial visual field stimulation of the right or left hemi-visual field [4]. In addition, any abnormal visual function can be quantitatively evaluated by the latency delay with or without amplitude attenuation [19, 20]. Therefore, VEFs can be used clinically to localize the visual cortex and to evaluate the visual function in patients with (1) either organic or functional brain diseases before surgical interventions such as craniotomy, endovascular, or radiosurgical procedures, and/or (2) suspected abnormal condition in a certain part of the visual system [4].

Regarding visual stimulation, checkerboard pattern-reversal stimulation is the most common procedure used to explore the function of the lower-level visual pathways [5] because local spatial frequency analyzers are presumably present in V1 [21]. The parameters of visual stimuli on VEFs should follow those used for conventional scalp VEP guidelines [5]. These parameters include the stimulus size or spatial frequency, their contrast, the size of the field of presentation, the mean luminance of the field and background, and the temporal frequency [2]. In particular, by selecting the appropriate pattern element size, one can predominantly stimulate the fovea or the peripheral retina. Smaller-sized patterns somewhere between 10 and 15′ of arc preferentially stimulate the fovea, while patterns subtending more than 30–40′ of arc stimulate both the fovea and extrafoveal region [2]. To distinguish and analyze the visual areas of the left and right hemispheres, partial visual stimulation such as the left- or right-half visual fields or the quadrant visual fields are suitable [4].

In healthy people, VEFs elicited by pattern-reversal stimulation are characterized by three major components with latencies of 75–90, 100–120, and 145–160 ms, which are termed M75, M100, and M145, respectively [2, 22–24] (Fig. 8.2). These components correspond to the N75, P100, and N145 components of pattern-reversal VEPs [2]. The equivalent current dipoles (ECDs) of three VEF peaks have been estimated in V1 [2, 22–24]. It is important to understand that VEFs on foveal stimulation are much smaller than those on peripheral stimulation because of dipole direction [26–28], whereas VEPs on foveal stimulation are much larger than on peripheral stimulation [2, 26].

There have been several VEF studies using checkerboard pattern stimulation in cases with structural lesions adjacent to the visual pathway, such as the optic chiasm and occipital lobe [19, 20, 29, 30]. For example, in patients with compressive

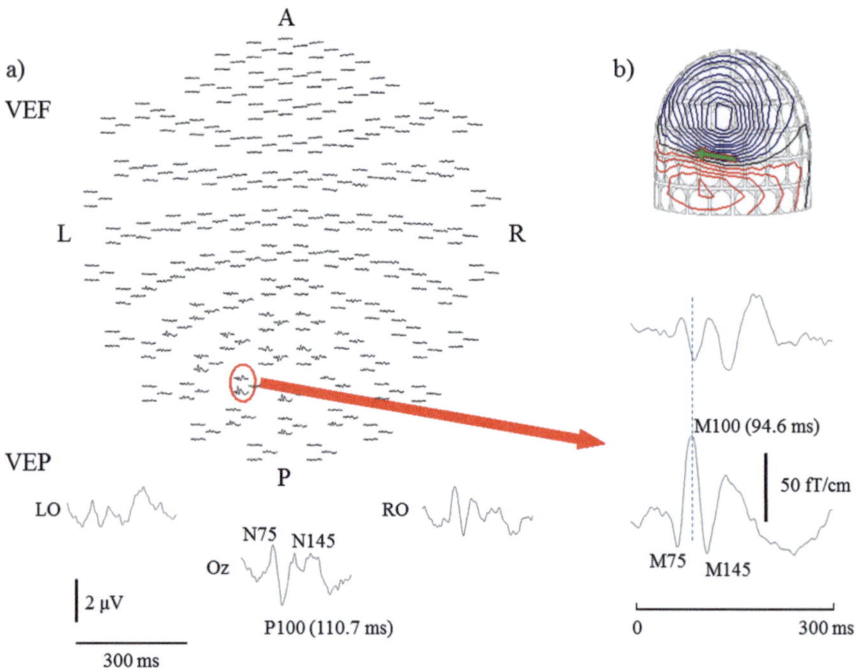

Fig. 8.2 Simultaneous recording of pattern-reversal VEFs and VEPs in a representative normal participant. In VEFs to the right hemifield stimulation, the M75, M100, and M145 waves are elicited in the left occipital area. However, VEPs show the right dominant occipital N75, P100, and N145 waves during the right hemifield stimulation (known as "paradoxical lateralization" [25]) (**a**). The contour map of magnetic fields at the peak of M100 demonstrates the dipole pattern in the occipital area (**b**). Abbreviations: *VEFs* visual evoked magnetic fields, *VEPs* visual evoked potentials

lesions of the optic chiasm, latencies of M75, M100, and M145 waves in temporal half-visual field stimulation were delayed more than those in nasal visual stimulation [29]. In patients with right lower quadrantanopsia after left parietal lobe lesionectomy, the M100 was obviously attenuated and delayed in the lesioned hemisphere [29]. In another study [30], VEFs for monocular hemifield pattern-reversal stimuli were recorded in patients with occipital lesions. VEFs were not detectable for the corresponding visual stimuli or the M100 latency was delayed in patients with homologous hemianopsia. In contrast, the M100 source was estimated on the calcarine fissure in patients with no visual deficits, even in patients with lesions adjacent to the V1 [30]. Thus, VEFs can provide a noninvasive method to objectively evaluate visual field deficit in patients with optic chiasm or occipital lesions. Accordingly, modulation of latency, amplitude, source localization, and source orientation of VEFs are useful for presurgical planning [19, 20, 30, 31].

When VEFs for isoluminant color stimuli were recorded, we can evaluate the function of the P-pathway selectively [2, 32]. In this condition, M120 was evoked, and ECDs of M120 were estimated in V1 [2, 32, 33]. M120/N120 (electrical equivalent of M120) abnormalities were reported in disorders showing the impairment of the P-pathway such as strabismic amblyopia [34], optic neuritis [35], and autism spectrum disorders [36]. Therefore, the exploration of M120/N120 abnormalities may detect subtle functional changes in V1 in several neurological and psychiatric disorders.

8.3.2 ERFs for Higher-Level Visual Pathway

For testing the higher-level visual pathway, ERF methods can be used. Similar to VEFs, ERFs offer an advantage over ERPs for localization of the anatomical sources of evoked brain activity.

8.3.2.1 Ventral Pathway

Category-specific stimuli (faces, words, and objects) are useful for evaluating the function of specialized processing modules of the fusiform gyrus within the ventral stream [3]. ERF and ERP studies have shown that visual processes are specialized for certain stimulus categories within the first 200 ms of stimulus presentation. Such fast and specialized brain processes have most consistently been indexed by an increased occipitotemporal M170/N170 component for faces and words compared with control stimuli (e.g., faces compared with non-face objects, words compared with non-orthographic strings of symbols or forms) [37–40]. In addition to amplitude differences, the lateralization of the M170 can differ by category; M170/N170 is right lateralized for faces, smaller and bilateral for objects, and as large for printed words in the left hemisphere [3, 41, 42]. ECDs of M170 for faces have been localized in fusiform gyri including FFA with right hemisphere dominance [43, 44] (Fig. 8.3). In contrast, ECDs of M170 for words are estimated in the left fusiform gyrus including VWFA [43, 45] (Fig. 8.4).

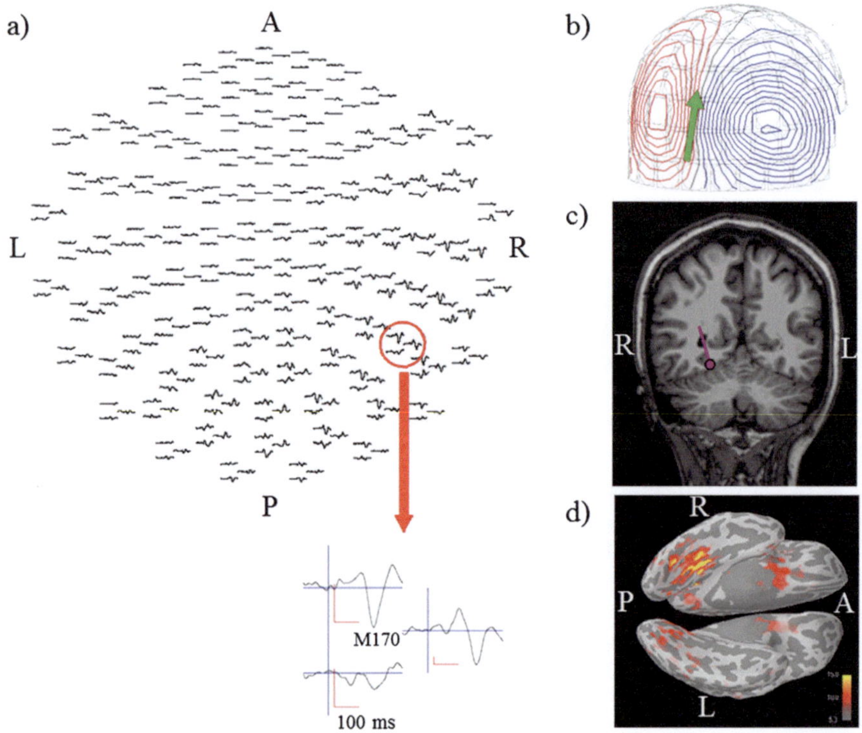

Fig. 8.3 ERFs for face stimuli in a representative normal participant. The full view of the MEG sensor array shows the right dominant occipitotemporal M170 (**a**). The contour map of magnetic fields at the peak of M170 demonstrates the dipole pattern in the right occipitotemporal area (**b**). The ECD is projected onto the participant's MRI and located in the right occipitotemporal area (**c**). The MNE solution provides source distribution maps at the peak of M170; dominant current sources are distributed over the right occipitotemporal area (**d**). Abbreviations in this and subsequent figures: *ERFs* event-related magnetic fields, *MEG* magnetoencephalography, *ECD* equivalent current dipole, *MNE* minimum norm estimate

Face stimuli are used when investigating disorders of face perception. Several ERF/ERP studies report the lack of or smaller M170/N170 in response to faces in acquired or congenital prosopagnosia (impairment of face recognition) [46–48]. Other ERF/ERP studies with congenital prosopagnosia have found attenuation or absence of face selectivity (no difference in amplitude between faces and other visual stimuli) in the M170/N170 response [49, 50]. Similarly, a functional magnetic resonance imaging (fMRI) study showed no differentiation in FFA activity between faces, houses, and other objects in congenital prosopagnosia [51]. Thus, the M170/N170 component for face stimuli is useful for evaluating FFA dysfunction. Additionally, in recent systematic review of face-specific ERFs and ERPs, abnormalities of M170/N170 responses were observed in various neurological or psychiatric disorders [52]. These included attention-deficit/hyperactivity disorder, alcohol dependence, Alzheimer's disease (AD), autism spectrum disorders, bipolar

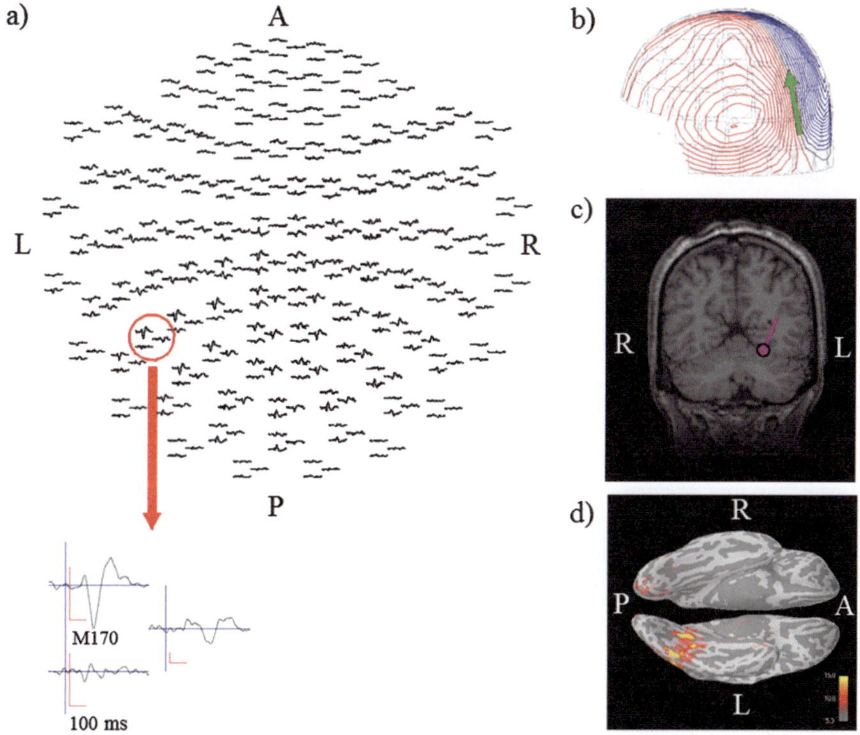

Fig. 8.4 ERFs for word stimuli in a representative normal participant. The full view of the MEG sensor array shows the left dominant occipitotemporal M170 (**a**). The contour map of magnetic fields at the peak of M170 demonstrates the dipole pattern in the left occipitotemporal area (**b**). The ECD is projected onto the participant's MRI and located in the left occipitotemporal area (**c**). The MNE solution provides source distribution maps at the peak of M170; dominant current sources are distributed over the left occipitotemporal area (**d**)

disorder, bulimia nervosa, fibromyalgia, Huntington's disease, major depressive disorder, Parkinson's disease, schizophrenia, and social phobia [52]. In schizophrenia, there was consistent evidence of smaller amplitudes for N170 to both emotional and nonemotional faces in a variety of tasks [52]. Accordingly, N170 is useful as a diagnostic aid in schizophrenia. A lack of N170 amplitude modulation to different emotional faces [53] and N170 insensitivity to face inversion [54] were also observed. In other disorders, smaller amplitudes or slower latencies of the M170/N170 component were frequently observed when using emotional stimuli. This suggests the impairment in facial affect processing rather than configural face processing. Therefore, M170/N170 abnormalities hold promise as diagnostic and treatment monitoring biomarkers for social dysfunction [52].

Word stimuli can be applied to reading disorders (dyslexia). Dyslexia is defined as a circumscribed difficulty in the acquisition of reading and writing skills despite adequate schooling and normal range of intelligence [55]. An ERF study reported that, unlike controls, adults with developmental dyslexia failed to show activation of the left inferior occipitotemporal cortex within 150 ms after the presentation of written words [56]. Supporting this view, fMRI studies recording participants with developmental dyslexia while they are reading have consistently reported underactivation of the left VWFA [57]. These ERFs and fMRI findings suggest a persistent deficit in VWFA specialization in patients with developmental dyslexia. Accordingly, the M170 for words is useful for evaluating reading disorders.

Yamasaki et al. [58] applied category-specific ERPs using faces, words, and objects to a unique case of multiple sclerosis with visual object agnosia and prosopagnosia without pure alexia [58]. MRI revealed white matter lesions in ventral parts of the temporo-occipital lobes including the fusiform and lingual gyri and inferior longitudinal fasciculi. Fluorodeoxyglucose-positron emission tomography also demonstrated decreased glucose metabolism in the ventral parts of the temporo-occipital lobes, with right dominance. In category-specific ERPs, the patient's N170 amplitude in response to faces and objects was significantly lower than that of normal controls, while N170 responses to words were normal. It is likely that this patient had a selective impairment of processing modules of objects and faces without that of words within the higher-level ventral stream. These electrophysiological findings were consistent with neuropsychological and neuroimaging examinations [58]. Therefore, category-specific ERFs/ERPs are useful to evaluate the functional impairment of subdivisions of the ventral pathway.

8.3.2.2 Dorsal Pathway

Motion stimuli can be useful for examining higher-level dorsal stream function. Visual motion stimuli can be characterized by the direction of motion: linear translation, rotation, expansion and contraction, and motion in depth. One particularly useful stimulus is the random dot kinematogram, in which the overall motion is extracted from a set of coherently and/or incoherently moving subunits. A distinction should also be made between real motion, apparent motion (stepwise dislocation), and illusory motion (motion aftereffects, etc.) [2]. Coherent motion stimuli using random dots have been widely used in psychophysical, electrophysiological, and neuroimaging studies to investigate global motion processing [3]. Coherent global motion, such as radial optic flow and horizontal motion, has been used in a number of studies [3]. Radial optic flow motion is a type of complex visual motion related to self-motion perception [59], while horizontal motion refers to simple unidirectional motion. As mentioned earlier, the dorsal stream is divided into two functional streams, the v-d stream and the d-d stream [60]. The d-d stream (SPL) is more closely related to horizontal motion processing, while the v-d stream (IPL) is important for optic flow motion processing [3, 61]. ERF/ERP studies reported that perception of these stimuli was associated with two major components (M170/N170, M200/P200) [3, 61] (Fig. 8.5). The occipitotemporal M170/N170 exhibited a V5/MT origin [3, 61] and was evoked by both types of stimuli. In contrast, the parietal M200/P200 was found to originate in IPL and was only elicited by optic flow motion stimuli [3, 61].

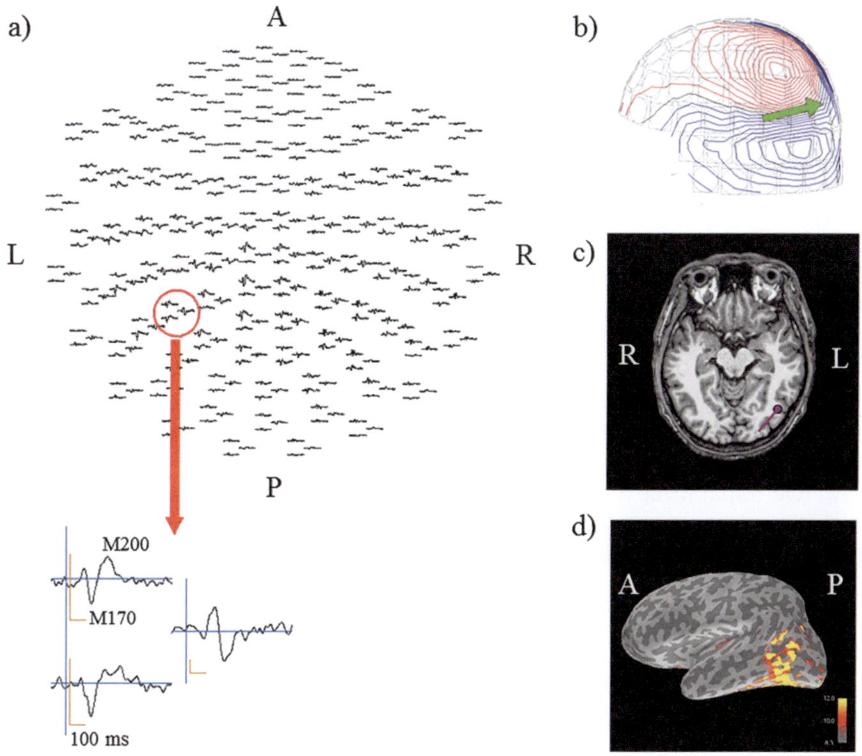

Fig. 8.5 ERFs for coherent optic flow motion stimuli in a representative normal participant. The full view of the MEG sensor array shows the occipito-temporoparietal M170 and M200 (**a**). The contour map of magnetic fields at the peak of M170 demonstrates the dipole pattern in the occipitotemporal area (**b**). The ECD is projected on to the subject's MRI and located in the occipitotemporal area (**c**). The MNE solution provides source distribution maps at the peak of M170; dominant current sources are distributed over the occipitotemporal area (**d**)

Disturbance of motion perception is caused by parietal lobe dysfunction in AD [62] as well as in stroke and brain trauma. To our knowledge, there have been no MEG studies on motion processing in patients with impaired motion perception including AD or mild cognitive impairment (MCI, the prodromal stage of AD). In contrast, several ERP studies using coherent motion stimuli have been reported in patients with AD or MCI [63–66]. Kavcic et al. [63] and Fernandez et al. [64] recorded ERPs in response to radial optic flow and horizontal motion in AD and healthy older controls. Patients with AD showed smaller N200 (corresponding to N170 in the present review) for both radial optic flow and horizontal motion stimuli compared with healthy controls. Abnormality in N200 for optic flow was well correlated with psychological tests such as navigational abilities [63]. These data suggest a deficit in extrastriate visual cortical motion processing in AD. Kubová and collaborators [62] recorded pattern-reversal VEP and motion ERPs (translational and radial movement) in patients with mild to moderate AD. Pattern-reversal

VEPs were comparable in AD patients and controls. However, there was a significantly smaller N2 (corresponding to N170 in the present review) ERP peak amplitude in response to motion stimuli (particularly, radially moving stimuli) in AD patients. This result implies a dysfunction in the higher-level dorsal cortical stream in AD. Yamasaki et al. [66] also evaluated motion perception in MCI and AD patients. The authors found no significant differences in N170 responses for both optic flow and horizontal motion stimuli between MCI patients and healthy controls. However, a significantly prolonged P200 latency for optic flow stimuli was observed in MCI patients compared with healthy controls. Therefore, within the dorsal stream, the v-d stream (IPL) related to optic flow perception, but not the d-d stream (SPL), is selectively impaired in MCI patients [66]. Furthermore, Yamasaki et al. [66] also investigated the dorsal stream function in patients with moderate AD. Patients with AD showed a prolongation of N170 for both optic flow and horizontal motion stimuli and of P200 for optic flow motion compared with healthy older and MCI patients. These results indicate that MCI patients exhibit a selective impairment of higher-level processing in the dorsal stream (v-d stream including IPL). Conversely, AD patients show an impairment in the distributed higher-level dorsal stream (both v-d and d-d streams including IPL, SPL, and V5/MT) [66]. In addition, N170 latency for both optic flow and horizontal motion and P200 latency and amplitude for optic flow were significantly correlated with the neuropsychological test. In particular, optic flow-specific P200 was the most highly correlated to the neuropsychological test [66]. Consequently, coherent motion stimuli can be useful in the examination of dorsal stream function.

8.4 Conclusions

In this review, we describe the clinical applications of VEFs and ERFs to patients with various types of visual dysfunction. We believe that the manipulation of visual stimulus parameters of VEFs and ERFs enables us to explore the function of distinct parallel visual pathways at different levels. Therefore, we propose that VEFs and ERFs with appropriate visual stimuli provide valuable and useful information for diagnosis, characterization, and therapeutic decisions in patients with visual dysfunction in various neurological and psychiatric disorders. However, ERFs have not been yet applied to clinical evaluation compared with VEFs. Accordingly, we hope that clinical application of ERFs as well as VEFs becomes more popular in the future.

Acknowledgments This study was supported in part by Grants-in-Aid for Scientific Research, No. 26350931 and No. 26560401 from the Ministry of Education, Culture, Sports, Science, and Technology in Japan.

References

1. Livingstone M, Hubel D. Segregation of form, color, movement, and depth: anatomy, physiology, and perception. Science. 1988;240:740–9.
2. Tobimatsu S, Celesia GG. Studies of human visual pathophysiology with visual evoked potentials. Clin Neurophysiol. 2006;117:1414–33.
3. Yamasaki T, Tobimatsu S. Electrophysiological assessment of the human visual system. In: Harris JM, Scott J, editors. Neuroscience research progress, visual cortex: anatomy, functions and injuries. New York: Nova Science Publishers; 2012. p. 37–67.
4. Hashimoto I, Kakigi R, Nagamine T, Nakasato N, Shiraishi H, Watanabe Y. Guideline for clinical application of magnetoencephalography. Jpn Clin Neurophysiol. 2005;33:231–52.
5. Burgess RC, Funke ME, Bowyer SM, Lewine JD, Kirsch HE, Bagić AI. American Clinical Magnetoencephalography Society Clinical Practice Guideline 2: presurgical functional brain mapping using magnetic evoked fields. J Clin Neurophysiol. 2011;28:355–61.
6. De Moraes CG. Anatomy of the visual pathways. J Glaucoma. 2013;22:S2–7.
7. Menjot de Champfleur N, Menjot de Champfleur S, Galanaud D, Leboucq N, Bonafé A. Imaging of the optic chiasm and retrochiasmal visual pathways. Diagn Interv Imaging. 2013;94:957–71.
8. Schiller PH, Logothetis NK, Charles ER. Role of the color-opponent and broad-band channels in vision. Vis Neurosci. 1990;5:321–46.
9. Kanwisher N, McDermott J, Chun MM. The fusiform face area: a module in human extrastriate cortex specialized for face perception. J Neurosci. 1997;17:4302–11.
10. Cohen L, Dehaene S, Naccache L, Lehéricy S, Dehaene-Lambertz G, Hénaff MA, et al. The visual word form area: spatial and temporal characterization of an initial stage of reading in normal subjects and posterior split-brain patients. Brain. 2000;123:291–307.
11. Damasio AR, Damasio H, Vanhoesen GW. Prosopagnosia-anatomic basis and behavioral mechanisms. Neurology. 1982;32:331–41.
12. Kleinschmidt A, Cohen L. The neural bases of prosopagnosia and pure alexia: recent insights from functional neuroimaging. Curr Opin Neurol. 2006;19:386–91.
13. DeRenzi E, Perani D, Carlesimo GA, Silveri MC, Fazio F. Prosopagnosia can be associated with damage confined to the right hemisphere-an MRI and PET study and a review of the literature. Neuropsychologia. 1994;32:893–902.
14. Damasio AR, Damasio H. The anatomic basis of pure alexia. Neurology. 1983;33:1573–83.
15. Rizzo M, Nawrot M, Zihl J. Motion and shape perception in cerebral akinetopsia. Brain. 1995;118:1105–27.
16. Zihl J, von Cramon D, Mai N. Selective disturbance of movement vision after bilateral brain damage. Brain. 1983;106:313–40.
17. Schenk T, Zihl J. Visual motion perception after brain damage. 1. Deficits in global motion perception. Neuropsychologia. 1997;35:1289–97.
18. Seki K, Nakasato N, Fujita S, Hatanaka K, Kawamura T, Kanno A, et al. Neuromagnetic evidence that the P100 component of the pattern reversal visual evoked response originates in the bottom of the calcarine fissure. Electroencephalogr Clin Neurophysiol. 1996;100:432–42.
19. Nakasato N, Seki K, Kawamura T, Ohtomo S, Kanno A, Fujita S, et al. Cortical mapping using an MRI-linked whole head MEG system and presurgical decision making. Electroencephalogr Clin Neurophysiol. 1996;S47:333–41.
20. Nakasato N, Yoshimoto T. Somatosensory, auditory, and visual evoked magnetic fields in patients with brain diseases. J Clin Neurophysiol. 2000;17:201–11.
21. De Valois KK, De Valois RL, Yund EW. Responses of striate cortex cells to grating and checkerboard patterns. J Physiol. 1979;291:483–505.
22. Hatanaka K, Nakasato N, Seki K, Kanno A, Mizoi K, Yoshimoto T. Striate cortical generators of the N75, P100 and N145 components localized by pattern reversal visual evoked magnetic fields. Tohoku J Exp Med. 1997;182:9–14.

23. Shigeto H, Tobimatsu S, Yamamoto T, Kobayashi T, Kato M. Visual evoked cortical magnetic responses to checkerboard pattern reversal stimulation: a study on the neural generators of N75, P100 and N145. J Neurol Sci. 1998;156:186–94.
24. Tabuchi H, Yokoyama T, Shimogawara M, Shiraki K, Nagasaka E, Miki T. Study of the visual evoked magnetic field with the M-sequence technique. Invest Ophthalmol Vis Sci. 2002;43:2045–54.
25. Cohn NB, Kircher J, Emmerson RY, Dustman RE. Pattern reversal evoked potentials: age, sex and hemispheric asymmetry. Electroencephalogr Clin Neurophysiol. 1985;62:399–405.
26. Brecelj J, Kakigi R, Koyama S, Hoshiyama M. Visual evoked magnetic responses to central and peripheral stimulation: simultaneous VEP recordings. Brain Topogr. 1998;10:227–37.
27. Nakamura M, Kakigi R, Okusa T, Hoshiyama M, Watanabe K. Effects of check size on pattern reversal visual evoked magnetic field and potential. Brain Res. 2000;872:77–86.
28. Chen WT, Ko YC, Liao KK, Hsieh JC, Yeh TC, Wu ZA, et al. Optimal check size and reversal rate to elicit pattern-reversal MEG responses. Can J Neurol Sci. 2005;32:218–24.
29. Hatanaka K, Nakasato N, Nagamatsu K, Iwasaki M, Kanno A, Ikeda H, et al. Modification of the pattern-evoked magnetic fields associated with the location of the lesion along the visual pathways. In: Proceedings of the 12th international conference on biomagnetism; 2001. p. 145–49.
30. Kanno A, Nakasato N, Fujiwara S, Kumabe T, Tominaga T. Visual-evoked magnetic fields for pattern-reversal stimulus in patients with occipital lesions. Clin Neurophysiol. 2008;119:e93.
31. Grover KM, Bowyer SM, Rock J, Rosenblum ML, Mason KM, Moran JE, et al. Retrospective review of MEG visual evoked hemifield responses prior to resection of temporo-parieto-occipital lesions. J Neuro-Oncol. 2006;77:161–6.
32. Shigeto H, Tobimatsu S, Yamamoto T, Kobayashi T, Kato M. Visual evoked cortical magnetic responses: effects of visual field and visual stimulation. Electroencephalogr Clin Neurophysiol. 1995;97:124–5.
33. Fylan F, Holliday IE, Singh KD, Anderson SJ, Harding GFA. Magnetoencephalographic investigation of human cortical area V1 using color stimuli. Neuroimage. 1997;6:47–57.
34. Anderson SJ, Holliday IE, Harding GFA. Assessment of cortical dysfunction in human strabismic amblyopia using magnetoencephalography (MEG). Vis Res. 1999;39:1723–38.
35. Tobimatsu S, Kato M. Multimodality visual evoked potentials in evaluating visual dysfunction in optic neuritis. Neurology. 1998;50:715–8.
36. Fujita T, Yamasaki T, Kamio Y, Hirose S, Tobimatsu S. Parvocellular pathway impairment in autism spectrum disorder: evidence from visual evoked potentials. Res Autism Spectr Disord. 2011;5:277–85.
37. Bentin S, Allison T, Puce A, Perez E, McCarthy G. Electrophysiological studies of face perception in humans. J Cogn Neurosci. 1996;8:551–65.
38. Itier RJ, Taylor MJ. N170 or N1? Spatiotemporal differences between object and face processing using ERPs. Cereb Cortex. 2004;14:132–42.
39. Liu J, Higuchi M, Marantz A, Kanwisher N. The selectivity of the occipitotemporal M170 for faces. Neuroreport. 2000;11:337–41.
40. Sams M, Hietanen JK, Hari R, Ilmoniemi RJ, Lounasmaa OV. Face-specific responses from the human inferior occipito-temporal cortex. Neurosci. 1997;77:49–55.
41. Rossion B, Joyce CA, Cottrell GW, Tarr MJ. Early lateralization and orientation tuning for face, word, and object processing in the visual cortex. Neuroimage. 2003;20:1609–24.
42. Maurer U, Rossion B, McCandliss BD. Category specificity in early perception: face and word N170 responses differ in both lateralization and habituation properties. Front Hum Neurosci. 2008;2:18.
43. Halgren E, Raij R, Marinkovic K, Jousmäki V, Hari R. Cognitive response profile of the human fusiform face area as determined by MEG. Cereb Cortex. 2000;10:69–81.
44. Allison T, Puce A, Spencer DD, McCarthy G. Electrophysiological studies of human face perception. I. Potentials generated in occipitotemporal cortex by face and non-face stimuli. Cereb Cortex. 1999;9:415–30.

45. Brem S, Bucher K, Halder P, Summers P, Dietrich T, Martin E, et al. Evidence for developmental changes in the visual word processing network beyond adolescence. Neuroimage. 2006;29:822–37.
46. Eimer M, McCarthy RA. Prosopagnosia and structural encoding of faces: evidence from event-related potentials. Neuroreport. 1999;10:255–9.
47. Righart R, de Gelder B. Impaired face and body perception in developmental prosopagnosia. Proc Natl Acad Sci. 2007;104:17234–8.
48. Dobel C, Putsche C, Zwitserlood P, Junghöfer M. Early left-hemispheric dysfunction of face processing in congenital prosopagnosia: an MEG study. PLoS One. 2008;3:e2326.
49. Harris AM, Duchaine BC, Nakayama K. Normal and abnormal face selectivity of the M170 response in developmental prosopagnosics. Neuropsychologia. 2005;43:2125–36.
50. Kress T, Daum I. Event-related potentials reflect impaired face recognition in patients with congenital prosopagnosia. Neurosci Lett. 2003;352:133–6.
51. Hadjikhani N, de Gelder B. Neural basis of prosopagnosia: an fMRI study. Hum Brain Mapp. 2002;16:176–82.
52. Feuerriegel D, Churches O, Hofmann J, Keage HAD. The N170 and face perception in psychiatric and neurological disorders: a systematic review. Clin Neurophysiol. 2014. doi:10.1016/j.clinph.2014.09.015.
53. Lynn SK, Salisbury DF. Attenuated modulation of the N170 ERP by facial expressions in schizophrenia. Clin EEG Neurosci. 2008;39:108–11.
54. Tsunoda T, Kanba S, Ueno T, Hirano Y, Hirano S, Maekawa T, et al. Altered face inversion effect and association between face N170 reduction and social dysfunction in patients with schizophrenia. Clin Neurophysiol. 2012;123:1762–8.
55. Peterson RL, Pennington BF. Developmental dyslexia. Lancet. 2012;379:1997–2007.
56. Helenius P, Tarkiainen A, Cornelissen P, Hansen PC, Salmelin R. Dissociation of normal feature analysis and deficient processing of letter-strings in dyslexic adults. Cereb Cortex. 1999;9:476–83.
57. Richlan F, Kronbichler M, Wimmer H. Functional abnormalities in the dyslexic brain: a quantitative meta-analysis of neuroimaging studies. Hum Brain Mapp. 2009;30:3299–308.
58. Yamasaki T, Taniwaki T, Tobimatsu S, Arakawa K, Kuba H, Maeda Y, et al. Electrophysiological correlates of associative visual agnosia lesioned in the ventral pathway. J Neurol Sci. 2004;221:53–60.
59. Gibson JJ. The perception of the visual world. Boston: Houghton Mifflin; 1950.
60. Rizzolatti G, Matelli M. Two different streams form the dorsal visual system: anatomy and functions. Exp Brain Res. 2003;153:146–57.
61. Yamasaki T, Muranaka H, Kaseda Y, Mimori Y, Tobimatsu S. Understanding the pathophysiology of Alzheimer's disease and mild cognitive impairment: a mini review on fMRI and ERP studies. Neurol Res Int. 2012;2012:719056.
62. Tsai PH, Mendez MF. Akinetopsia in the posterior cortical variant of Alzheimer disease. Neurology. 2009;73:731–2.
63. Kavcic V, Fernandez R, Logan D, Duffy CJ. Neurophysiological and perceptual correlates of navigational impairment in Alzheimer's disease. Brain. 2006;129:736–46.
64. Fernandez R, Kavcic V, Duffy CJ. Neurophysiologic analyses of low- and high-level visual processing in Alzheimer disease. Neurology. 2007;68:2066–76.
65. Kubová Z, Kremláček J, Vališ M, Langrová J, Szanyi J, Vít F, et al. Visual evoked potentials to pattern, motion and cognitive stimuli in Alzheimer's disease. Doc Ophtalmol. 2010;121:37–49.
66. Yamasaki T, Goto Y, Ohyagi Y, Monji A, Munetsuna S, Minohara M, et al. Selective impairment of optic flow perception in amnestic mild cognitive impairment: evidence from event related potentials. J Alzheimers Dis. 2012;28:695–708.

Part VI

Epilepsy

Childhood Epilepsy

9

Hideaki Shiraishi

Abstract

Magnetoencephalography (MEG) is one of the best ways to analyze neural function. In particular, MEG is valuable for assessing brain activity in children with epilepsy, because it is noninvasive and can be used multiple times for the same patient, thus enabling changes in epileptogenicity to be monitored as a child's growth.

Conventional MEG analysis is not always able to define the epileptogenic area. To this end, the single-dipole analysis tools can resolve localized epileptic MEG discharges and demonstrate equivalent current dipoles (ECDs) in cerebral cortex. However, diffuse or multifocal epileptic activities are not suitable for such ECD analyses, because the formula underpinning the single-dipole method assumes a circumscribed epileptogenic area. In this chapter, we discuss an alternative MEG tool whereby spatial filtering analysis is used for widespread or multifocal epileptogenic areas. Furthermore, we performed time-frequency analysis on a patient with symptomatic localization-related epilepsy who showed rhythmic activities as subclinical electrical discharges.

In conclusion, the single-dipole method, spatial filtering analysis, and time-frequency analysis could successfully resolve an epileptogenic area in patients with epilepsy. Thus, MEG analysis is potentially useful for presurgical evaluation or the diagnosis of epileptic syndromes for almost every patient with epilepsy.

Keywords

MEG • Epilepsy • Short-time Fourier transform • Dynamic statistical parametric mapping • Time-frequency analysis

H. Shiraishi, M.D., Ph.D. (✉)
Department of Pediatrics, Graduate School of Medicine, Hokkaido University, North 15, West 7, Kita-ku, Sapporo, Hokkaido 060-8638, Japan
e-mail: siraisi@med.hokudai.ac.jp

S. Tobimatsu, R. Kakigi (eds.), *Clinical Applications of Magnetoencephalography*, DOI 10.1007/978-4-431-55729-6_9

9.1 Introduction

Clinical applications of magnetoencephalography (MEG) should be expanded in the field of child epilepsy, first, because MEG can demonstrate valuable information about epileptogenicity noninvasively and, second, because a patient can undergo MEG multiple times to monitor changes in epileptogenicity during the child's growth.

MEG analysis is now established as an indispensable presurgical tool to evaluate candidates for epileptic surgery. In addition, MEG can also generate valuable information for the precise diagnosis of epileptic syndrome, though this application necessitates further development of appropriate methods of analysis to evaluate various kinds of epileptic activities.

In this chapter, we introduce a system of evaluating epileptic current in patients with epilepsy and further demonstrate the valuable role of MEG in this clinical field.

9.2 Methods

9.2.1 MEG Data Acquisition

MEG data were recorded using 204-channel helmet-shaped gradiometers (Vector View, Elekta AB, Stockholm, Sweden) at a 600-Hz sampling rate and then digitally filtered with a band-pass from 3 to 30 Hz for offline analysis. Segments containing abnormal paroxysms were identified manually.

9.2.2 Approach for Appropriate Analyses of Epileptic Current

ECD analysis is applicable for patients with localized epileptic foci, although this analysis is widely used and is available via preinstalled software in almost every MEG machine. In the first step of a MEG analysis or trial evaluation for epileptic spikes, ECD analysis is widely used because for the first screening of epileptic patients, we have to concentrate on the whole-brain MEG activity in a single view to select the most appropriate analysis method(s) for each patient. For localizing epileptiform activity, ECD analysis is applicable. Diffuse activity is not applicable for ECD analysis but for spatial filtering analysis, such as dynamic statistical parametric mapping (dSPM). Low-voltage fast activity, such as subclinical discharge or ictal activity, is not applicable for ECD analysis, since the signal to noise ratio is low. Time-frequency analysis can calculate the major gravity of distribution of the low-voltage oscillation and demonstrate the propagation of rhythmic activity.

9.2.3 Equivalent Current Dipole

Individual spikes were aligned on the basis of the peak latency and then analyzed. The distribution of brain activity generating the spikes was determined using a source estimation approach, the equivalent current dipole (ECD) model. This model is appropriate when the underlying brain activity is focal, i.e., restricted to a relatively small region of the brain.

ECDs were calculated with xfit software (Elekta Neuromag Oy, Helsinki, Finland) using the single-dipole model. The conductivity geometry of the head was assumed to be spherical and symmetrical. Dipoles were calculated for each time-point measurement (every 2.5 ms) within 100 ms of each MEG spike. Results from all sensors were analyzed, with no regions of interest selected. The initial location for the iterative ECD fit was assumed to be under the sensor with the largest signal, and the ECD with the best goodness of fit (GOF) was selected as being representative of that particular MEG signal spike. The GOF is a measure of how well the ECD model explains the measured signals, and a dipole fit was accepted when the GOF was greater than 70 %. To visualize anatomical locations, the ECDs were superimposed on the magnetic resonance imaging (MRI) generated for each patient.

9.3 Direct Impact for Diagnosis of Epileptic Syndrome

Case 1: A 22-year-old boy with parietal lobe epilepsy (PLE). His seizures first occurred at the age of 8 years, beginning with sensory auras (tingling in the right hand) and evolving to right hemi-convulsion with his face and eyes deviated to the right side. His seizures have been precipitated by touch on the right side of his body and have occurred daily even with multiple antiepileptic drugs (AEDs). Initial scalp electroencephalography (EEG) screening detected no significant epileptiform discharges, and the patient was referred to our hospital. On performing MEG, we demonstrated remarkable spikes in the left parietal region with corresponding EEG spikes in the same area and ECDs accumulated in the postcentral gyrus (Fig. 9.1a). According to the clustering ECD location, this patient had a cortical lesion with a high signal on T2-weighted imaging (T2WI) in the vicinity of the central gyrus by MEG-guided MRI scanning (Fig. 9.1b). [18F]-Fluorodeoxyglucose positron emission tomography (FDG-PET) also showed glucose hypometabolism at the left postcentral gyrus. He could not attend junior high school for 3 years due to

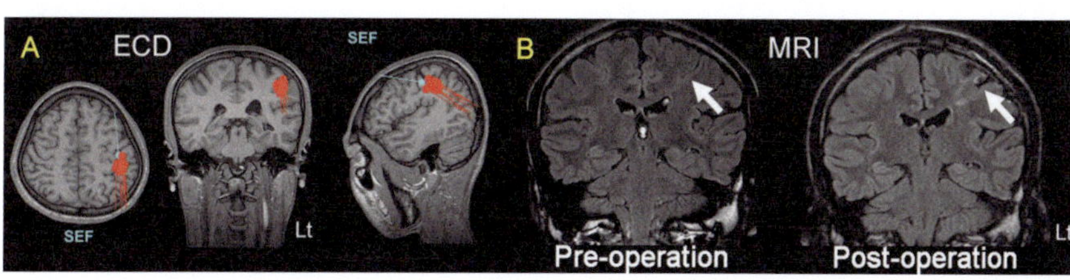

Fig. 9.1 (**a**) Equivalent current dipoles (ECDs) of MEG spikes were located in the posterior bank of the left central gyrus. The ECD of the somatosensory evoked field (SEF) localized the central gyrus. (**b**) MRI with FLAIR image showed a high signal at the bottom of the central gyrus and transmantle sign. Partial resection of the high-signal lesion was undertaken

recurrent complex partial seizures and secondary generalized seizures before the surgery. Furthermore, he was depressed by his disabled situation and prescribed antidepressant. Partial resection at the lesion stopped his complex partial seizure, but he retained abnormal sensation in the right upper arm (Fig. 9.1b). After his surgery, however, the patient showed improved mood and was able to attend and graduate from high school. His antidepressant prescription could be stopped, and he is currently working as a technologist in a company.

Case 2: A 17-year-old boy with atypical benign partial epilepsy (ABPE) in childhood. ABPE is characterized as childhood epilepsy with central-temporal EEG spikes and continuous spike and waves during sleep (CSWS). Patients with ABPE have multiple seizure types: focal motor seizures, atypical absence seizures, and myoclonic astatic seizures including epileptic negative myoclonus (ENM) but not tonic seizures [1]. Case 2 experienced his first seizure, a generalized tonic-clonic seizure, during sleep at the age of 3 years. His seizures occurred every 2–3 months, and he was treated with carbamazepine (CBZ) initially and up to the age of 7 years, when ENM occurred and the seizure frequency increased to more than 100 times every day. He was referred to our hospital at the age of 7 years. At this time, EEG showed CSWS, and MEG demonstrated concentrated spike sources at the peri-sylvian and peri-rolandic area (Fig. 9.2a). These findings and the patient's clinical course were consistent with the diagnostic criteria of ABPE [2]. His drug treatment was changed, with the CBZ replaced with ethosuximide (ESM), after which his ENM decreased dramatically. By the age of 14 years, his medication was stopped altogether, and the MEG spikes located unilaterally had disappeared by the age of 16 years (Fig. 9.2b, c). He currently shows normal intelligence and attends regular high school.

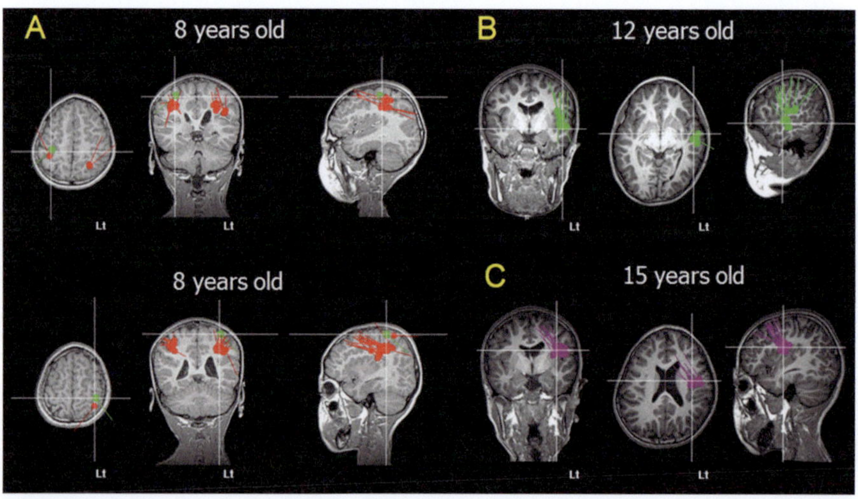

Fig. 9.2 (**a**) MEG showed concentrated MEG spike sources at the peri-sylvian and peri-rolandic areas bilaterally at the age of 8 years. (**b**) MEG spikes were located at the left side unilaterally at the age of 12 years. (**c**) MEG spikes were located at the left side unilaterally at the age of 15 years

9.3.1 Short-Time Fourier Transform Analysis

Short-time Fourier transform (STFT) analysis may be used to reveal the distribution of MEG polyspikes [15]. The MATLAB (MathWorks, Natick, USA) program was used to execute the STFT for the MEG signals in this study. Each signal was divided into small sequential frames, and fast Fourier transformation (FFT) applied to each frame.

In the present study, the STFT was implemented using a 256-point window at 426.7 ms per window (i.e., 256 points × 1000 ms/600 Hz). The window was shifted every four points, corresponding to 6.7 ms (i.e., 1000 ms/600 Hz × 4 points). The fast Fourier transform method was applied to each window. This process was repeated for all representative signals, and the time-frequency distributions were displayed as graphs.

Fast Fourier transform was performed for frequencies in the ranges of 3–30 Hz, 30–50 Hz, and 50–100 Hz. A signal's spectrum was considered to be aberrant when it was isolated from the background frequency spectrum on the graph. Such aberrant frequency spectra were superimposed onto the 3D-MRI reconstruction.

9.4 Direct Impact for Diagnosis of Epileptic Syndrome

Case 3: A 26-year-old woman with daily seizures. At the age of 2 months, she began to have spasms with bilateral upper limb contraction and head nodding, and her EEG showed hypsarrhythmia. She was diagnosed with West syndrome. After ACTH therapy, her seizures were initially resolved; however, they relapsed after a while. At the age of 2 years, she showed tonic posturing of the upper limbs and myoclonic seizures that were refractory to various AEDs (valproate, CBZ, clonazepam, phenobarbital, acetazolamide, and nitrazepam) as well as ketogenic diets. At the age of 17 years, her seizures became epileptic spasms with series of transform and tonic seizures. These seizures were symmetric from clinical findings, but her head deviated to the left occasionally. Her interictal EEG showed paroxysmal fast activities (PFAs) and bilateral frontal dominant diffuse spikes. Ictal EEG showed diffuse desynchronization in her seizures. At 19 years of age, 99mTc-ECD-SPECT showed hypoperfusion in the right frontal and temporal lobes dominantly. She was diagnosed with unclassifiable epilepsy.

At the age of 17 years, this patient underwent MEG analysis with PFA captured during the recordings. Although ECDs were located in bilateral hemispheres, most were scattered in the right parietal area (Fig. 9.3c). The MEG showed PFA corresponding to the EEG PFA in the right central-parietal-temporal area. Analysis of the PFA by STFT showed a significant power spectrum in the range from 10 to 25 Hz. The 3D-MRI moving image showed generation of PFA in the right angular gyrus that propagated contiguously to the postcentral gyrus and superior parietal lobule (Fig. 9.3a).

Ictal EEG and MEG of epileptic spasms were obtained simultaneously. STFT analysis before the onset of a clinical seizure showed a specific aberrant oscillation band of 10–18 Hz, although the EEG showed no specific findings. The 3D-MRI

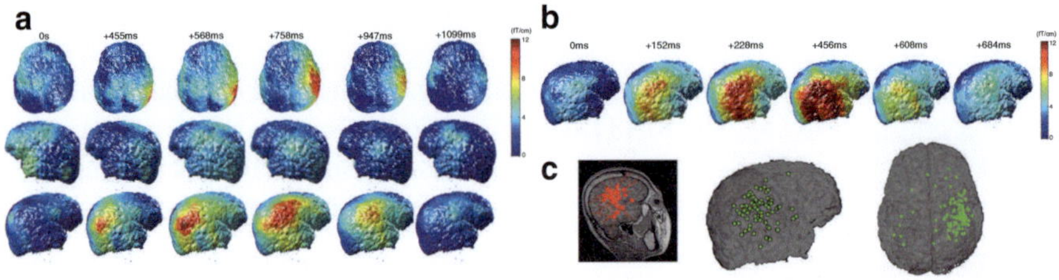

Fig. 9.3 (**a**) Analysis of PFA by STFT showed a significant power spectrum in the range from 10 to 25 Hz. The 3D-MRI moving image showed that PFA was generated in the right angular gyrus and propagated contiguously to the postcentral gyrus and superior parietal lobule. (**b**) In the ictal period, STFT analysis before the onset of a clinical seizure showed a specific, aberrant, 10–18-Hz oscillation band. The 3D-MRI moving image indicated that the specific oscillation band was generated in the right inferior parietal lobule. (**c**) Although ECDs were located in bilateral hemispheres, most ECDs were scattered in the right parietal area

moving image indicated that the specific oscillation band was generated in the right inferior parietal lobule, while analysis by STFT during the onset of a clinical seizure showed no specific aberrant oscillation. The high-power area of the PFA overlapped with that of the ictal paroxysmal discharge in the MEG (Fig. 9.3b).

The case was diagnosed and treated as cryptogenic-generalized epilepsy, suspected to be Lennox-Gastaut syndrome, and STFT analysis of the MEG demonstrated cryptogenic-localized epileptogenic foci in the right frontal and temporal lobes. STFT analysis also demonstrated the mechanism of the MEG rhythmic activity: the ictal discharges, providing the most useful information for the analysis of epileptogenic foci [3–5].

9.4.1 Dynamic Statistical Parametric Mapping (dSPM)

The dSPM method [6] is based on a noise-normalized minimum L2-norm estimate. For such analysis, we used an anatomically constrained distributed source model consisting of current dipole vectors located at each element; this model assumes the sources are located in the cerebral cortex. The cortical surface was segmented from the high-resolution MRI using FreeSurfer software [7, 8] and subsampled to about 2500 elements per hemisphere. For this estimate, the forward solution was calculated using a boundary element model (BEM) [9, 10] with conductivity boundaries determined from the segmented MRI. Only the inner surface of the skull is needed for MEG [10, 11].

The dSPM approach used to estimate the time course of activity at each cortical location is based on the generalized least-square or weighted minimum-norm solution [12, 13]; here, however, the estimate is normalized for noise sensitivity, thus providing a statistical parametric map [6] and reducing variation in the point-spread function between locations [14]. Simulations have suggested that the spatial resolution is 15 mm or better [6, 14]. Maps were calculated at 2.5-ms intervals. The significance of modulation at each site was calculated using an F-test [6, 15]. These statistical maps differ from maps of estimated source strengths, since the estimated noise variance is not constant across different cortical locations. However, since the same noise covariance estimates are used at all time points for a given cortical location, source strength at a given location over time is directly proportional to the statistical map. The current approach thus provides dSPM of cortical activity, similar to the statistical maps typically generated using fMRI or PET data but with a millisecond temporal resolution.

9.5 Direct Impact for Diagnosis of Epileptic Syndrome

Case 4: A 21-year-old male patient with symptomatic/cryptogenic localization-related epilepsy with unclassifiable epileptic focus. He experienced drug-resistant recurrent seizures with impairment of consciousness from infancy. His EEG showed a bilateral diffuse spike and wave complex (Fig. 9.4). The paroxysm occurred frequently and was sometimes represented as rhythmic activity: subclinical

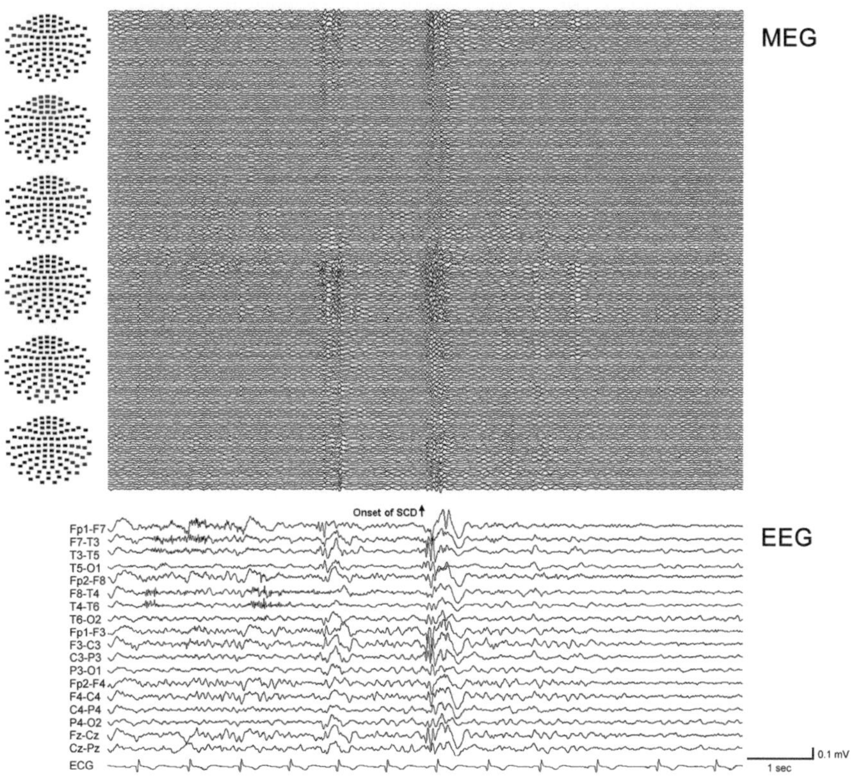

Fig. 9.4 Case 4, EEG and MEG. This patient's EEG showed a bilateral diffuse spike and wave complex. Paroxysm occurred frequently and sometimes represented as rhythmic activity: subclinical discharge. The MEG also showed bilateral diffuse spike activity

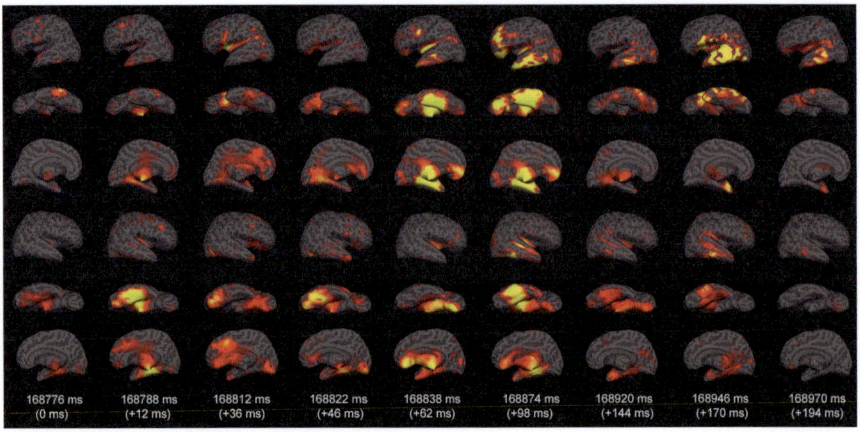

Fig. 9.5 Case 4, dSPM. dSPM demonstrated the change in spike activities at milliseconds resolution. The initial activity was generated in the left fusiform gyrus, propagated to the right cingulate gyrus and right fusiform gyrus (+12–46 ms), and moved to the left fusiform gyrus, middle and inferior temporal gyrus, and left superior frontal gyrus (+62–194 ms)

discharge. As for the EEG, his MEG showed bilateral diffuse spike activity (Fig. 9.4). ECD analysis could not resolve the epileptogenic focus, because the activity was not sufficiently localized to define the ideal single spike model. dSPM demonstrated the change in spike activities with milliseconds resolution: the initial activity was generated in the left fusiform gyrus, propagated to the right cingulate gyrus and right fusiform gyrus (+12–46 ms), and then moved to the left fusiform gyrus, middle and inferior temporal gyrus, and left superior frontal gyrus (+62–194 ms) (Fig. 9.5).

Due to the significant advantage of MEG over EEG in spatial resolution, MEG can show the evolution of epileptic current during the seizure. Furthermore, MEG study can be performed repeatedly, since it can be done noninvasively. This advantage is especially important for children.

Case 5: A 14-year-old patient with Lennox-Gastaut syndrome. He had tonic seizures and atypical absence seizures. His EEG and MEG showed diffuse slow spike and wave complex (Fig. 9.6). Gliosis was found at the marginal area circulated by the middle and posterior cerebral arteries on diagnostic MRI. This lesion was probably related to asphyxia during the perinatal period. dSPM of the epileptiform discharges estimated the largest activation occurred during the spiking period in the right inferior frontal and superior temporal gyri and then suddenly propagated widely to the frontal and temporal lobes. The distribution of activity for the diffuse spike rhythm was wider than that for the diffuse spike and wave complex (Fig. 9.7a, c).

By using spatial filtering analysis, MEG studies are suitable for both localization-related and generalized epilepsy. MEG also has a significant advantage over EEG in spatial resolution, thus providing additional useful information in the diagnosis of epileptic syndromes.

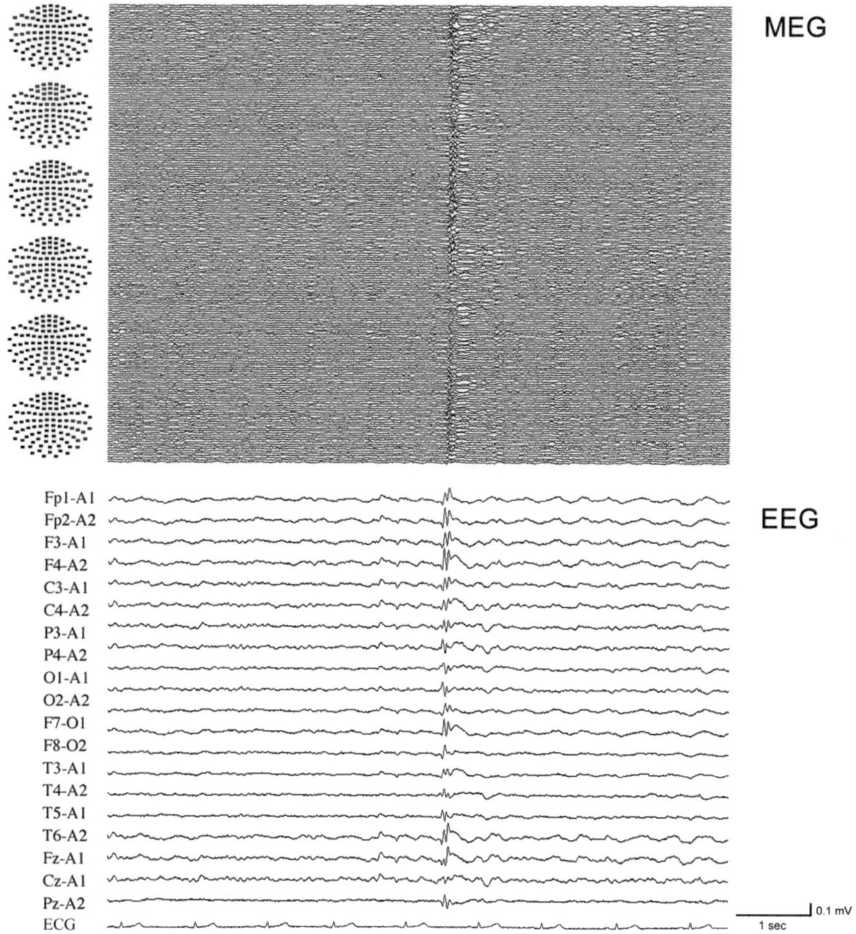

Fig. 9.6 Case 5, EEG and MEG. This patient's EEG and MEG showed a diffuse slow spike and wave complex

9.6 For the Global Application of MEG for Every Patient with Epilepsy

We performed MEG analyses for 109 cases with epilepsy: 85 cases with symptomatic localization-related epilepsy (SLRE), five cases with symptomatic generalized epilepsy (SGE), 15 cases with idiopathic localization-related epilepsy (ILRE), two cases with idiopathic generalized epilepsy (IGE), and two cases with undetermined epilepsy (UDE). We initially performed single-dipole analysis for all cases, and the findings from 83 cases (65 cases with SLRE, 12 cases with ILRE, three cases with SGE, a case with IGE, two cases with UDE) were estimated as reasonable when

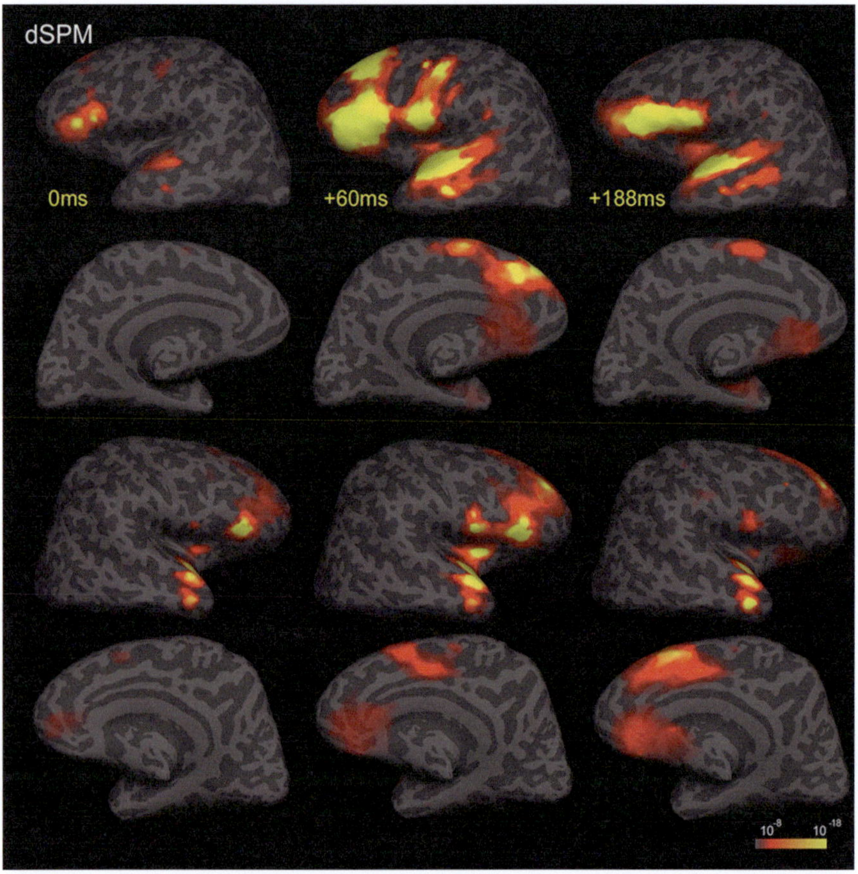

Fig. 9.7 Case 5, dSPM. dSPM of epileptiform discharges estimated the largest activation during the spiking period in the right inferior frontal and superior temporal gyri that then suddenly propagated widely to the frontal and temporal lobes. The distribution of activity for the diffuse spike rhythm was wider than that for the diffuse spike and wave complex

compared to their seizure manifestation and other electroclinical findings. We then used dSPM for the remaining cases, and 16 (ten cases with SLRE, three cases with ILRE, two cases with SGE, a case with IGE) could be evaluated reasonably. Finally, we applied SFT for the remaining ten cases, with reasonable findings achieved.

By using a combination of the single-dipole method, spatial filtering analysis, and time-frequency analysis, we successfully demonstrated the distribution of epileptogenic areas in patients with epilepsy. These findings suggested that MEG analysis could be useful for almost every patient with epilepsy during the presurgical evaluation and diagnosis of epileptic syndromes.

References

1. Aicardi J, Chevrie JJ. Atypical benign partial epilepsy of childhood. Dev Med Child Neurol. 1982;24:281–92.
2. Shiraishi H, Haginoya K, Nakagawa E, Saitoh S, Kaneko Y, Nakasato N, et al. Magnetoencephalography localizing spike sources of atypical benign partial epilepsy. Brain Dev. 2014;36:21–7.
3. Yagyu K, Takeuchi F, Shiraishi H, Nakane S, Sueda K, Asahina N, et al. The applications of time-frequency analyses to ictal magnetoencephalography in neocortical epilepsy. Epilepsy Res. 2010;90:199–206.
4. Sueda K, Takeuchi F, Shiraishi H, Nakane S, Asahina N, Kohsaka S, et al. MEG time-frequency analyses for pre- and post-surgical evaluation of patients with epileptic rhythmic fast activity. Epilepsy Res. 2010;88:100–7.
5. Sueda K, Takeuchi F, Shiraishi H, Nakane S, Sakurai K, Yagyu K, et al. Magnetoencephalographic analysis of paroxysmal fast activity in patients with epileptic spasms. Epilepsy Res. 2013;104:68–77.
6. Dale AM, Liu AK, Fischl BR, Buckner RL, Belliveau JW, Lewine JD, et al. Dynamic statistical parametric mapping: combining fMRI and MEG for high-resolution imaging of cortical activity. Neuron. 2000;26:55–67.
7. Dale AM, Fischl B, Sereno MI. Cortical surface-based analysis. I. Segmentation and surface reconstruction. Neuroimage. 1999;9:179–94.
8. Fischl B, Sereno MI, Tootell RB, Dale AM. High-resolution intersubject averaging and a coordinate system for the cortical surface. Hum Brain Mapp. 1999;8:272–84.
9. Oostendorp TF, van Oosterom A. Source parameter estimation in inhomogeneous volume conductors of arbitrary shape. IEEE Trans Biomed Eng. 1989;36:382–91.
10. Hämäläinen MS, Sarvas J. Realistic conductivity geometry model of the human head for interpretation of neuromagnetic data. IEEE Trans Biomed Eng. 1989;36:165–71.
11. Meijs JW, Peters MJ. The EEG and MEG, using a model of eccentric spheres to describe the head. IEEE Trans Biomed Eng. 1987;34:913–20.
12. Hämäläinen MS, Ilmoniemi RJ. Interpreting magnetic fields of the brain: minimum norm estimates. Med Biol Eng Comput. 1994;32:35–42.
13. Dale AM, Sereno MI. Improved localizadon of cortical activity by combining EEG and MEG with MRI cortical surface reconstruction: a linear approach. J Cogn Neurosci. 1993;5:162–76.
14. Liu AK, Dale AM, Belliveau JW. Monte Carlo simulation studies of EEG and MEG localization accuracy. Hum Brain Mapp. 2002;16:47–62.
15. Dhond RP, Buckner RL, Dale AM, Marinkovic K, Halgren E. Spatiotemporal maps of brain activity underlying word generation and their modification during repetition priming. J Neurosci. 2001;21:3564–71.

Adult Epilepsy

10

Kazutaka Jin and Nobukazu Nakasato

Abstract

Both electroencephalography (EEG) and magnetoencephalography (MEG) measure the same underlying brain activities. The greatest advantage of MEG is that source estimation techniques are easier to apply for MEG than for EEG. Magnetic source imaging (MSI) of interictal spikes as part of presurgical evaluations is one of the most successful clinical applications of MEG. MSI provides additional information to conventional presurgical evaluations by other non-invasive modalities in some patients with intractable epilepsy, especially those with neocortical/extratemporal lobe epilepsy or epilepsy with normal magnetic resonance imaging (MRI). MSI is recommended for the following situations: (1) no clear hypothesis regarding ictal onset, (2) insular onset suspected, (3) interhemispheric onset (especially frontal) suspected, (4) mesial temporal onset suspected without clear evidence of hippocampal sclerosis on MRI, (5) intrasylvian onset suspected, (6) multiple or very large epileptogenic lesion on MRI, and (7) planned intracranial EEG (especially previous craniotomy including revision epilepsy surgery). MSI can provide guidance for additional electrode coverage for intracranial EEG and the extent of the resection area when planning surgery. A single tight cluster of MEG spike dipoles is well correlated with ictal onset zone. Complete resection of the MEG focus often results in seizure freedom after surgery.

Keywords

Temporal lobe epilepsy • Extratemporal lobe epilepsy • Magnetic source imaging • Presurgical evaluation

K. Jin • N. Nakasato (✉)
Department of Epileptology, Tohoku University Graduate School of Medicine, 2-1 Seiryo-machi, Aoba-ku, Sendai 980-8575, Japan
e-mail: nkst@med.tohoku.ac.jp

S. Tobimatsu, R. Kakigi (eds.), *Clinical Applications of Magnetoencephalography*,
DOI 10.1007/978-4-431-55729-6_10

10.1 Introduction

Epilepsy is a disorder of the brain characterized by an enduring predisposition to generate epileptic seizures and by the neurobiological, cognitive, psychological, and social consequences of this condition. An epileptic seizure is defined as a transient occurrence of signs and/or symptoms due to abnormal excessive or synchronous neuronal activity in the brain [1]. Electroencephalography (EEG) is the classical test used to measure and record such electrical activity. Magnetoencephalography (MEG) is a more recent test that can detect the magnetic fields induced by neuronal activity in the brain [2]. MEG has some advantages over EEG. The major advantage of MEG is that source estimation techniques are easier to apply for MEG than for EEG. In addition, the magnetic fields are not distorted by the inhomogeneous conductivity of brain tissue, and the volume conduction effect is often negligible. Consequently, MEG provides higher reliability and accuracy of the estimated dipoles. Magnetic source imaging (MSI) of interictal spikes in the presurgical evaluation of epilepsy is one of the most successful clinical applications of MEG.

This chapter focuses on the clinical application of MEG for adult patients with epilepsy, especially focal epilepsy.

10.2 Comparison Between EEG and MEG

EEG and MEG both measure the same underlying brain activities [2]. The detectability of discharges is dependent on the orientation of the current dipoles. MEG has higher sensitivity to tangential sources, whereas EEG is better at detecting radial sources [3]. Completely tangential or radial spikes are very rare in the clinical setting of patients with epilepsy, and epileptic spikes usually contain both components. Therefore, spikes can be simultaneously captured by both EEG and MEG in most patients. However, the number of epileptic spikes may vary between these modalities [4, 5]. Therefore, EEG and MEG are complementary in the detection of interictal spikes. Simultaneous EEG and MEG recording also influences the spike identification in each modality, and combined data from both modalities increase the sensitivity and specificity of the spikes.

A few systematic studies have compared EEG and MEG spike source localizations obtained with simultaneous recordings. Early studies with a small number of MEG sensors showed that both EEG and MEG dipole modeling gave good concordance with electrocorticography findings [6, 7]. However, MEG was considered to achieve superior localizing accuracy compared to EEG. Recent studies with whole-head systems showed that one modality was likely to be better than the other, but which modality depended on individual patients [8–12]. Comparison of source localization results with EEG and MEG using three spherical shells and a boundary element method volume conductor model found that the effect of the volume conductor model was different in each modality but was apparently small enough to neglect in the clinical setting (differences

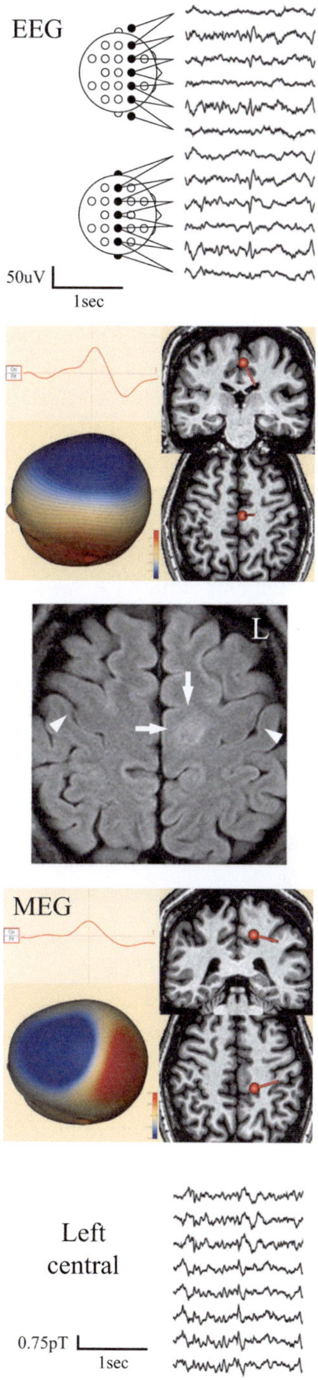

Fig. 10.1 EEG and MEG source localization of interictal spikes in 41-year-old female with intractable epilepsy due to cortical dysplasia in the dorsal peri-rolandic region. *Upper* and *lower rows*, raw waveforms of interictal spikes on EEG (*upper row*) and MEG (*lower row*). Source waveforms of averaged spikes, surface map, and single equivalent current dipoles (ECDs) on T1-weighted MR images using EEG and MEG spikes. *Middle row*, axial fluid-attenuated inversion recovery (FLAIR) MR images showing a focal hyperintense lesion (*arrows*). Central sulcus is indicated by the *arrowhead*. Both EEG and MEG spike sources were close to the MR imaging lesion. Raw EEG waveform of interictal spikes showing paradoxical lateralization

<5–20 mm) [13]. We also reported that both EEG and MEG are equally useful to estimate averaged spike sources arising from a single focal epileptogenic lesion in the dorsal peri-rolandic area (Fig. 10.1) [14]. In addition, the clinical value of MEG is often compared with that of scalp EEG in terms of costs and benefits.

10.3 Temporal Lobe Epilepsy (TLE)

The effectiveness of MEG is limited in mesial TLE compared to neocortical epilepsy because MEG spike localization does not indicate the epileptogenic focus in the mesial temporal region [15, 16]. Typical anterior and middle temporal spikes/sharp waves in scalp EEG or MEG are known to be generated in the anterior and lateral temporal neocortex and do not arise directly from the mesial temporal structures, based on the results of simultaneous recordings with intracranial EEG. MSI can identify which compartments of the temporal lobe are involved in epileptic discharges and may be helpful for the noninvasive classification of subtypes of TLE.

Patients with TLE are classified into the following three subgroups: anterior horizontal, anterior vertical, and posterior vertical types (Fig. 10.2) [11, 17, 18]. Spikes localized in the anterior temporal region suggest the diagnosis of mesial TLE. These spikes are classified into spikes in the temporal tip with horizontal orientation to the temporal lobe axis (anterior horizontal type) and spikes in the superior or basal temporal cortex with vertical orientation (anterior vertical type). On the other hand, spikes localized vertically in the posterior temporal region (posterior vertical type) are seen frequently in patients with lateral TLE (neocortical TLE).

10.4 Extra-TLE

MEG and intracranial EEG show concordance in spike localization in patients with extra-TLE [6, 7, 19–24]. Both detection and localizing accuracy of MEG spikes can be better in the extratemporal region than in the temporal regions [22, 24]. Resection of the irritative zone defined by MEG is correlated with better postoperative seizure outcome [20, 25]. In particular, a single tight cluster of MEG spike dipoles is well correlated with ictal onset zone, and complete resection of the area leads to better seizure outcome [26]. On the other hand, multiple clusters of MEG spikes suggest multiple or extensive epileptogenic zones, which should be evaluated by intracranial EEG before planning resective surgery [27]. MEG is useful in diagnosing insular epilepsy/opercular epilepsy because tangential dipoles generated in the insular/opercular region are more easily detected by MEG than by scalp EEG. MEG spikes in the insula can be seen in 60–100 % of patients with insular epilepsy detected by other means [28–30].

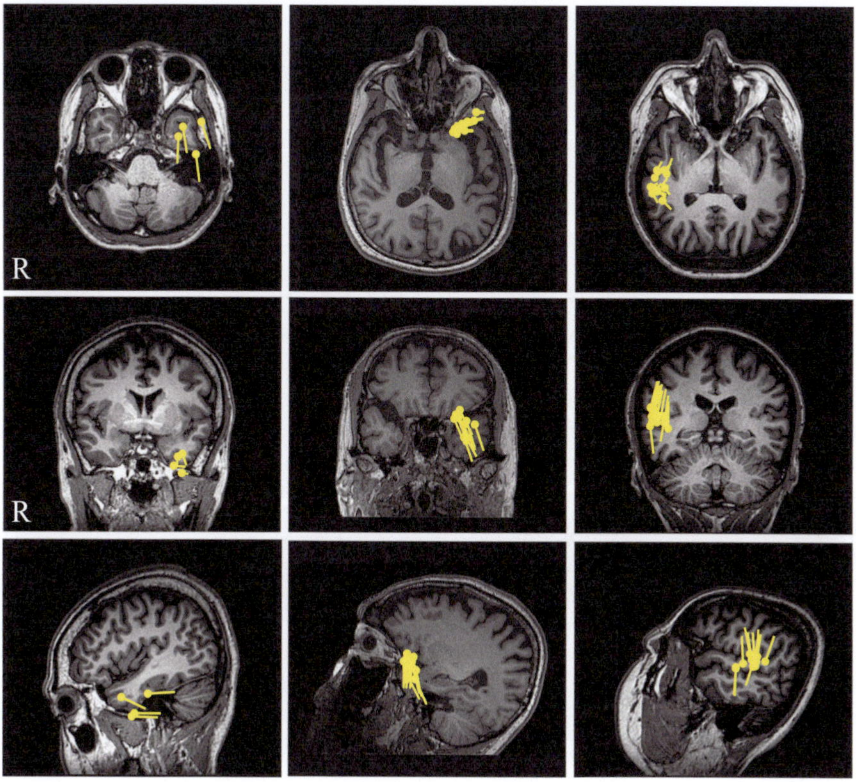

Fig. 10.2 Typical example of each subtype of equivalent current dipole (ECD) patterns in patients with temporal lobe epilepsy. Single ECD model was used for source estimation of epileptic spikes. *Circles* and *bars* indicate locations and orientations of dipoles. *Left column*, typical example of anterior horizontal type in 30-year-old female with left hippocampal sclerosis. ECDs of her left temporal MEG spikes were localized in the temporal tip with a horizontal orientation to the temporal lobe axis. *Middle column*, typical example of anterior vertical type in 67-year-old male with left hippocampal sclerosis. ECDs of his left temporal MEG spikes were localized in the tip of superior temporal cortex with a vertical orientation. *Right column*, typical example of posterior vertical type in 42-year-old female with right temporal lobe epilepsy. Her MRI was normal, but FDG-PET showed hypometabolism in right temporal lobe. ECDs of her right temporal MEG spikes were localized in the posterior temporal cortex with a vertical orientation

10.5 Presurgical Evaluations

MEG can be used as a method of presurgical evaluation. A few large studies have evaluated the additional value of MEG findings on conventional other modalities. A retrospective review of the presurgical evaluations of 104 epilepsy patients to evaluate the contribution of MEG relative to the other modalities classified the MEG findings into "additional information affecting decision about treatment and/or invasive recordings" in 11 %, "additional information" in 24 %, "confirmation" in 54 %, "no contribution (no spikes)" in 10 %, and "contradiction" in 2 %. Thus, MSI provided additional clues as to the location of epileptogenic activity in 35 % of these operated cases [23].

A prospective study evaluated the impact of MSI results on planning of intracranial electrode placement in 77 patients. MSI indicated the need for additional electrode coverage in 18 (23 %) of 77 patients with intracranial EEG. In seven (39 %) of these 18 patients, ictal onset patterns on intracranial EEG involved the additional electrodes indicated by MSI. MSI spike localization increases the chance that the ictal onset zone is sampled when patients undergo intracranial EEG for presurgical evaluations [31].

A similar prospective study assessed the clinical added value of MSI in the presurgical evaluation of 70 patients with refractory focal epilepsy, including 21 patients with extratemporal epilepsy, 38 with temporal epilepsy, and 11 with undetermined localization epilepsy. MSI showed interictal epileptiform discharges in 52 patients (74.5 %) and changed the initial management in 15 patients (21 %). MSI-related changes were significantly more frequent in patients with extratemporal or undetermined localization epilepsy compared with patients with temporal epilepsy [32].

These studies imply that MEG may be crucial in estimating the epileptogenic zone of some surgical candidates, especially patients with presumed extratemporal or undetermined localization epilepsy. MSI can provide additional useful information in surgical decision-making based on the information collected from other modalities such as video-EEG monitoring, magnetic resonance imaging (MRI), and 2-deoxy-2[F-18]fluoro-D-glucose positron emission tomography (FDG-PET) [23, 31, 32].

Which patients should undergo MEG as a part of presurgical evaluations? This question is not easy to answer and should be decided based on the individual situation of a patient. However, recommendations were reported as follows: (1) no clear hypothesis regarding ictal onset, (2) insular onset suspected (see Sect. 10.4), (3) interhemispheric onset (especially frontal) suspected, (4) mesial temporal onset suspected without clear evidence of hippocampal sclerosis on MRI, (5) intrasylvian onset suspected, (6) multiple or very large epileptogenic lesion on MRI, and (7) planned intracranial EEG (especially previous craniotomy including revision epilepsy surgery) [33].

Patients with no abnormalities on MRI are always challenging while considering surgical indications. If both MRI and FDG-PET show no abnormalities, MEG should be performed to obtain any localizing information as additional presurgical evaluations. In patients with no abnormalities on MRI, MEG can indicate the location of the irritative zone and guide the correct placement of intracranial electrodes for better surgical planning or even obviate the need for invasive recording. The distribution of MEG spike dipoles can provide information about the extent of the ictal onset zone confirmed by intracranial EEG [34–36]. EEG source localization is also useful for presurgical evaluation of patients with no abnormalities on MRI [37]. Please see Sect. 10.2 for a comparison between EEG and MEG source localization. In some patients without abnormalities on initial

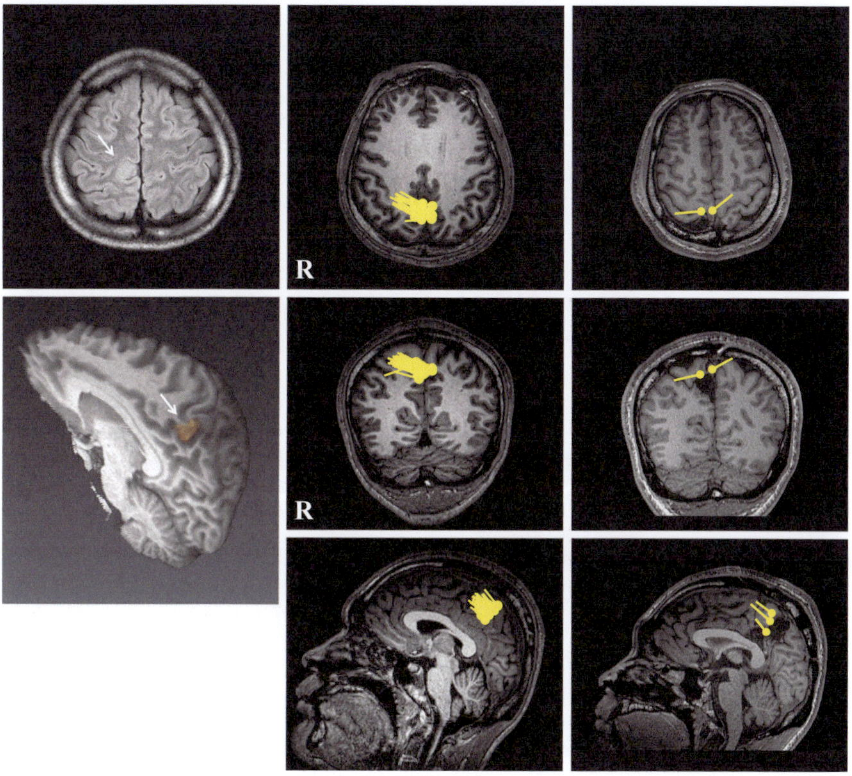

Fig. 10.3 Magnetic source imaging of interictal spikes in 20-year-old male with intractable epilepsy due to cortical dysplasia in the right paracentral lobule. Preoperative fluid-attenuated inversion recovery (FLAIR) MR images showed an abnormal hyperintense lesion in the right paracentral lobule (*left column*). Preoperative equivalent current dipoles (ECDs) of his MEG spikes were localized in the parietal interhemispheric area with orientation to the right side, suggesting spike sources in the right medial parietal cortex close to the MRI lesion (*middle column*). He underwent tailored cortical resection including both the MRI lesion and the MEG spike area, which led to significant improvement of his seizures. But, he had residual seizures after the surgery. Postoperative ECDs of his MEG spikes were localized just anterior to the resection cavity (*right column*)

MRI, repeat MRI with high-resolution imaging and surface coil can reveal subtle MRI abnormalities after obtaining localizing information from MSI [38, 39].

MEG has advantages for spike source localization of postoperative patients with previous craniotomy, because magnetic fields are not distorted by skull defects and other conductivity inhomogeneities (Fig. 10.3) [40, 41].

10.6 Ictal MEG

Although MEG is less suitable for long-term monitoring to obtain ictal recordings than EEG, ictal activity is occasionally recorded during routine MEG study by chance. Source localization of ictal MEG activities was reported to be concordant with intracranial ictal localization [42–44]. Some reports showed that ictal onset activity on MEG was localized closer to the ictal onset zone on intracranial EEG than interictal activity on MEG [45–47]. Patients usually undergo body movements during seizures, resulting in massive artifacts which obscure ictal activities. However, recording of only the initial part of seizures would be useful, which could indicate the precise localization of the ictal onset zone.

10.7 Source Modeling Algorithm Other Than Equivalent Current Dipole (ECD) Modeling

An ECD model is most commonly used to estimate the distribution of interictal epileptiform discharges and correlates well with the ictal onset zone as identified by intracranial EEG. However, the ECD model cannot be applied to activities arising from an extended area or multiple areas [48, 49]. Thus, other methods for estimating the distribution of activities are needed, such as scanning methods including multiple signal classification, current density methods including minimum-norm estimate, and beam-forming methods including synthetic aperture magnetometry [34, 50, 51]. A recent study described the use of multiple MEG source estimation techniques and demonstrated that all algorithms had similar rates of concordance with intracranial EEG [51]. The clinical significance of the results provided by methods other than ECD is still under discussion, because the algorithm is a "black box."

10.8 Conclusion

Both EEG and MEG measure the same underlying brain activities. The greatest advantage of MEG is that source estimation techniques are easier to apply for MEG than for EEG. MSI of interictal spikes as part of presurgical evaluations is one of most successful clinical applications of MEG. MSI provides additional information to conventional presurgical evaluations by other noninvasive modalities in some patients with intractable epilepsy, especially those with neocortical epilepsy/extra-TLE or epilepsy with normal MRI. MSI provides guidance for additional electrode coverage for intracranial EEG and the extent of the resection area when planning surgery. Complete resection of the MEG focus often results in seizure freedom after surgery.

References

1. Fisher RS, van Emde Boas W, Blume W, Elger C, Genton P, Lee P, et al. Epileptic seizures and epilepsy: definitions proposed by the International League Against Epilepsy (ILAE) and the International Bureau for Epilepsy (IBE). Epilepsia. 2005;46:470–2. doi:10.1111/j.0013-9580.2005.66104.x.
2. Barth DS. The neurophysiological basis of epileptiform magnetic fields and localization of neocortical sources. J Clin Neurophysiol. 1993;10:99–107.
3. Hillebrand A, Barnes GR. A quantitative assessment of the sensitivity of whole-head MEG to activity in the adult human cortex. Neuroimage. 2002;16:638–50. doi:10.1006/nimg.2002.1102.
4. Iwasaki M, Pestana E, Burgess RC, Lüders HO, Shamoto H, Nakasato N. Detection of epileptiform activity by human interpreters: blinded comparison between electroencephalography and magnetoencephalography. Epilepsia. 2005;46:59–68. doi:10.1111/j.0013-9580.2005.21104.x.
5. Lin YY, Shih YH, Hsieh JC, Yu HY, Yiu CH, Wong TT, et al. Magnetoencephalographic yield of interictal spikes in temporal lobe epilepsy. Comparison with scalp EEG recordings. Neuroimage. 2003;19:1115–26. doi:10.1016/S1053-8119(03)00181-2.
6. Ko DY, Kufta C, Scaffidi D, Sato S. Source localization determined by magnetoencephalography and electroencephalography in temporal lobe epilepsy: comparison with electrocorticography: technical case report. Neurosurgery. 1998;42:414–21; discussion 421–2.
7. Nakasato N, Levesque MF, Barth DS, Baumgartner C, Rogers RL, Sutherling WW. Comparisons of MEG, EEG, and ECoG source localization in neocortical partial epilepsy in humans. Electroencephalogr Clin Neurophysiol. 1994;91:171–8. doi:10.1016/0013-4694(94)90067-1.
8. Bast T, Oezkan O, Rona S, Stippich C, Seitz A, Rupp A, et al. EEG and MEG source analysis of single and averaged interictal spikes reveals intrinsic epileptogenicity in focal cortical dysplasia. Epilepsia. 2004;45:621–31. doi:10.1111/j.0013-9580.2004.56503.x.
9. Bast T, Ramantani G, Boppel T, Metzke T, Ozkan O, Stippich C, et al. Source analysis of interictal spikes in polymicrogyria: loss of relevant cortical fissures requires simultaneous EEG to avoid MEG misinterpretation. Neuroimage. 2005;25:1232–41. doi:10.1016/j.neuroimage.2004.12.059.
10. Leijten FS, Huiskamp GJ, Hilgersom I, Van Huffelen AC. High-resolution source imaging in mesiotemporal lobe epilepsy: a comparison between MEG and simultaneous EEG. J Clin Neurophysiol. 2003;20:227–38.
11. Pataraia E, Lindinger G, Deecke L, Mayer D, Baumgartner C. Combined MEG/EEG analysis of the interictal spike complex in mesial temporal lobe epilepsy. Neuroimage. 2005;24:607–14. doi:10.1016/j.neuroimage.2004.09.031.
12. Yoshinaga H, Nakahori T, Ohtsuka Y, Oka E, Kitamura Y, Kiriyama H, et al. Benefit of simultaneous recording of EEG and MEG in dipole localization. Epilepsia. 2002;43:924–8. doi:10.1016/j.neuroimage.2004.09.031.
13. Scheler G, Fischer MJ, Genow A, Hummel C, Rampp S, Paulini A, et al. Spatial relationship of source localizations in patients with focal epilepsy: comparison of MEG and EEG with a three spherical shells and a boundary element volume conductor model. Hum Brain Mapp. 2007;28:315–22. doi:10.1002/hbm.20277.
14. Itabashi H, Jin K, Iwasaki M, Okumura E, Kanno A, Kato K, et al. Electro- and magneto-encephalographic spike source localization of small focal cortical dysplasia in the dorsal peri-rolandic region. Clin Neurophysiol. 2014;125:2358–63. doi:10.1016/j.clinph.2014.02.028.
15. Rose DF, Sato S, Smith PD, Porter RJ, Theodore WH, Friauf W, et al. Localization of magnetic interictal discharges in temporal lobe epilepsy. Ann Neurol. 1987;22:348–54. doi:10.1016/j.clinph.2014.02.028.
16. Sutherling WW, Barth DS. Neocortical propagation in temporal lobe spike foci on magneto-encephalography and electroencephalography. Ann Neurol. 1989;25:373–81. doi:10.1002/ana.410250409.

17. Baumgartner C, Pataraia E, Lindinger G, Deecke L. Neuromagnetic recordings in temporal lobe epilepsy. J Clin Neurophysiol. 2000;17:177–89.
18. Iwasaki M, Nakasato N, Shamoto H, Nagamatsu K, Kanno A, Hatanaka K, et al. Surgical implications of neuromagnetic spike localization in temporal lobe epilepsy. Epilepsia. 2002;43:415–24. doi:10.1046/j.1528-1157.2002.30801.x.
19. Knowlton RC, Laxer KD, Aminoff MJ, Roberts TP, Wong ST, Rowley HA. Magnetoencephalography in partial epilepsy: clinical yield and localization accuracy. Ann Neurol. 1997;42:622–31. doi:10.1002/ana.410420413.
20. Mamelak AN, Lopez N, Akhtari M, Sutherling WW. Magnetoencephalography-directed surgery in patients with neocortical epilepsy. J Neurosurg. 2002;97:865–73.
21. Schwartz DP, Badier JM, Vignal JP, Toulouse P, Scarabin JM, Chauvel P. Non-supervised spatio-temporal analysis of interictal magnetic spikes: comparison with intracerebral recordings. Clin Neurophysiol. 2003;114:438–49. doi:10.1016/S1388-2457(02)00413-3.
22. Stefan H, Hummel C, Hopfengärtner R, Pauli E, Tilz C, Ganslandt O, et al. Magnetoencephalography in extratemporal epilepsy. J Clin Neurophysiol. 2000;17:190–200.
23. Stefan H, Hummel C, Scheler G, Genow A, Druschky K, Tilz C, et al. Magnetic brain source imaging of focal epileptic activity: a synopsis of 455 cases. Brain. 2003;126:2396–405. doi:10.1093/brain/awg239.
24. Wheless JW, Willmore LJ, Breier JI, Kataki M, Smith JR, King DW, et al. A comparison of magnetoencephalography, MRI, and V-EEG in patients evaluated for epilepsy surgery. Epilepsia. 1999;40:931–41. doi:10.1111/j.1528-1157.1999.tb00800.x.
25. Fischer MJ, Scheler G, Stefan H. Utilization of magnetoencephalography results to obtain favourable outcomes in epilepsy surgery. Brain. 2005;128:153–7. doi:10.1093/brain/awh333.
26. Stefan H, Wu X, Buchfelder M, Rampp S, Kasper B, Hopfengärtner R, et al. MEG in frontal lobe epilepsies: localization and postoperative outcome. Epilepsia. 2011;52:2233–8. doi:10.1111/j.1528-1167.2011.03265.x.
27. Oishi M, Kameyama S, Masuda H, Tohyama J, Kanazawa O, Sasagawa M, et al. Single and multiple clusters of magnetoencephalographic dipoles in neocortical epilepsy: significance in characterizing the epileptogenic zone. Epilepsia. 2006;47:355–64. doi:10.1111/j.1528-1167.2006.00428.x.
28. Kakisaka Y, Iwasaki M, Alexopoulos AV, Enatsu R, Jin K, Wang ZI, et al. Magnetoencephalography in fronto-parietal opercular epilepsy. Epilepsy Res. 2012;102:71–7. doi:10.1016/j.eplepsyres.2012.05.003.
29. Mohamed IS, Gibbs SA, Robert M, Bouthillier A, Leroux JM, Khoa Nguyen D. The utility of magnetoencephalography in the presurgical evaluation of refractory insular epilepsy. Epilepsia. 2013;54:1950–9. doi:10.1111/epi.12376.
30. Park HM, Nakasato N, Tominaga T. Localization of abnormal discharges causing insular epilepsy by magnetoencephalography. Tohoku J Exp Med. 2012;226:207–11. doi:10.1620/tjem.226.207.
31. Knowlton RC, Razdan SN, Limdi N, Elgavish RA, Killen J, Blount J, et al. Effect of epilepsy magnetic source imaging on intracranial electrode placement. Ann Neurol. 2009;65:716–23. doi:10.1002/ana.21660.
32. De Tiège X, Carrette E, Legros B, Vonck K, Op de Beeck M, Bourguignon M, et al. Clinical added value of magnetic source imaging in the presurgical evaluation of refractory focal epilepsy. J Neurol Neurosurg Psychiatry. 2012;83:417–23. doi:10.1136/jnnp-2011-301166.
33. Kharkar S, Knowlton R. Magnetoencephalography in the presurgical evaluation of epilepsy. Epilepsy Behav. 2014;pii:S1525-5050(14)00648-9. doi:10.1016/j.yebeh.2014.11.029.
34. Zhang R, Wu T, Wang Y, Liu H, Zou Y, Liu W, et al. Interictal magnetoencephalographic findings related with surgical outcomes in lesional and nonlesional neocortical epilepsy. Seizure. 2011;20:692–700. doi:10.1016/j.seizure.2011.06.021.
35. Schneider F, Wang ZI, Alexopoulos AV, Almubarak S, Kakisaka Y, Jin K, et al. Magnetic source imaging and ictal SPECT in MRI-negative neocortical epilepsies: additional value and comparison with intracranial EEG. Epilepsia. 2013;54:359–69. doi:10.1111/epi.12004.

36. Wu XT, Rampp S, Buchfelder M, Kuwert T, Blümcke I, Dörfler A, et al. Interictal magnetoencephalography used in magnetic resonance imaging-negative patients with epilepsy. Acta Neurol Scand. 2013;127:274–80. doi:10.1111/j.1600-0404.2012.01712.x.
37. Brodbeck V, Spinelli L, Lascano AM, Pollo C, Schaller K, Vargas MI, et al. Electrical source imaging for presurgical focus localization in epilepsy patients with normal MRI. Epilepsia. 2010;51:583–91. doi:10.1111/j.1528-1167.2010.02521.x.
38. Moore KR, Funke ME, Constantino T, Katzman GL, Lewine JD. Magnetoencephalographically directed review of high-spatial-resolution surface-coil MR images improves lesion detection in patients with extratemporal epilepsy. Radiology. 2002;225:880–7. doi:10.1148/radiol.2253011597.
39. Wilenius J, Medvedovsky M, Gaily E, Metsähonkala L, Mäkelä JP, Paetau A, et al. Interictal MEG reveals focal cortical dysplasias: special focus on patients with no visible MRI lesions. Epilepsy Res. 2013;105:337–48. doi:10.1016/j.eplepsyres.2013.02.023.
40. Mohamed IS, Otsubo H, Ochi A, Elliott I, Donner E, Chuang S, et al. Utility of magnetoencephalography in the evaluation of recurrent seizures after epilepsy surgery. Epilepsia. 2007;48:2150–9. doi:10.1111/j.1528-1167.2007.01271.x.
41. Yoshinaga H, Kobayashi K, Hoshida T, Kinugasa K, Ohtuska Y. Magnetoencephalogram in a postoperative case with a large skull defect. Pediatr Neurol. 2008;39:48–51. doi:10.1016/j.pediatrneurol.2008.03.010.
42. Tilz C, Hummel C, Kettenmann B, Stefan H. Ictal onset localization of epileptic seizures by magnetoencephalography. Acta Neurol Scand. 2002;106:190–5. doi:10.1034/j.1600-0404.2002.02047.x.
43. Assaf BA, Karkar KM, Laxer KD, Garcia PA, Austin EJ, Barbaro NM, et al. Ictal magnetoencephalography in temporal and extratemporal lobe epilepsy. Epilepsia. 2003;44:1320–7. doi:10.1046/j.1528-1157.2003.14303.x.
44. Tang L, Mantle M, Ferrari P, Schiffbauer H, Rowley HA, Barbaro NM, et al. Consistency of interictal and ictal onset localization using magnetoencephalography in patients with partial epilepsy. J Neurosurg. 2003;98:837–45.
45. Eliashiv DS, Elsas SM, Squires K, Fried I, Engel Jr J. Ictal magnetic source imaging as a localizing tool in partial epilepsy. Neurology. 2002;59:1600–10. doi:10.1212/01.WNL.0000032493.83875.0B.
46. Fujiwara H, Greiner HM, Hemasilpin N, Lee KH, Holland-Bouley K, Arthur T, et al. Ictal MEG onset source localization compared to intracranial EEG and outcome: improved epilepsy presurgical evaluation in pediatrics. Epilepsy Res. 2012;99:214–24. doi:10.1016/j.eplepsyres.2011.11.007.
47. Medvedovsky M, Taulu S, Gaily E, Metsähonkala EL, Mäkelä JP, Ekstein D, et al. Sensitivity and specificity of seizure-onset zone estimation by ictal magnetoencephalography. Epilepsia. 2012;53:1649–57. doi:10.1111/j.1528-1167.2012.03574.x.
48. Ebersole JS. Noninvasive localization of epileptogenic foci by EEG source modeling. Epilepsia. 2000;41 Suppl 3:S24–33. doi:10.1111/j.1528-1157.2000.tb01531.x.
49. Yoshinaga H, Nakahori T, Ohtsuka Y, Oka E, Kitamura Y, Kiriyama H, et al. Benefit of simultaneous recording of EEG and MEG in dipole localization. Epilepsia. 2002;43:924–8. doi:10.1046/j.1528-1157.2002.42901.x.
50. de Gooijer-van de Groep KL, Leijten FS, Ferrier CH, Huiskamp GJ. Inverse modeling in magnetic source imaging: comparison of MUSIC, SAM(g2), and sLORETA to interictal intracranial EEG. Hum Brain Mapp. 2013;34:2032–44. doi:10.1002/hbm.22049.
51. Tenney JR, Fujiwara H, Horn PS, Rose DF. Comparison of magnetic source estimation to intracranial EEG, resection area, and seizure outcome. Epilepsia. 2014;55:1854–63. doi:10.1111/epi.12822.

Part VII

Neurological Disorders

Cerebrovascular Diseases

11

Naohiro Tsuyuguchi

Abstract

Measuring local cerebral blood flow and metabolism by various mapping methods, such as PET (positron emission tomography), SPECT (single-photon emission computed tomography), perfusion computed tomography, MRI, and so on, helps us to evaluate detailed functions of brain areas containing a focal ischemic lesion, but does not necessarily represent neural activities of the areas. Scalp electroencephalography (EEG), reflecting volume-conducted neural activities, demonstrates that slow wave activity is dominant in an acute ischemic cerebral region, but this technique presents major problems with the lack of objective indices for brain functions and low spatial resolution. Magnetoencephalography (MEG), an important new method in neuroscience to directly detect neural activities with high spatial resolution, has been applied in stroke patients. This chapter mainly describes the relation between magnetic responses and cerebral ischemic changes from several stroke-related manuscripts.

Some papers stressed the clinical usefulness of MEG, for example, that slow wave activities occur on the affected cerebral hemisphere and somatosensory evoked potential becomes indicator of brain plasticity. However, other neurophysiological signal changes after stroke are various and not consistent. The usage of MEG for assessing neural activities in an ischemic brain area has not been fully established as yet. Therefore, more objective analysis of MEG findings in ischemic conditions is needed for future development.

Keywords

Magnetoencephalography • Stroke • Cerebrovascular disease • Ischemia • Brain plasticity

N. Tsuyuguchi (✉)
Graduate School of Medicine, Department of Neurosurgery, Osaka City University, 1-4-3, Asahimachi, Abeno, Osaka 545-8585, Japan
e-mail: ntsuyuguchi@gmail.com

S. Tobimatsu, R. Kakigi (eds.), *Clinical Applications of Magnetoencephalography*,
DOI 10.1007/978-4-431-55729-6_11

11.1 Introduction

There are various correlations between electrophysiological and hemodynamic responses in stroke, epilepsy, brain tumor, dementia, rehabilitation process, etc.

There are two main clinical approaches for cerebral vascular disease, especially stroke. The first is measurement of cerebrovascular reserve capacity, evaluation of which is very important to treat cerebral infarction. The other approach involves determination of brain plasticity, which is an indicator of rehabilitation after stroke [1].

Cerebral blood flow and metabolism can be measured by nuclear medicine studies, such as single-photon emission computed tomography (SPECT) and positron emission tomography (PET), as well as by perfusion CT, MRI, and near-infrared spectroscopy (NIRS). These methods have been established for diagnosing and assessing cerebrovascular disease. Some reports have indicated that brain plasticity can be evaluated by diffusion-weighted MRI or fMRI [2].

However, hemodynamics in cerebrovascular disease as assessed by these imaging studies only indirectly reflects brain function.

On the other hand, electrophysiological studies like electroencephalography (EEG) more directly reflect neural activity. It is well known that the slow wave activity in EEG occurs in ischemic areas of the brain. On the other hand, scalp-recorded EEGs are greatly affected by the skull itself. Electrophysiological signal in small infarct lesions may be buried within normal signal findings in surrounding areas and go undetected. Thus, routine EEGs are problematic because of low spatial resolution and a lack of objective indices for quantitative measurement of *brain function*. Magnetoencephalography (MEG) quantitatively measures magnetic activity in the brain and may overcome some of these limitations of EEG. This chapter mainly describes the relations between magnetic responses and cerebral ischemic changes.

MEG is now performed with devices that can cover the entire head, thus providing improved spatial resolution. MEG was first used in clinical research to assess brain function and search for epileptic foci preoperatively in patients undergoing neurosurgery. The usefulness of MEG for brain function evaluation in other diseases has also been frequently reported, but its clinical use in diseases other than epilepsy has been sparse, so it is still unclear what level of scientific evidence exists for its utility outside of epilepsy.

Therefore, trends in clinical research of MEG in ischemic cerebral disease were examined in a literature search to evaluate the level of scientific evidence for using MEG in clinical evaluation.

11.2 Methods

A PubMed database literature search (http://www.ncbi.nlm.nih.gov/pubmed/) was conducted with the keywords (stroke or cerebral ischemia) and (MEG or magnetoencephalography) for publications dated between January 1990 and July 2014. A total of 62 papers were retrieved from this search regarding MEG based on the titles, 28 of these papers were reviewed based on their abstract contents, and then, focusing on the level of evidence, 14 papers that measured cerebral blood flow (CBF) or used controlled comparisons were selected and reviewed.

Although it was difficult to directly compare these reports due to the nonuniformity of presentation and varied ischemic conditions, some common results are described.

11.3 Results 1

A total of 62 papers about MEG and cerebral ischemia were retrieved from the literature search. These were then limited to original manuscripts, and 28 papers were selected based on their abstract contents. These included reports about diagnostic criteria for cerebral ischemia and functional recovery/neuroplasticity after stroke. They were further divided into reports about spontaneous cerebral magnetic fields and evoked cerebral magnetic fields.

Evaluation of the use of MEG in functional recovery after stroke is reported separately, and some reports about the evaluation of ischemia are described.

11.3.1 Diagnostic Criteria for Cerebral Ischemia in MEG

1. Spontaneous magnetic fields: slow waves occur after stroke

   Affected side
   - In areas surrounding lesions after stroke, MEG shows a decrease of high-frequency components (gamma band) and an increase of low-frequency delta (theta band) [3, 4].
   - Increased theta waves are seen in the temporoparietal area of the affected side [5] and correlate with misery perfusion [6].
   - Slow waves are seen on the affected side in transient ischemic attack (TIA) patients [7].
   - The power of delta waves is correlated with the National Institute of Health Stroke Scale (NIHSS) [8].
   - Slow wave components are correlated with decreased *N*-acetylaspartate (NAA) on MR spectroscopy [9].
   - Slow wave activity during re-buildup in areas of impaired circulation in moyamoya disease is seen in deep cortical sulci [10, 11].

   Unaffected side
   - Increased slow waves are also seen on the unaffected side [4, 12].
   - Increased slow waves (delta) on the unaffected side may be clinically correlated with symptom improvement [2].

2. Evoked magnetic fields

   SEF (somatosensory evoked fields)
   - In stroke cases, asymmetry of localization of equivalent current dipoles (ECDs) in N20m of the affected side [13], prolonged latency [14, 15], abnormal waveforms [16], and decreased [12, 17] or increased ECD strength [11, 15] are seen. This asymmetry is correlated with improvement of clinical symptoms [12, 18].
   - The SEF ECD component at N20m is correlated with the NIHSS [8].
   - In parallel with decreased CBF, N20m dipole moment decreases and P30m increases [19].

   MEF (motor evoked fields)
   - Shifts in source position and latencies are seen [20].
   - Beta band event-related desynchronization (ERD) is seen in motor areas and the ipsilateral hemisphere [12].
   - The amplitude of motor-related direct current (DC) signal (infra-slow MEG signal) decreases in the affected hemisphere [21].

   AEF (auditory evoked fields)
   - The temporal lobe response on the affected side is decreased [22].

11.3.2 Functional Recovery, Plasticity, and Indices of Reorganization

1. Spontaneous magnetic fields (slow waves)

   - Slow waves on the affected side decrease with improved circulation after surgical treatment [6, 23].
   - Normalization of slow waves leads to a clinical improvement in symptoms [12].
   - Revascularization improves CBF and decreases theta bands, but there are no changes in delta bands [23]. In another report, delta waves near the lesion did not change during follow-up, and there was no correlation with clinical symptoms [3].
   - In TIAs, normalization of slow waves from the somatosensory cortex may be an index of short-term functional recovery after stroke [24].
   - In cases of aphasia due to stroke, delta waves (1–4 Hz) were seen near the lesion, and a decrease of delta waves after speech therapy was associated with good outcomes [25].
   - Delta waves on the unaffected side and gamma waves on the affected side may be indices of functional recovery.

2. Evoked magnetic fields

- Improved SEF latency correlates with sensory improvement [14].
- Source power localized to S1 (SEF) and to M1 with finger tapping showed a correlation with sensory and motor function improvement [26].

11.4 Results 2

Of the abovementioned 28 papers, 14 with high evidence levels, including measurement of CBF or comparisons with normal controls, were further reviewed. Table 11.1 describes their content in detail.

11.4.1 Target Diseases and Comparisons

Most papers discussed internal carotid artery or middle cerebral artery occlusion or stenosis. One paper dealt with TIAs, and one paper was about moyamoya disease. There were normal controls in 11 papers.

11.4.2 Analysis Methods

Spontaneous magnetic fields were measured in six studies, evoked cerebral magnetic fields were measured in six studies, and both spontaneous and evoked cerebral magnetic fields were measured in one study. Among papers dealing with evoked cerebral magnetic fields, four used median nerve stimulation, one used tactile finger stimulation [15], one used auditory stimulation [4], and two used motor-related magnetic fields [27, 28].

Most papers on evoked magnetic fields used a single equivalent current dipole (ECD) method, but one used a spatial filtering technique (synthetic aperture magnetometry, SAM) [28]. Measurement of spontaneous magnetic fields included analysis of the waveforms themselves, analysis of magnetic field distribution using ECDs, analysis using a spatial filtering technique (standardized low-resolution brain electromagnetic tomography, sLORETA) [23], and analysis using power spectral density (PSD) [4].

One paper showed the relation between direct current (DC)-MEG signal and near-infrared spectroscopy (NIRS) for a finger movement task [27].

Table 11.1 Summary of some meritorious manuscripts

No.	Reference no.	Authors	Journal	Title	PubMed PMID	Clinical diagnosis	Case number	Age
1	4	Tecchio F, Zappasodi F, Pasqualetti P, Tombini M, Salustri C, Oliviero A, Pizzella V, Vernieri F, Rossini PM	Neuroimage. 2005 28:72–83	Rhythmic brain activity at rest from rolandic areas in acute mono-hemispheric stroke: a magnetoencephalographic study	16023869	Unilateral MCA ischemia	32	30–86
2	5	Seki S, Nakasato N, Ohtomo S, Kanno A, Shimizu H, Tominaga T	Neuroimage. 2005 25:502–10	Neuromagnetic measurement of unilateral temporo-parietal theta rhythm in patients with internal carotid artery occlusive disease	15784429	ICA occlusive disease	48	40–76
3	6	Ohtomo S, Nakasato N, Shimizu H, Seki S, Kanno A, Kumabe T, Tominaga T	Clin Neurophysiol. 2009 120:1227–34	Temporo-parietal theta activity correlates with misery perfusion in arterial occlusive disease	19539523	ICA or MCA occlusive disease	56	40–75
4	8	Assenza G, Zappasodi F, Squitti R, Altamura C, Ventriglia M, Ercolani M, Quattrocchi CC, Lupoi D, Passarelli F, Vernieri F, Rossini PM, Tecchio F	Neuroimage. 2009 44:1267–73	Neuronal functionality assessed by magnetoencephalography is related to oxidative stress system in acute ischemic stroke	19010427	MCA ischemia	18	Mean: 73
5	11	Qiao F, Kuroda S, Kamada K, Houkin K, Iwasaki Y	Childs Nerv Syst. 2003 19: 760–764	Source localization of the re-build up phenomenon in pediatric moyamoya disease-a dipole distribution analysis using MEG and SPECT		Moyamoya disease	4	8–16

6	16	Forss N, Hietanen M, Salonen O, Hari R	Brain. 1999 122:1889–99	Modified activation of somatosensory cortical network in patients with right-hemisphere stroke	10506091	Right MCA infarction	6	45–65
7	17	Tsutada T, Ikeda H, Tsuyuguchi N, Hattori H, Shimogawara M, Shimada H, Miki T	J Neurol Sci. 2002 198:51–61	Detecting functional asymmetries through the dipole moment of magnetoencephalography		Various lesion	34	37–39
8	19	Bundo M, Inao S, Nakamura A, Kato T, Ito K, Tadokoro M, Kabeya R, Sugimoto T, Kajita Y, Yoshida J	Stroke. 2002 33:61–6	Changes of neural activity correlate with the severity of cortical ischemia in patients with unilateral major cerebral artery occlusion	11779890	"ICA or MCA occlusive disease no apparent clinical symptom"	7	18–75
9	22	Toyoda K, Ibayashi S, Yamamoto T, Kuwabara Y, Fujishima M	J Neurol Neurosurg Psychiatry. 1998 64:777–84	Auditory evoked neuromagnetic response in cerebrovascular diseases: a preliminary study	2170114	"Major branch stenosis and occlusion no apparent infarct area in temporal lobe"	24	43–75
10	28	Oshino S, Kato A, Hirata M, Kishima H, Saitoh Y, Fujinaka T, Yoshimine T	Stroke. 2008 39:2769–75	Ipsilateral motor-related hyperactivity in patients with cerebral occlusive vascular disease	18635836	"Occlusive cerebrovascular disease (arteliosclerosis:28 cases、others:10 cases)with right handedness and no apparent motor weakness	38	55–79

(continued)

Table 11.1 (continued)

No.	Reference no.	Authors	Journal	Title	PubMed PMID	Clinical diagnosis	Case number	Age
11	23	Sakamoto S, Tanaka H, Tsuyuguchi N, Terakawa Y, Ohata K, Inoue Y, Miki Y, HaraM, Takahashi Y, Nitta K, Sawa H, Satone A, Ide W, Hashimoto I, Kamada H	Neuroimage. 2010 49:488–97	Quantitative imaging of spontaneous neuromagnetic activity for assessing cerebral ischemia using sLORETA-qm	19632340	"ICA stenosis: 4 ICA occlusion:1"	5	60–77
12	24	Stippich C, Kassubek J, Kober H, Sörös P, Vieth JB	Neuroreport. 2000 11:3309–13	Time course of focal slow wave activity in transient ischemic attacks and transient global amnesia as measured by magnetoencephalography	11059893	"TIA: 6 cases TGA: 2 cases"	8	44–82
13	27	Leistner S, Sander-Thoemmes T, Wabnitz H, Moeller M, Wachs M, Curio G, Macdonald R, Trahms L, Mackert BM	Biomed Tech (Berl) 56:85–90	Non-invasive simultaneous recording of neuronal and vascular signals in subacute ischemic stroke	21299378	Subcortical infarction: 4cases	4	44–77
14	21	Forss N, Mustanoja S, Roiha K, Kirveskari E, Mäkelä JP, Salonen O, Tatlisumak T, Kaste M	Hum Brain Mapp. 2012 33:534–41	Activation in parietal operculum parallels motor recovery in stroke	21425393	MCA infarction	18	46–85

No.	Control number	MEG and device	Evoked field/ spontaneous field	Other modality	Task Mesurements	Analysis	Result	Clinical approach
1	14	28 channel	SF	None	"MEG was performed between 2 and 10 days after stroke onset 0.48–250Hz sampling:1000Hz 3 min examination"	"Power spectral density (PSD) spectral entropy"	The perilesional increase of the low frequency, 2–3.5Hz 4–7.5Hz, is detect whithin the AH rolandic areas, the same effect is present in the UH. This phenomenon may show interhemispheric diaschisis. An increased of the toital power and a reduction of the spectral entropy were detected in the AH. This effect suggests a high synchrony of local neuronal activity, areduction of the intra-cortical inhibitory networks efficiency and an increase of neuronal excitability	C
2	27	"Neuromag 204 planar-type gradiometer"	SF	None	"0.03–100Hz sampling 300Hz awake 6 min examination"	Wave form	6–8Hz temporoparietal theta wave (in 14 of 48 patients)	C
3		"Neuromag 204 planar-type gradiometer"	SF	XeCT CBF, CVR	"0.03–100Hz sampling 500Hz awake 10 min examination"	"Fourier analysis ECD analysis"	"The presence of TPTA (6–8Hz) was detected in 14 patients, and significantly correlated with both reduced rCBF and reduced rCVR. After anastomosis, TPTA disappeared in 7 of the 10 studied patients. The area of leptpmeningeal collateral correlated with the region of TPTA"	C

(continued)

Table 11.1 (continued)

No.	Control number	MEG and device	Evoked field/ spontaneous field	Other modality	Task Mesurements	Analysis	Result	Clinical approach
4	20	28 channel	"SF(bilateral rolandic area 3min,eye open state) EF (median nerve stimulation)"	None	"SEF:0.48–250 Hz sampling rate 1000 Hz"	"NIHSS for ishemia evaluation ECD analysis Power Spectral Density"	"NIHSS has positive correlation with delta power and negative correlation with N20mECD stremgth. Serum transferrin also correlated with affected hemisphere N20m ECD strength and inversely with its delta power"	C
5		"204 channel Vectorview"	SF	"Xe-SPECT, IMP-SPECT acetazolamide loading"	"High pass filter 0.1Hz sampling 600Hz awake 10 min examination"	"1–6Hz filter extract 50ms data at slow wave activity caliculate ECDs at each 2ms interval"	"During hyper ventilation, slow wave activity on rebuild-up was detected in the area of poor response for acetazolamide loading, especially, deep sulcus region. After surgical anastomosis, rebuild-up was disappeared"	C
6	7	122 channel Neuromag	EF(median nerve stimulation)	None	"0.03–320Hz accumulation: 200 times"	"Neurological evaluation for ischemia ECD analysis"	AH SI showed abnormal response, and AH S2 responce decreased. UH S2 responce showed no change	C
7	22	"160 channel MEG Vision"	EF (median nerve stimulation)	SPECT	"3–200Hz accumulation: 400 times"	ECD analysis	ECD moment showed latelarity between AH and UH	C

8		"37 channel BTI"	EF (median nerve stimulation)	"H2O^{15}-PET CBF study"	"Sampling rate: 1041.7 Hz accumulation: 200 times"	"ECD analysis Comparison CBF with the amplitude and latensy of dipole moment at 20ms, 30ms, 40ms"	"The rCBF in the primary sensory area and N20 m were significantly reduced and the second component (P30 m) was significantly augmented in AH as compared with UH. P30m activation may indicate the compensatory change or the inhibitory system"	C
9	12	"37 channel BTI"	EF(AEF)	PET	"Auditory stimulation 1Hz tone and burst Orgogozo score, functional independence score, two points discrimination test MEG was performed within 1 week, at 3months, at one year after stroke onset"	Evaluation for responces at P50m, N100m	"In nine cases, the accurate magnetic sources of P50m or N100m were not identified. Cases with abnormal P50m responses had decreased supratemporal and hemispheric blood flow"	C
10	8	"64 channel CTF"	EF(motor related evoked field)	None	"Clench and un clench during 10 s 60 times index finger tapping"	"SAM analysis for event related synchronization at βband (13–30Hz) comparison between right and left t volume"	"Abnormal ipsilateral dominant distribution of beta ERD was observed significantly more often during contralesional hand grasping in patients with atherosclerotic vascular lesion. This may indicate the inhibitory system of bilateral cerebral hemishere connectivity"	C

(continued)

Table 11.1 (continued)

No.	Control number	MEG and device	Evoked field/ spontaneous field	Other modality	Task Mesurements	Analysis	Result	Clinical approach
11	5	"160 channel MEG Vision"	SF	"$O_2{}^{15gas}$-PET IMP-SPECT"	"0.16–200Hz sampling 500Hz awake 10 min examination"	"sLORETA extract 45 s data from each band, 0.3–2, 2–4, 4–6 6–8Hz comaprison between pre and post operative activated position, signal intensity of slow wave"	"In all 5 patients, slow waves in every frequency band were distributed in the area of cerebrovascular insufficiency. Slow-wave intensities in theta bands decreased postoperatively along with improvements in CBF and metabolism, whereas delta bands showed no significant differences between pre- and postoperatively"	C
12	8	"27 × 2 channel BTI"	SF	None	MEG was performed at 4 days, at 11 days, between 31 and 68 days after stroke onset	ECD analysis	"Slow wave (2–6Hz) appeared in ipsilateral sensorymoter for for TIA (6 cases) and bilateral temporal for TGA (2 cases). These wave disappeared within 11 days after onset"	P
13	4	49 channel	EF (motor related evoked field)	NIRS	"0.16–200Hz sampling 500Hz finger movement task (30-s period movemnt and 30-s rest period, total 30-min)"	Motor relatede DC-field, simultaneous to the NIRS	The motor-related activation shows a trend towards stronger infraslow neuronal signals over the unaffected hemispheres compared with the affected hemisphere	P

14	18	"Neuromag 306 planar-type gradiometer"	EF (tactile finger stimulation)	None	"0.03–308Hz sampling 941Hz MEG was performed at 1–7 days, at 3–4 weeks, ane at 3 months after stroke onset"	"ECD analysis signal space separetion method"	The S1-SEF (contralateral primary somatosensory evoked field) amplitude or latency did not correlate with any of the functional outcome scores. The cPO-SEF (contralateral parietal opercula somatosensory evoked field) amplitude correlated with hand function in the acute phase and during recovery	P

TPTA: Temporo-parietal theta activity
CBF
CBV
NIHSS: National Institute of Health Stroke Scale
AH: affected hemisohere
UH: unaffected hemisphere
N20m: the strength of the initial component of somatosensory evoked magnetic field
P30m: the second component
C: serebral metabolism and circulation relatede matter
P: neuronal plasticity related matter

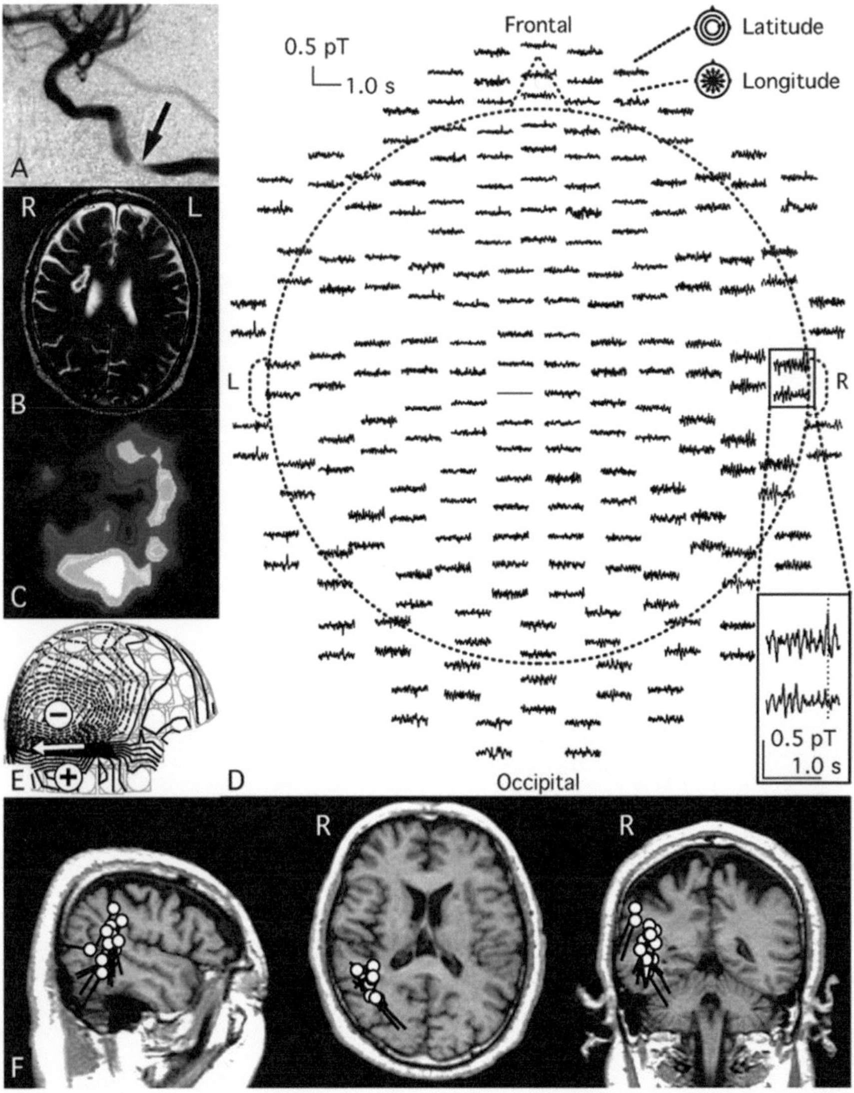

Fig. 11.1 A 54-year-old male with temporoparietal theta activity (TPTA) detected by MEG. (**a**) Digital subtraction angiogram showing stenosis (*arrow*) of the right internal carotid artery. (**b**) T2-weighted MR image showing a right striatocapsular infarct lesion. (**c**) SPECT scan showing flow reduction in the territory of the right middle cerebral artery. (**d**) MEG waveforms, detected in the awake condition with the eyes closed using the latitudinal and longitudinal tangential derivatives of 204 planar-type gradiometers, showing TPTA in the right hemisphere. (**e**) Isofield map at a typical peak of TPTA (*broken line* in the inlet square of D), showing a single dipole pattern. The *arrow* shows the approximate location and orientation of the equivalent current dipole (ECD) of TPTA over the right temporal area. (**f**) ECDs at ten similar peaks of TPTA to (**d**) and (**e**) projected onto three orthogonal MR images. *Circles* and *bars* indicate the ECD location and orientation, respectively (Ohtomo et al. [15])

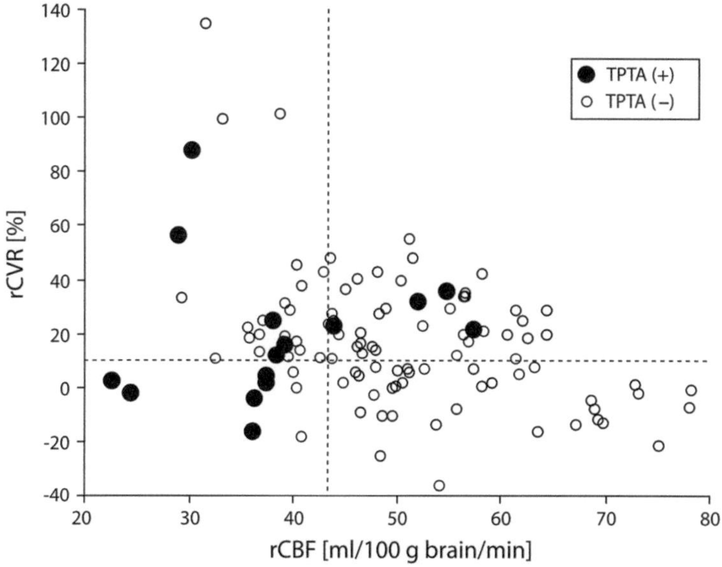

Fig. 11.2 Resting regional cerebral blood flow (rCBF) and regional cerebrovascular reactivity (rCVR) in the middle cerebral artery territory of 112 hemispheres in 56 patients with or without temporoparietal theta activity (TPTA). *Dashed vertical* and *horizontal lines* indicate the criteria for reduced rCBF (43.4 ml/100 g brain/min) and reduced rCVR (10 %) as defined in normal subjects in our institute. Note that TPTA was detected in six of nine hemispheres in the patients with both reduced rCBF and reduced rCVR (Ohtomo et al. [15])

11.4.3 Cerebrovascular Measurements

Six papers described measurement of CBF, including five that measured cerebrovascular reserve capacity. Three papers discussed areas of decreased cerebrovascular reserve (penumbra). They mainly described the correlation with areas of slow wave appearance and discussed postoperative improvement. Ohtomo and coworkers [6] demonstrated typical temporoparietal theta activity (TPTA) in ischemic lesion in Fig. 11.1 and the relation between cerebral circulation and TPHA in Fig. 11.2. Sakamoto and coworkers [23] showed the changes in slow wave activity between pre- and postsurgical anastomosis after stroke in Fig. 11.3.

11.5 General Remarks

Low-frequency activity often appears near ischemic foci and in an ischemic hemisphere, and these waves also tend to occur in the temporoparietal area [5]. In addition, with improved cerebral blood flow (CBF), slow waves, particularly theta band activity, usually disappear [6, 23]. However, this phenomenon is not

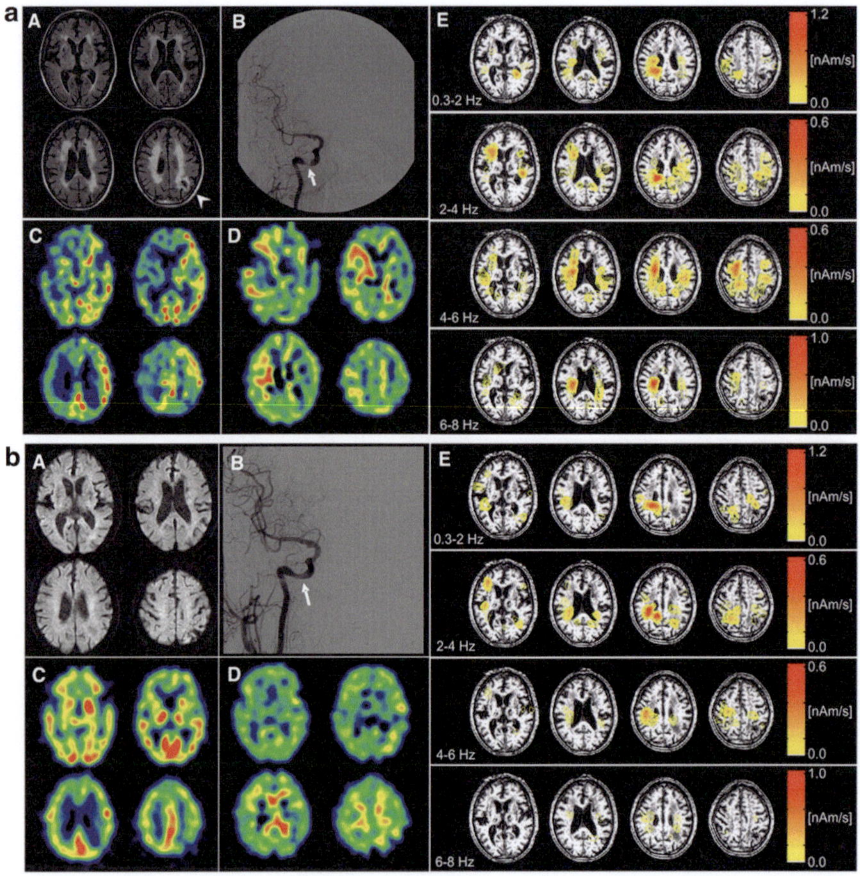

Fig. 11.3 (**a**) A 70-year-old man with the severe stenosis of right internal carotid artery showed slow wave activity in preoperative state by sLORETA of MEG. (*A*) MRI (FLAIR) showed no severe cerebral infarction, excluding old infarct lesion in the left parietooccipital area (*arrowhead*) and multiple lacunar infarctions in bilateral basal ganglia and periventricular white matter. (*B*) Cerebral angiography showed N90 % stenosis of the C5 segment of the right internal carotid artery (*arrow*). (*C*, *D*) 15O-gas PET showed reduced CBF (C, 32.5 ml/100 g/min) and increased OEF (D, 73.3 %) in the right cerebral cortex. (*E*) Quantitative imaging of cerebral neuromagnetic fields using sLORETA-qm in four frequency bands showed distribution and increased intensities of slow waves dominantly in the parietooccipital area of the right cerebral hemisphere. Distribution of slow waves was also recognized in the left parietooccipital area and periventricular white matter corresponding with old infarct lesion (Sakamoto et al. [19]). (**b**) Postoperative investigations in same case. (*A*) MRI (DWI) showed no additional infarction in any area of cerebrum. (*B*) Cerebral angiography showed recovery of stenosis after CAS of the right internal carotid artery (*arrow*). (*C*, *D*) 15O-gas PET showed increased CBF (C, 38.0 ml/100 g/min) and decreased OEF (D, 47.1 %) in the right cerebral cortex. (*E*) Quantitative imaging of cerebral neuromagnetic fields with sLORETA-qm in each frequency band showed decreased intensities at 2–4 Hz, 4–6 Hz, and 6–8 Hz in the parietooccipital area of the right cerebral hemisphere, although no apparent change of intensity at 0.3–2 Hz was observed compared with preoperative data (Sakamoto et al. [19])

necessarily seen in all cases, so using changes in slow waves as an objective index of a penumbra can be unreliable.

Many investigators have reported a decreased SEF N20m response to median nerve stimulation in an ischemic hemisphere, but it is difficult to determine whether this signifies complete ischemia or a penumbra.

MEG reflects activity in the sulcal cortex, whereas EEG also reflects activity in the sulcal and gyral cortex activity. On the other hand, MEG uses large number of channels (> 100 sensors on the whole head), so MEG can detect brain function over a wider area. Furthermore, its spatial resolution is much better than that of EEG. MEG directly detects magnetic fields because it is unaffected by the skin and bones and is, therefore, quantitatively superior to EEG. Moreover, MEG and MRI images can be fused, making it convenient for clinical application, such as neurosurgical navigation techniques.

MEG is a noninvasive technique to quantify neural activity, but it is not yet an established method for evaluating cerebral ischemia. The use of new analytical methods, including spatial filtering, not only measuring spontaneous magnetic fields but also objectively assessing neural activity when patients perform simple tasks [28], is a novel approach using cerebral magnetic fields to evaluate ischemic cerebrovascular disease.

11.6 Summary

1. Slow waves appear on the affected side in patients with stroke.
2. Slow waves are seen in areas of decreased blood perfusion (often the temporoparietal area) in patients with cerebrovascular occlusion.
3. The first peak of the somatosensory evoked field (SEF) in the ischemic hemisphere may decrease, may disappear, or be abnormal.

11.7 Conclusion

This section examined trends in clinical research and reviewed the scientific evidence, based on a literature search, for the use of MEG in evaluating ischemic cerebrovascular disease. MEG could not absolutely evaluate brain plasticity and circulation. Although it is complete as an alternative device to other modalities, a few papers reported a relatively high level of evidence for using MEG in diagnosis and treatment planning. MEG may be superior to scalp EEG in assessing ischemic changes in neural function.

MEG may also be useful to objectively evaluate neurological function, other than from the perspective of cerebral blood flow and metabolism, in patients with ischemic cerebrovascular disease.

References

1. Jang SH. Recovery mechanisms of somatosensory function in stroke patients: implications of brain imaging studies. Neurosci Bull. 2013;29:366–72.
2. Zappasodi F, Tombini M, Milazzo D, et al. Delta dipole density and strength in acute monohemispheric stroke. Neurosci Lett. 2007;416:310–4.
3. Butz M, Gross J, Timmermann L, et al. Perilesional pathological oscillatory activity in the magnetoencephalogram of patients with cortical brain lesions. Neurosci Lett. 2004;355:93–6.
4. Tecchio F, Zappasodi F, Pasqualetti P, et al. Rhythmic brain activity at rest from rolandic areas in acute mono-hemispheric stroke: a magnetoencephalographic study. Neuroimage. 2005;28:72–83.
5. Seki S, Nakasato N, Ohtomo S, et al. Neuromagnetic measurement of unilateral temporo-parietal theta rhythm in patients with internal carotid artery occlusive disease. Neuroimage. 2005;25:502–10.
6. Ohtomo S, Nakasato N, Shimizu H, et al. Temporo-parietal theta activity correlates with misery perfusion in arterial occlusive disease. Clin Neurophysiol. 2009;120:1227–34.
7. Vieth J, Kober H, Weise E, et al. Functional 3D localization of cerebrovascular accidents by magnetoencephalography (MEG). Neurol Res. 1992;14:132–4.
8. Assenza G, Zappasodi F, Squitti R, et al. Neuronal functionality assessed by magnetoencephalography is related to oxidative stress system in acute ischemic stroke. Neuroimage. 2009;44:1267–73.
9. Kamada K, Saguer M, Moller M, et al. Functional and metabolic analysis of cerebral ischemia using magnetoencephalography and proton magnetic resonance spectroscopy. Ann Neurol. 1997;42:554–63.
10. Nakasato N, Kanno A, Tominaga T. Magnetoencephalography (MEG): its application to moyamoya disease. In: Buying-Kyu Cho, Tominaga T, editors. Moyamoya disease update. Tokyo: Springer Japan; 2010. p. 220–4.
11. Qiao F, Kuroda S, Kamada K, et al. Source localization of the re-build up phenomenon in pediatric moyamoya disease-a dipole distribution analysis using MEG and SPECT. Childs Nerv Syst. 2003;19:760–4.
12. Tecchio F, Zappasodi F, Tombini M, et al. Brain plasticity in recovery from stroke: an MEG assessment. Neuroimage. 2006;32:1326–34.
13. Altamura C, Torquati K, Zappasodi F, et al. fMRI-vs-MEG evaluation of post-stroke inter-hemispheric asymmetries in primary sensorimotor hand areas. Exp Neurol. 2007;204:631–9.
14. Gallien P, Aghulon C, Durufle A, et al. Magnetoencephalography in stroke: a 1-year follow-up study. Eur J Neurol. 2003;10:373–82.
15. Rossini PM, Tecchio F, Pizzella V, et al. Interhemispheric differences of sensory hand areas after monohemispheric stroke: MEG/MRI integrative study. Neuroimage. 2001;14:474–85.
16. Forss N, Hietanen M, Salonen O, et al. Modified activation of somatosensory cortical network in patients with right-hemisphere stroke. Brain. 1999;122:1889–99.
17. Tsutada T, Ikeda H, Tsuyuguchi N, et al. Detecting functional asymmetries through the dipole moment of magnetoencephalography. J Neurol Sci. 2002;198:51–6.
18. Tecchio F, Zappasodi F, Tombini M, et al. Interhemispheric asymmetry of primary hand representation and recovery after stroke: a MEG study. Neuroimage. 2007;36:1057–64.
19. Bundo M, Inao S, Nakamura A, et al. Changes of neural activity correlate with the severity of cortical ischemia in patients with unilateral major cerebral artery occlusion. Stroke. 2002;33:61–6.
20. Kotani K, Kinomoto Y, Yamada M, et al. Spatiotemporal patterns of movement-related fields in stroke patients. Neurol Clin Neurophysiol. 2004;2004:63.
21. Forss N, Mustanoja S, Roiha K, Kirveskari E, Mäkelä JP, Salonen O, Tatlisumak T, Kaste M. Activation in parietal operculum parallels motor recovery in stroke. Hum Brain Mapp. 2012;33:534–41.

22. Toyoda K, Ibayashi S, Yamamoto T, et al. Auditory evoked neuromagnetic response in cerebrovascular diseases: a preliminary study. J Neurol Neurosurg Psychiatry. 1998;64:777–84.
23. Sakamoto S, Tanaka H, Tsuyuguchi N, et al. Quantitative imaging of spontaneous neuromagnetic activity for assessing cerebral ischemia using sLORETA-qm. Neuroimage. 2010;49:488–97.
24. Stippich C, Kassubek J, Kober H, et al. Time course of focal slow wave activity in transient ischemic attacks and transient global amnesia as measured by magnetoencephalography. Neuroreport. 2000;11:3309–13.
25. Meinzer M, Elbert T, Wienbruch C, et al. Intensive language training enhances brain plasticity in chronic aphasia. BMC Biol. 2004;2:20.
26. Huang M, Davis LE, Aine C, et al. MEG response to median nerve stimulation correlates with recovery of sensory and motor function after stroke. Clin Neurophysiol. 2004;115:820–33.
27. Leistner S, Sander-Thoemmes T, Wabnitz H, Moeller M, Wachs M, Curio G, Macdonald R, Trahms L, Mackert BM. Non-invasive simultaneous recording of neuronal and vascular signals in subacute ischemic stroke. Biomed Tech (Berl). 2011;56(2):85–90.
28. Oshino S, Kato A, Hirata M, et al. Ipsilateral motor-related hyperactivity in patients with cerebral occlusive vascular disease. Stroke. 2008;39:2769–75.

Neurodegenerative Disorders

12

Isamu Ozaki and Isao Hashimoto

Abstract

Scientific evidence of the efficacy of magnetoencephalography (MEG) is currently lacking except in diagnosing epilepsy. In the present study, we performed a review of clinical MEG studies on neurodegenerative disorders using a website bibliographic survey. We searched MEDLINE for MEG papers on neurodegenerative disorders published before December 2014 using the following keywords: a representative diagnosis such as amyotrophic lateral sclerosis (ALS) and magnetoencephalography or MEG. We further narrowed the search to 30 papers based on the levels of evidence and abstract contents; 3 papers on ALS, 18 papers on Parkinson disease, and 9 papers on multiple sclerosis were included in the final review. Levels of evidence were classified as follows: grade I, no paper; grade II, 19 papers; grade III, 2 papers; and grade IV, 9 papers. The majority of studies were conducted with a small number of patients. However, MEG has the advantage of being able to detect spontaneous activity in small brain regions and to measure functional network activity between multiple brain areas or coherent activity between deep brain nuclei and distinct cortical areas. Accordingly, MEG allows the assessment of functional changes in certain diseases and provides novel insights into disease-specific pathophysiology, such as in Parkinson disease.

Keywords

Amyotrophic lateral sclerosis • Parkinson disease • Multiple sclerosis

---

I. Ozaki (✉)
Faculty of Health Sciences, Aomori University of Health and Welfare, 58-1 Mase, Hamadate, Aomori 030-8505, Japan
e-mail: isamu@auhw.ac.jp

I. Hashimoto
Kanazawa Institutes of Technology, Tokyo 102-0083, Japan

S. Tobimatsu, R. Kakigi (eds.), *Clinical Applications of Magnetoencephalography*,
DOI 10.1007/978-4-431-55729-6_12

12.1 Introduction

Although magnetoencephalography (MEG) provides noninvasive information regarding the localization of epileptic foci in patients with epilepsy, scientific evidence of the utility of MEG in diagnosing various other neurologic diseases, defining disease status, and predicting disease progression or prognosis remains lacking. To evaluate clinical utility of MEG in patients diagnosed as having neurodegenerative diseases, we reviewed clinical MEG studies on neurodegenerative disorders, including amyotrophic lateral sclerosis (ALS) or motor neuron disease, multiple sclerosis (MS), Parkinson disease (PD), and spinocerebellar degeneration, based on a website bibliographic survey. We identified MEG studies on neurodegenerative disorders published before December 2014 by searching MEDLINE using the following keywords: a representative diagnosis such as ALS and magnetoencephalography or MEG. We further narrowed the search to 30 papers based on the levels of evidence and abstract contents; 3 papers on ALS, 18 papers on PD, and 9 papers on MS were included in the final review. The evidence level and recommendations were judged in each paper based on published criteria, as shown in Table 12.1 [1]. In this chapter, we provide a brief overview of the clinical and pathophysiological features of each abovementioned neurodegenerative disease and the current status of MEG research related to each disease. We further discuss the present capabilities and future possibilities of MEG in relation to neurodegenerative diseases.

12.2 Magnetoencephalography (MEG) in Amyotrophic Lateral Sclerosis (ALS)

12.2.1 Clinical and Pathophysiological Features of ALS

Amyotrophic lateral sclerosis (ALS), often referred to as Lou Gehrig's disease, is a progressive neurodegenerative disease that affects both upper motor neurons in the brain and lower motor neurons in the spinal cord. *Amyotrophy* means neurogenic

Table 12.1 Grades of recommendation and levels of evidence

Grade	Level	Type of evidence
A	Ia	Evidence obtained from meta-analysis of randomized-controlled trials
	Ib	Evidence obtained from at least one randomized-controlled trial
B	IIa	Evidence obtained from at least one well-designed controlled study without randomization
	IIb	Evidence obtained from at least one other type of well-designed quasi-experimental study
	III	Evidence obtained from well-designed nonexperimental descriptive studies, such as comparative studies, cohort, and case–control studies
C	IV	Evidence obtained from expert committee reports or opinions and/or clinical experiences of respected authorities

atrophy of muscle; *lateral sclerosis* refers to the firmness of affected spinal cords reported by pathologists at autopsy [2]. Lateral sclerosis results from the proliferation of astrocytes and the scarring of the lateral columns of the spinal cord in response to degeneration of the corticospinal tracts. Early symptoms of ALS include increasing muscle weakness that predominantly involves the arms and legs, speech, swallowing, or breathing. The disease begins focally and then spreads relentlessly. However, some motor neurons innervating the ocular muscles are spared, and sensation and bladder functions also are spared during the course of the disease. Whereas approximately 10 % of ALS cases are inherited as dominant traits and approximately 25 % of inherited ALS patients result from mutations in the gene encoding superoxide dismutase 1 (SOD1), the cause of sporadic ALS remains unknown [2]. Cognitive impairment in ALS was considered uncommon until recently. Overt frontotemporal dementia (FTD) occurs in approximately 15 % of ALS patients, but up to 50 % of ALS patients are classified as impaired if measured by neuropsychological tests [3].

12.2.2 MEG Study in Patients with ALS

A summary of three MEG studies of ALS is presented in Table 12.2. A number of divergent approaches to the study of MEG in ALS are observed. Pekkonen et al. [4] reported that the P50m and N100m responses or MMNm of auditory-evoked fields, the magnetic counter part of mismatch negativity potentials, are augmented in ALS patients with severe bulbar signs, indicating that auditory processing underlying stimulus detection, and subsequent memory-based comparison processes are abnormal in ALS. Boyajian et al. [5] performed single-dipole analysis of focal delta–theta activity in ALS patients and demonstrated the presence of slow-wave bursts generated from the frontal, temporal, and parietal cortices but not occipital areas, indicating widespread cortical dysfunction in patients with ALS. By performing MEG before and after swallowing in normal controls and ALS patients, Teismann et al. [6] demonstrated event-related desynchronization (ERD) in beta and low gamma bands in bilateral sensorimotor areas in control subjects; however, the ERD response was predominantly on the right side in ALS patients with difficulty in swallowing. This right hemispheric predominance in the activation of the primary motor cortex during volitional swallowing may be the only sign of cortical plasticity in dysphagic ALS patients.

Table 12.2 MEG study in patients with amyotrophic lateral sclerosis

No	Number of patients (gender, mean age, range)	Number of control subjects (gender, mean age, range)	MEG apparatus	MEG modality (evoked/spontaneous)	Recording method	Analysis	Main results	Levels of evidence	Grades of recommendation	Authors	Year	Journal
1	10, ALS having bulbar signs without dementia (F 6, M4; 63; 55–78 years)	10 (65; 59–76 years)	122Ch (61sites, planar gradiometer), Neuromag (Neuroimag Ltd)	Auditory evoked	Auditory-evoked fields and MMN fields were elicited by standard 700 Hz tones with two different ISIs (0.5 s and 2.5 s).	The amplitudes or latencies of P50m, N100m, and MMNm were compared.	The amplitudes of the P50m, N100m, and MMNm were larger in patients with ALS than in control subjects. The P50m latencies were shorter in patients with ALS.	IIb	B	Pekkonen et al.	2004	Clin Neurophysiol
2	7, ALS without dementia (F 3, M4; 41–77 years)	8 (39–83 years)	148Ch, Magnes 2500 radial magnetometer and/or axial gradiometer (4D Neuroimaging)	Spontaneous	Localized dipolar sources of focal delta–theta (1–7 Hz) discharges (slow-wave activity) were computed from spontaneous MEG (0.1–100 Hz) recorded for 15 min.	Regional slow-wave activity was obtained as dipole numbers/min (single-dipole method). Localization was based on a subset of 37 channels surrounding the region producing the focal slow-wave activity.	The slow-wave activity was produced in all brain regions except for occipital regions in ALS patients. The slow-wave dipole density in the cingulate gyrus was correlated with the disability score of upper limbs. In control subjects, the slow-wave activity was not found.	IIb	B	Boyajian et al.	2008	Am J Phys Med Rehabil
3	14, bulbar–onset ALS (5F, 9M; 58.9; 44–74 years)	7 (3F, 4M; 57.6; 41–71 years)	275Ch (Omega 275, CTF Systems Inc., Canada)	Swallowing evoked	EMGs of submental muscles and spontaneous MEG were recorded during a self-paced swallowing.	MEGs before and after swallowing were analyzed by means of time-frequency analysis and synthetic aperture magnetometry (SAM).	Event-related desynchronization (ERD) in beta band and low gamma band was found in bilateral sensorimotor areas in control subjects. The ERD response was predominant on the right side in ALS patients with difficulty in swallowing.	IIb	B	Teismann et al.	2011	PLoS One

12.3 Magnetoencephalography (MEG) in Parkinson Disease

12.3.1 Clinical and Pathophysiological Features of Parkinson Disease

Parkinson disease (PD) is a progressive neurodegenerative disorder involving varying combinations of bradykinesia, rest tremor, rigidity, and loss of postural adjustment. The pathological hallmark of idiopathic PD is the degeneration of the dopaminergic cells in the substantia nigra pars compacta that project to the striatum and a number of other basal ganglia nuclei. Dopaminergic hypofunction in cortico-basal ganglia circuits is thought to underlie the majority of motor disturbances observed in PD as dopamine replacement or dopamine receptor agonist administration has demonstrated efficacy in reducing motor disturbances. Several hypotheses have been proposed to account for the mechanisms underlying the pathogenesis of motor disturbances in PD. Of these hypotheses, the "rate model" [7] or standard "antagonist balance model" [8] posits that the pathophysiology of bradykinesia or hypokinesia is as follows. Loss of dopaminergic input to the striatum gives rise to increased activity of the indirect pathway; the striatum to the external segment of the globus pallidus (GPe) to the subthalamic nucleus (STN) to the internal segment of the globus pallidus (GPi), and decreased activity in the direct pathway; and the striatum to the GPi [7]. Both of these changes would lead to a net increase in the activity of neurons in the GPi and substantia nigra pars reticulata. This increase in basal ganglia output would then result in increased inhibition of thalamocortical and midbrain tegmental neurons and account for the hypokinetic features of PD [7]. A modification of the standard model, known as the "center–surround model," states that the two pathways interact in a center–surround organization similar to that described in the visual system [9]. In this model, desired movement is normally achieved via activation of the direct pathway, and undesired movement due to competing motor programs is suppressed via activation of the indirect pathway that causes surrounding inhibition [9]. In PD, STN hyperactivity leads to excessive inhibition of all movements, both desired and undesired, leading to akinesia and bradykinesia, whereas STN hypoactivity results in decreased suppression of undesired movements and its florid expression in the form of hemiballism [8, 9]. Another emerging hypothesis regarding the pathogenesis of PD is the "abnormal firing pattern model [7, 8]." Studies employing microelectrode recordings from MPTP parkinsonian primates and PD patients have demonstrated abnormal firing pattern in the indirect pathway, i.e., increased oscillatory and synchronized activity in GPi, GPe, and STN neurons [7, 8]. Abnormal neural oscillations in the low frequency range of 5–8 Hz may contribute to rest tremor [7, 8]. In addition, abnormal oscillation and increased synchronization at the 15–30 Hz frequency range (beta band) may either block the normal flow of information through the basal ganglia or be associated with a loss of neuronal selectivity, resulting in the slowing or prevention of movements [7, 8]. High-frequency deep brain stimulation (DBS) has been shown to improve motor deficits in PD patients by suppressing oscillatory beta activity of the basal ganglia.

Recently, an increasing number of studies have examined PD-related non-motor symptoms, such as hyposmia, autonomic dysfunctions, mood disorders, sleep disorders, sensory disorders, and cognitive deficits [10]. In accordance with these clinical symptoms, accumulated pathological studies of sporadic or idiopathic PD patients have provided evidence that PD involves a multisystem degenerative process, possibly initiated by the migration of pathogens to the brain from the stomach or nose [11], involving not only the dopaminergic but also the noradrenergic, serotoninergic, cholinergic, and other systems [10]. During the progression of PD, Lewy bodies have been shown to extend from brainstem areas to multiple cortical areas, leading to the onset of dementia [10]. Therefore, it is now accepted that the majority of PD patients develop cognitive deficits with prolonged disease duration in contradiction to the original description of PD by James Parkinson.

12.3.2 MEG Studies in Patients with Parkinson Disease

A summary of 18 MEG studies of PD is presented in Table 12.3. MEG studies of PD include a number of different approaches dependent on the subject of interest. Eight articles were published by the VU University Medical Center in Amsterdam [12–19]. In earlier studies, the frequency spectrum of spontaneous MEG, or odor stimulus-conditioned MEG, was analyzed in sensor space of ten cortical regions covering the bilateral frontal, central, temporal, parietal, and occipital areas [12, 13]. As a result of sensor-based synchronization likelihood (SL) analysis of odor stimulus-conditioned MEG, Boesveldt et al. [13] found a decrease in beta (β) local SL and an increase in delta (δ) interhemispheric SL in controls but not patients with PD (for further details, see Table 12.3). Gómez et al. [14] also investigated sensor-space functional connectivity in PD patients by obtaining Lempel–Ziv complexity (LZC) values calculated by channel-by-channel analysis and demonstrated lower LZC values in PD patients compared to controls. Recently, Olde Dubbelink et al. [15–18] published four papers focusing on changes in functional connectivity and the possible contribution of such changes to cognitive decline in PD patients. First, the relationship between sensor-space frequency spectral power and cognitive function was assessed with decreased cognitive performance shown to be associated with increased delta (0.5–4 Hz) and theta (4–8 Hz) power, as well as decreased in alpha1 (8–10 Hz), alpha2 (10–13 Hz), and gamma (30–48 Hz) power. Increased motor impairment was found to be associated with increased theta power only [15]. Next, using a beamforming approach to measure brain activity in 78 cortical regions, a source-space analysis was performed to assess frequency spectral power and the phase lag index (PLI), as a measure of functional connectivity, with PD patients and controls examined twice with an interval of 4 years [16]. In patients with PD, longitudinal analyses over a 4-year period revealed decreased alpha1 and alpha2 band connectivity in multiple seed regions associated with motor or cognitive deterioration (see Table 12.3) [16]. In a further study with the same PD patients and controls, Olde Dubbelink et al. [17] applied minimum spanning tree (MST) network analyses as another measure of

Table 12.3 MEG in patients with Parkinson disease

No	Number of patients (gender, mean age, range)	Number of control subjects (gender, mean age, range)	MEG apparatus	MEG modality (evoked/ spontaneous)	Recording method	Analysis	Main results	Levels of evidence	Grades of recommendation	Authors	Year	Journal
1	70 (F 30, M40; 62.1 ± 6.8) including 18 de novo Parkinson (F 6, M12; 59.4 ± 7.9)	21 (F 10, M11; 59.4 ± 7.3)	151Ch, axial gradiometer system (CTF)	Spontaneous	Spontaneous MEGs of 3 epochs for 13 s were digitally filtered into standard frequency bands, and sensor-space (artifact-free 141 Ch data) analyses were performed for bilateral frontal, central, parietal, temporal, occipital regions.	Relative spectral power was calculated into the delta (0.5–4 Hz), theta (4–8 Hz), low alpha (8–10 Hz), high alpha (10–13 Hz), beta (13–30 Hz), and gamma (30–48 Hz) bands. Analysis of covariance (ANCOVA) was used to evaluate changes in each spectral power in de novo PD and to analyze the relation of spectral power to disease duration, disease stage, disease severity and effect of dose of dopaminomimetics within the whole group of PD patients.	ANCOVA testing showed that patients with de novo PD have higher power in the theta and low alpha frequency bands and lower power in the beta and gamma frequency bands than controls. In patients with PD, relative spectral power in low alpha band decreased with increasing disease duration in the right temporal and right occipital regions. Spectral power values in any frequency band were unrelated with disease severity or dose of dopaminomimetics. After intake of dopaminomimetics, a slight modulatory effect on spectral power was found; decreases of relative power in right frontal theta, left occipital beta, and left temporal gamma, and an increase in right parietal gamma power.	IIb	B	Stoffers et al.	2007	Brain

(continued)

Table 12.3 (continued)

No	Number of patients (gender, mean age, range)	Number of control subjects (gender, mean age, range)	MEG apparatus	MEG modality (evoked/ spontaneous)	Recording method	Analysis	Main results	Levels of evidence	Grades of recommendation	Authors	Year	Journal
2	20 (F 8, M 12; 61.5; 50–73)	21 (F 12, M 9; 56.3; 49–73)	151Ch, axial gradiometer system (CTF)	Spontaneous, olfactory evoked	Spontaneous or olfactory stimulus-evoked MEGs of around 30 s were digitally filtered into standard frequency bands, and sensor-space (artifact-free 141 Ch data) analyses were performed for bilateral frontal, central, parietal, temporal, occipital regions.	Relative spectral power was calculated into the delta (1–4 Hz), theta (4–8 Hz), alpha 1 (8–10 Hz), alpha 2 (10–13 Hz), beta (13–30), and gamma (30–48 Hz) bands. Changes in overall spectral power for the odor stimulus compared to the rest condition were evaluated. Sensor-based synchronization likelihood (SL) was computed.	For the odor stimulus condition, a power decrease in beta band and a power increase in theta band were found in bilateral central regions and right temporal region in controls. A significant decrease in alpha 1 band was found in bilateral central and parietal regions and the left temporal region in patients with PD. As to analysis of SL, for the stimulus condition in controls, a decrease in beta local SL and an increase in delta interhemispheric SL were found. Control subjects showed an increase in alpha2 intrahemispheric SL and a decrease in the beta band, whereas patients with PD showed the opposite pattern.	IIb	B	Boesveldt et al.	2009	Hum Brain Mapp

3	18 (F 6, M12; 60.0 ± 8.0), untreated Parkinson	20 (F 9, M11; 59.4 ± 7.5)	151Ch, axial gradiometer system (CTF)	Spontaneous	Spontaneous MEGs of artifact-free 12 epochs of 4 s were recorded. A complexity of MEG data for bilateral frontal, central, parietal, temporal, occipital regions was calculated.	The Lempel–Ziv complexity (LZC) was obtained by channel-by-channel analysis, and regional LZC values for 10 cortical regions were compared between the patients with PD and controls.	PD patients displayed lower LZC values than control subjects for all the MEG channels both with the binary and the three-symbol sequence conversion. There were significant differences in regional LZC values between PD patients and controls subjects, suggesting that the complexity for spontaneous MEG is lower in PD patients than in control subjects.	IIb	B	Gómez et al.	2011	Ann Biomed Eng
4	49 (F 18, M31; 61.4 ± 6.39) at baseline, 70 patients [12]	14 (F 4, M10; 60.0 ± 8.55) at baseline, 21 controls [12]	151Ch, axial gradiometer system (CTF)	Spontaneous	MEGs were recorded in an eye-closed resting-state condition for 5 min and digitally filtered into standard frequency bands. Sensor-space (artifact-free 139 Ch data) analyses were performed for bilateral frontal, central, parietal, temporal, occipital regions.	MEG recordings were performed twice at an approximate 4-year interval. Global relative spectral power density averaged over all channels was obtained for controls and PD patients. Relative spectral power was also calculated into the delta (0.5–4 Hz), theta (4–8 Hz), alpha 1 (8–10 Hz), alpha 2 (10–13 Hz), beta (13–30 Hz), and gamma (30–48 Hz) bands. The relationship between cognitive function and spectral power was also analyzed.	In control subjects, global relative spectral power density averaged over all channels was unchanged. In contrast to healthy controls, PD patients showed a slowing of the dominant peak frequency. In PD patients, decreasing cognitive performance was associated with increases in delta and theta power, as well as decreases in alpha1, alpha2, and gamma power, whereas increasing motor impairment was associated with a theta power increase only.	IIb	B	Olde Dubbelink et al.	2013	Neurobiol Aging

(continued)

Table 12.3 (continued)

No	Number of patients (gender, mean age, range)	Number of control subjects (gender, mean age, range)	MEG apparatus	MEG modality (evoked/ spontaneous)	Recording method	Analysis	Main results	Levels of evidence	Grades of recommendation	Authors	Year	Journal
5	43 (F 15, M28; 61.5 ± 6.45) including 12 de novo Parkinson at baseline, 70 patients [12]	14 (F 4, M10; 60.0 ± 8.55) at baseline, 21 controls [12]	151Ch, axial gradiometer system (CTF)	Spontaneous	MEGs were recorded in an eye-closed resting-state condition for 5 min. Five epochs for 13 s were digitally filtered into standard frequency bands. Source-space analyses were performed for 78 cortical regions.	MEG recordings were performed twice at an approximate 4-year interval. Relative spectral power was calculated into the delta (δ), theta (θ), low alpha (α), high alpha (α), beta (β), and gamma (γ) bands. For analysis of functional connectivity, the phase lag index (PLI) was obtained from 78 cortical regions and averaged for parahippocampal, inferior and middle temporal, temporal pole, orbitofrontal, precuneus, anterior	In PD patient group, longitudinal analyses over a 4-year period revealed decreases in alpha1 and alpha2 band connectivity for multiple seed regions that were associated with motor or cognitive deterioration. Untreated PD patients had lower parahippocampal and temporal delta band connectivity and higher temporal alpha1 band connectivity compared to controls.	IIb	B	Olde Dubbelink et al.	2013	Neuroimage Clin

						cingulate, and middle frontal regions.						
6	43 (F 15, M28; 61.5 ± 6.45) including 12 de novo Parkinson at baseline, 70 patients [12]	14 (F 4, M10; 60.0 ± 8.55) at baseline, 21 controls [12]	151Ch, axial gradiometer system (CTF)	Spontaneous	MEGs were recorded in an eye-closed resting-state condition for 5 min. Five epochs for 13 s were digitally filtered into standard frequency bands. Source-space analyses were performed for 78 cortical regions.	MEG recordings were performed twice at an approximate 4-year interval. Relative spectral power was calculated into the delta (0.5–4 Hz), theta (4–8 Hz), low alpha (8–10 Hz), high alpha (10–13 Hz), beta (13–30 Hz), and gamma (30–48) bands. For analysis of functional connectivity, the phase lag index (PLI) was obtained from 78 cortical regions. Minimum spanning tree (MST) network analyses were added.	Brain networks in early-stage untreated PD patients displayed lower local clustering with preserved path length in the delta frequency band in comparison to controls. MSTs of untreated PD patients revealed lower leaf number and lower tree hierarchy in the alpha2 frequency band when compared to controls. Longitudinal analysis over a 4-year period in PD patients showed a progressive decrease in local clustering in multiple frequency bands together with a decrease in path length in the alpha2 frequency band.	IIb	B	Olde Dubbelink et al.	2014	Brain
7	63 (F 24, M39) at baseline, 70 patients [12] 19 patients (F 4, M15;66.0 ± 5.19) out of 63 converted to PD-related dementia (PDD). 44 patients (F 20,	(–)	151Ch, axial gradiometer system (CTF)	Spontaneous	MEGs were recorded in an eye-closed resting-state condition for 5 min. Five epochs for 13 s were digitally filtered into standard frequency bands.	Baseline cognitive assessments and MEG recordings were analyzed in relation to PD-related dementia (PDD) conversion over a 7-year period. Relative spectral power was calculated into the delta (0.5–4 Hz), theta	Of the neurophysiologic markers, beta power less than median was the strongest PDD predictor, followed by peak frequency less than median and theta power more than median. Of the cognitive test battery,	IIb	B	Olde Dubbelink et al.	2014	Neurology

(continued)

Table 12.3 (continued)

No	Number of patients (gender, mean age, range)	Number of control subjects (gender, mean age, range)	MEG apparatus	MEG modality (evoked/ spontaneous)	Recording method	Analysis	Main results	Levels of evidence	Grades of recommendation	Authors	Year	Journal
	M 24;60.9 ± 6.48) did not				Source-space analyses were performed for 78 cortical regions.	(4–8 Hz), low alpha (8–10 Hz), high alpha (10–13 Hz), beta (13–30 Hz), and gamma (30–48 Hz) bands for 78 cortical regions. Among the cognitive and neurophysiologic markers, predictive factors for converting PDD were evaluated.	performance on a posterior (pattern recognition memory score less than median) and a fronto-executive (spatial span scoreless than median) task most strongly predicted dementia conversion. The combination of impaired fronto-executive task performance and low beta power was associated with the highest dementia risk.					
8	11 (F 3, M8; 61.0 ± 15.5)	11 (F 3, M8; 62.2 ± 8.4)	151Ch (third-order synthetic gradiometers) (CTF)	Movement-related magnetic fields	MEGs were recorded during a rhythmic squeezing movement of the right hand, under self-paced condition or regularly cued condition.	Left M1 sources during right hand movement were estimated by synthetic aperture magnetometry (SAM) beamformers. These activities were assessed via the signals' event- related field, power, and phase uniformity.	PD patients displayed more power in the alpha band and less in the beta band during both rest and motor activity, as compared to controls. The result implied a slowing of oscillatory activity of the primary motor cortex in PD.	IIb	B	Vardy et al.	2011	Clin Neurophysiol

9	10 de nove PD (F 3, M7; 43–72) 10 with medication (F 3, M7; 44–69)	10 (age matched)	306Ch(102sites, planar gradiometer), Neuromag (Eleckta Oy)	Spontaneous and movement related	Spontaneous MEGs and electromyograms (EMG)from the extensor digitorum forearm muscle of the more severely affected side were simultaneously recorded for 1 min during muscle contraction period and rest period, respectively.	Motor task was one minute isometric contraction of the forearm of the more severely affected side at about 20 % of maximal contraction strength. Four epochs of rest and motor task conditions were analyzed. Spectral power and coherence between bilateral S1/M1and EMG were calculated. For source localization, using a spatial filter algorithm and a realistic head model of each individual, the analysis tool dynamic imaging of coherent sources (DICS) was applied.	In control subjects, beta frequency power of the hemisphere contralateral to isometric contraction was lower as compared to the ipsilateral side. This pattern appeared to be attenuated in de novo PD patients and reversed in medicated PD patients. Contralateral beta power was correlated with motor impairment during isometric contraction but not during rest. DICS analysis showed that cortical sources coherent with EMG from the extensor digitorum forearm muscle at beta frequency were localized within the primary motor cortex corresponding to Brodmann area 4.	IIb	B	Pollok et al.	2012	J Physiol
10	10 (F 4, M6; 60.0 ± 3.7)	11 (28.9 ± 2.4)	122Ch (61sites, planar gradiometer system), Neuromag (Neuroimag Ltd)	Spontaneous and movement related	Spontaneous MEGs and electromyograms (EMG)from the extensor digitorum communis (EDC) muscle of the tremor-dominant site were simultaneously recorded for	Power and coherence spectra between MEG and EMG were calculated. For source localization, using a spatial filter algorithm and a realistic head model of each individual, the analysis tool dynamic imaging of coherent sources (DICS) was applied.	The EMG power spectral analysis during off (resting tremor period) revealed discernible peaks at tremor frequency (4.8 Hz) and at double the tremor frequency (9.4 Hz). DICS analysis showed the coherent activity in the contralateral	IIb	B	Pollok et al.	2013	Mov Disord

(continued)

Table 12.3 (continued)

No	Number of patients (gender, mean age, range)	Number of control subjects (gender, mean age, range)	MEG apparatus	MEG modality (evoked/ spontaneous)	Recording method	Analysis	Main results	Levels of evidence	Grades of recommendation	Authors	Year	Journal
					5 min during off and on, respectively.		primary sensorimotor cortex (M1/S1). Using M1/S1 as reference region, cerebro–cerebral coherence analysis revealed several other areas oscillating at double the tremor frequency: premotor cortex, supplementary motor area, secondary somatosensory cortex, posterior parietal cortex, and thalamus contralateral to the EMG recorded site, and ipsilateral cerebellum in all patients. Coherence strength of each cerebro–cerebral coupling was decreased with improvement of tremor. The results were similar in healthy subjects mimicking resting tremor.					

11	13 (F 2, M10; 53.6;40–61). 13 patients who underwent subthalamic nucleus deep brain stimulation electrode implantation prior to deep brain stimulation (DBS) therapy. 12 patients underwent bilateral operation	(–)	275 Ch (CTF/VSM MedTech, Vancouver, Canada)	Spontaneous	Subthalamic nucleus (STN) electrode local field potentials (LFP) and MEGs were simultaneously recorded between 2 and 6 days postoperatively.	For each patient, coherence was calculated at the sensor level between each STN-LFP channel and each MEG channel. To locate coherent cortical sources spatially, the dynamic imaging of coherent sources (DICS) beamforming method was used to calculate coherence between each STN-LFP channel and a 3D grid of points representing potential sources within the brain.	Two spatially and spectrally separated cortico-subthalamic networks were identified. A temporoparietal-brainstem network was coherent with the STN in the alpha (7–13 Hz) band, while a predominantly frontal network was coherent in the beta (15–35 Hz) band. Dopaminergic medication modulated the resting beta network, by increasing beta coherence between the STN region and ipsilateral prefrontal cortex.	IV	B	Litvak et al.	2011	Brain
12	17 (F 6, M11; 55 ± 7; 40–66). 17 patients who underwent subthalamic nucleus deep brain stimulation electrode implantation prior to deep brain stimulation (DBS) therapy. 16 patients underwent bilateral operation	(–)	275 Ch (CTF/VSM MedTech, Vancouver, Canada)	Spontaneous and movement related	Subthalamic nucleus (STN) electrode local field potentials (LFP), MEGs, and EMGs of the right and left first interosseous muscles were simultaneously recorded between 2 and 7 days postoperatively.	Source localization of the primary motor hand area (M1) contralateral to the movement was performed using the dynamic imaging of coherent sources (DICS) beamforming method. Spectral power in the contralateral M1 and STN and the coherence between the two structures were evaluated during synchronous or sequential finger movements. Source time series were estimated with a linearly constrained minimum variance (LCMV) beamformer.	There were discrete peaks in M1 and STN power at 60–90 Hz and at 300–400 Hz. All these power peaks increased with either synchronous or sequential finger movement and levodopa treatment. Only STN activity at 60–90 Hz was coherent with activity in M1. Directionality analysis showed that STN gamma activity at 60–90 Hz tended to drive gamma activity in M1.	IV	B	Litvak et al.	2012	J Neurosci

(continued)

Table 12.3 (continued)

No	Number of patients (gender, mean age, range)	Number of control subjects (gender, mean age, range)	MEG apparatus	MEG modality (evoked/ spontaneous)	Recording method	Analysis	Main results	Levels of evidence	Grades of recommendation	Authors	Year	Journal
13	9 (F 3, M6; 64;47–75). The patients who underwent subthalamic nucleus deep brain stimulation electrode implantation bilaterally prior to deep brain stimulation (DBS) therapy. Data from one patient was discarded	(–)	306Ch(102sites, planar gradiometer + magnetometer), Neuromag (Eleckta Oy)	Spontaneous	Subthalamic nucleus (STN) electrode local field potentials (LFP) and MEGs were simultaneously recorded the day after electrode implantation.	For each patient, the five MEG sensors showing the highest mean coherence between STN-LFP were identified, and their coherence spectra were averaged. To locate coherent cortical sources spatially, the dynamic imaging of coherent sources (DICS) beamforming method was used to calculate coherence between each STN-LFP channel and a 3D grid of points representing potential sources within the brain.	Significant coherence between STN-LFPs and MEG sensors, lateralized to the ipsilateral side with respect to the STN, was found in the alpha, low beta, and high beta band in all subjects. Alpha source maxima were mostly situated in the temporal cortex and the sensorimotor cortices ipsilateral to the STN. Low beta 12–20 Hz and high beta 20–35 Hz sources maxima were clustered in the sensorimotor cortex and premotor cortex ipsilateral to the STN.	IV	B	Hirschmann et al.	2011	Neuroimage

14	10 (F 4, M6; 64.1; 53–75). The patients who underwent subthalamic nucleus deep brain stimulation electrode implantation bilaterally prior to deep brain stimulation (DBS) therapy	(–)	306Ch(102sites, planar gradiometer), Neuromag (Eleckta Oy) MEG signals of 204 gradiometer channels were used for the analysis	Spontaneous and movement related	Subthalamic nucleus (STN) electrode local field potentials (LFP), MEGs, and EMG of the extensor digitorum communis and flexor digitorum superficialis muscles of both upper limbs were simultaneously recorded the day after electrode implantation.	For three recording conditions of resting, holding the hand, and moving task (opening and closing one's fist), two cortical regions, M1 and superior temporal gyrus (STG), were investigated in terms of coherence between the STN-LFP or EMG channel. To locate coherent cortical sources spatially, the dynamic imaging of coherent sources (DICS) beamforming method was used.	The beta coherence between STN-LFP and M1 was identified during motor task and decreased after levodopa medication. M1–muscular coherence in alpha and beta frequency bands was decreased by movement but unchanged by medication. STG showed strong alpha band coherence with STN in all experimental conditions, and STG-STN coherence was neither modulated by administration of levodopa nor by motor task.	IV	B	Hirschmann et al.	2013	Neuroimage

(continued)

Table 12.3 (continued)

No	Number of patients (gender, mean age, range)	Number of control subjects (gender, mean age, range)	MEG apparatus	MEG modality (evoked/ spontaneous)	Recording method	Analysis	Main results	Levels of evidence	Grades of recommendation	Authors	Year	Journal
15	11 (M 11; 64.5; 53–74). The patients who underwent subthalamic nucleus deep brain stimulation electrode implantation prior to deep brain stimulation (DBS) therapy. 7 patients were bilaterally treated	(–)	306Ch(102sites, planar gradiometer), Neuromag (Eleckta Oy) MEG signals of 204 gradiometer channels were used for the analysis	Spontaneous and movement related	Subthalamic nucleus (STN) electrode local field potentials (LFP), MEGs, and EMG of the extensor digitorum communis and flexor digitorum superficialis muscles of both upper limbs were simultaneously recorded the day after electrode implantation.	Epochs containing spontaneous rest tremor and tremor-free epochs were analyzed. A set of 24 gradiometers covering sensorimotor cortex contralateral to the tremulous limb was selected a priori as MEG sensors of interest to investigate cerebro–muscular coherence. To locate coherent cortical sources spatially, the dynamic imaging of coherent sources (DICS) beamforming method was used.	For the sensor level analysis, prior to sudden increases of tremor amplitude, MEG beta power decreased. When tremor amplitude increased, MEG power at tremor frequency increased. For the source level analysis, increases in cerebral synchronization were observed in a rest tremor network including subthalamic nucleus, primary motor, premotor, and posterior parietal cortex contralateral to the tremulous limb. Analysis of the imaginary part of coherency revealed tremor-dependent coupling between these cortical areas at tremor frequency and double the tremor frequency. The tremor-associated increase in STN-cortex coherence was positively correlated with the tremor-associated increase in muscle activity.	IV	B	Hirschmann et al.	2013	Brain

16	12(F 6, M6; 62; 49–75) The patients who underwent subthalamic nucleus deep brain stimulation electrode implantation bilaterally for deep brain stimulation (DBS) therapy	(–)	306Ch(102sites, planar gradiometer + magnetometer), Neuromag (Eleckta Oy)	Auditory, somatosensory, and visual evoked	Auditory or somatosensory-evoked magnetic fields were compared between DBS ON and off state.	1-kHz sinusoidal 50-ms tone pips, electrical shocks to the median nerve bilaterally, and visual checkerboard stimuli were randomly given with irregular interstimulus interval (ISI) (the mean ISI, 5.5 s). N100m responses of auditory-evoked fields and N20m or P60m of somatosensory-evoked fields were analyzed.	The ipsilateral auditory N100m responses in the right hemisphere were augmented by 10 % following DBS ON. Contralateral N100m responses and somatosensory P60m responses also had a tendency to increase when bilateral DBS was on.	IV	B	Airaksinen et al.	2011	Hum Brain Mapp
17	11(F 6, M5; 61; 49–75) The patients who underwent subthalamic nucleus deep brain stimulation electrode implantation bilaterally for deep brain stimulation (DBS) therapy	(–)	306Ch(102sites, planar gradiometer + magnetometer), Neuromag (Eleckta Oy)	Spontaneous	MEGs when eyes closed or opened were compared between DBS ON and off state.	The spatiotemporal signal space separation (tSSS) method was applied to reduce the artifact by DBS when recording MEGs. Spontaneous MEG activity was recorded for 3 min when the patient's eyes were closed and 5 min with eyes open.	When DBS was turned on, the mean source strengths in the 6–10 Hz range (mu rhythm) and in the lower and higher beta ranges over the pericentral cortical regions decreased nonsignificantly. During DBS on, UPDRS (Unified Parkinson's Disease Rating Scale) motor disability rigidity scores correlated with 6–10 Hz and 12–20 Hz somatomotor source strengths when eyes were open. During DBS off, UPDRS action tremor scores correlated with the pericentral 6–10 Hz and 12–20 Hz and occipital alpha source strength when eyes were open.	IV	B	Airaksinen et al.	2012	Clin Neurophysiol

(continued)

Table 12.3 (continued)

No	Number of patients (gender, mean age, range)	Number of control subjects (gender, mean age, range)	MEG apparatus	MEG modality (evoked/ spontaneous)	Recording method	Analysis	Main results	Levels of evidence	Grades of recommendation	Authors	Year	Journal
18	19(F 4, M 15; 57; 36–71) The patients who underwent subthalamic nucleus deep brain stimulation electrode implantation bilaterally for deep brain stimulation (DBS) therapy	(–)	306Ch(102sites, planar gradiometer + magnetometer), Neuromag (Eleckta Oy)	Spontaneous and movement related	Spontaneous MEGs and electromyograms (EMG) from the extensor carpi radialis longus muscle were simultaneously recorded for 1 min during wrist extension period and 20 s rest period, respectively. Effects of DBS on and off were investigated.	Power spectral density values of MEG and rectified EMG and the MEG-EMG coherence spectra were calculated for each patient both in the DBS on and off conditions. The sensor level analysis for 4–6 Hz, 6–13 Hz, and 13–25 Hz frequency bands was performed for a selection of 15 gradiometer pairs over the sensorimotor cortex contralateral to the activated hand.	Corticomotor coherence (CMC) peaks for 13–25 Hz were found in 15 out of 19 patients. CMC peaks for 6–13 Hz also were found in 15 patients. CMC peaks for 4–6 Hz were found in 5 patients. The effect of DBS on these corticomotor coherence peaks was variable among patients.	IV	B	Airaksinen et al.	2014	Clin Neurophysiol

functional connectivity and found lower leaf number and lower tree hierarchy in the alpha2 (10–13 Hz) frequency band in PD patients compared to controls [17]. A 4-year longitudinal analysis in PD patients demonstrated progressive decreases in local clustering in multiple frequency bands concurrently with decreases in path length in the alpha2 (10–13 Hz) frequency band [17]. More recently, Olde Dubbelink et al. [18] analyzed the results of MEG testing and cognitive functions in PD patients over a 7-year period and evaluated predictive factors for the development of PD-related dementia (PDD). The authors concluded that the combination of impaired fronto-executive task performance and low beta power was associated with the highest dementia risk [18]. Vardy et al. [19] assessed primary motor cortex activity during rest and rhythmic movement in PD patients. This study reported that PD patients displayed more power in the alpha (7–11 Hz) band and less power in the beta (13–30 Hz) band during both rest and motor activity compared to controls, indicating that the slowing of neural activity is a structural, systemic phenomenon in PD that progresses over time [19].

Several MEG studies from other institutes have used different methods to investigate the pathophysiology of movement disorders in PD. One approach for understanding the mechanisms of tremor or hypokinesia is the simultaneous recording of spontaneous MEG and electromyograms (EMG) from the hand muscle of the more severely affected side in PD. Pollok et al. measured coherent activity between the primary motor hand area (M1) and the hand muscle EMG or between multiple motor-related brain areas during muscle contraction or rest [20] with or without the administration of the study drug [21]. In control subjects, the beta frequency power of M1 contralateral to movement was decreased during muscle contraction compared to the ipsilateral side [20]. This pattern appeared to be attenuated in de novo PD patients and fully reversed in medicated PD patients [20]. During the period without drug administration in PD patients, EMG power spectrum analysis revealed discernible peaks at the tremor frequency (4.8 Hz) and double the tremor frequency (9.4 Hz) [21]. Coherent activity at double the tremor frequency was found in the contralateral primary sensorimotor cortex (M1/S1) and several other areas: the premotor cortex (PMC), supplementary motor area (SMA), secondary somatosensory cortex (S2), posterior parietal cortex (PPC), and thalamus contralateral to the EMG recorded site, and ipsilateral cerebellum [21]. The coherence strength of each cerebro–cerebral coupling was seen to decrease with improvements in tremor during drug administration (Table 12.3) [21].

Deep brain stimulation (DBS) neurosurgery allows the assessment of the interactions between populations of neurons in the human cerebral cortex and basal ganglia in PD patients. Another approach for understanding the pathogenesis of motor disturbances in PD is the scrutinization of cortico-basal ganglia network activity. There have been five papers that have used simultaneous recording of MEG and local field potentials (LFP) of the STN to map cortico-STN coherence. Litvak et al. [22] identified two spatially and spectrally separated cortico-STN networks: a temporoparietal-STN network in the alpha (7–13 Hz) band and a predominantly frontal-STN network in the beta (15–35 Hz) band. Dopaminergic medications were shown to modulate the resting beta network by increasing beta coherence between the STN region and ipsilateral prefrontal cortex. Litvak et al. [23] further investigated cortico-STN coherence in PD patients during the performance of synchronous and sequential finger movements or during the administration of the dopamine prodrug, levodopa. Discrete peaks in M1 and STN power were observed at 60–90 Hz and at 300–400 Hz. All power peaks increased with either synchronous or sequential finger movement and levodopa treatment. Only STN activity at 60–90 Hz was coherent with activity in M1. Based on directionality analysis, STN gamma activity at 60–90 Hz was found to contribute to gamma activity in M1 [23]. Hirschmann et al. also studied cortico-STN coherence using simultaneous MEG and STN-LFP recording [24–26]. During rest, coherent activity in the low (12–20 Hz) and high (20–35 Hz) beta range was observed in the sensorimotor and premotor cortex on the ipsilateral side to STN-LFP recording [24]. Coherence in the alpha range (7–12 Hz) was observed at various locations in the ipsilateral temporal lobe [24]. During the motor task, beta coherence between STN-LFP and M1 was identified and seen to decrease following the administration of levodopa [25]. M1-muscular coherence in alpha and beta frequency bands was decreased by movement but was unchanged by medication [25]. This study also observed strong alpha band coherence between the superior temporal gyrus (STG) and STN in all experimental conditions and that motor tasks and the administration of levodopa had no effect on STG-STN coherence [25]. Hirschmann et al. further investigated cortical activity coherent to EMG spectral power produced by tremor [26]. Increases in cerebral synchronization at tremor frequencies were observed in a rest tremor network that included the STN, M1, premotor, and posterior parietal cortex contralateral to the tremulous limb [26]. Analysis of the imaginary part of coherency revealed tremor-dependent coupling between these cortical areas at tremor frequency and double the tremor frequency [26].

An alternative approach to the investigation of PD pathophysiology was undertaken by Airaksinen et al. [27–29] in a study comparing brain responses with or without DBS. In this study, a spatiotemporal signal space separation (tSSS) method was used to reduce DBS artifacts during MEG recording. When assessing auditory-evoked fields, the ipsilateral auditory N100m responses in the right hemisphere were found to be augmented by 10 % during DBS [27]. A trend toward increased contralateral N100m responses and somatosensory P60m responses was observed in response to bilateral DBS [27]. Spontaneous MEG during DBS demonstrated a nonsignificant decrease in mean source strength at the 6–10 Hz range (mu rhythm) and lower and higher beta ranges over pericentral cortical regions [28]. During DBS, the severity of rigidity correlated with 6–10 Hz and 12–20 Hz somatomotor source strengths when patients had their eyes open [28] (Table 12.3). When DBS was not being applied, action tremor severity correlated with pericentral 6–10 Hz and 12–20 Hz and occipital alpha source strength when patients had their eyes open [28]. By recording MEG during rest and movement, Airaksinen et al. [29] demonstrated sensor-space corticomotor coherence peaks at 13–25 Hz in 15 out of 19 PD patients. In addition, corticomotor coherence peaks at 6–13 Hz and 4–6 Hz were observed in 15 and 5 patients, respectively. The effect of DBS on corticomotor coherence peaks was variable among individual patients [29].

12.4 Magnetoencephalography (MEG) in Multiple Sclerosis (MS)

12.4.1 Clinical and Pathophysiological Features of Multiple Sclerosis (MS)

Multiple sclerosis (MS) is a common chronic demyelinating disease of the central nervous system (CNS) with a highly variable clinical course. MS is characterized by the spatial and temporal progression of lesions affecting the brain, spinal cord, and/or optic nerves [30]. Exacerbations and remissions occur frequently. The symptoms and signs of MS usually indicate the presence of more than one lesion, and some of them may be transient. Multiple sclerosis is a clinical diagnosis that requires appropriate expertise to confirm the spatial and temporal progression of CNS lesions and exclude other possible diseases [30]. Magnetic resonance imaging (MRI) is essential for the diagnosing of MS as it allows the visualization of MS plaques representing white matter inflammation, demyelination, and glial scarring (sclerosis). The identification of oligoclonal immunoglobulin G bands in cerebrospinal fluid may support the diagnosis of MS. Although autoimmune processes are thought to underlie the pathogenesis of MS, there are currently no serological tests with utility in diagnosing MS [30]. For the 10–20 years before MRI was introduced as a diagnostic tool for MS, evoked potentials were used as an important diagnostic tool for the detection of clinically silent CNS lesions. Currently, evoked potentials are considered less sensitive than MRI. However, visual-evoked potentials are still considered to have utility in providing evidence of optic nerve demyelination through the demonstration of markedly delayed P100 wave of normal amplitude [30].

12.4.2 MEG Studies in Patients with Multiple Sclerosis

A summary of nine MEG studies of MS is presented in Table 12.4. Two different approaches to the use of MEG to study MS were identified. Three of the nine papers focus on cortical somatosensory network activity following electrical finger stimulation or median nerve stimulation. Tecchio et al. [31] identified source activity representing the thumb and little finger at around 24 ms poststimulus and estimated sensory cortical connectivity as the phase locking between these source activities in the gamma frequency range. In this study, an altered pattern of the intracortical connectivity index was observed in MS patients (see Table 12.4) compared to controls. Dell'Acqua et al. [32] examined the profiles of M20 and M30 responses following median nerve stimulation in MS patients. Although the latency and signal strength for M20 were not affected, the analysis of M30 responses demonstrated prolonged latency, decreased signal strength, and asymmetry of right and left M30 dipole locations [32]. Hagiwara et al. [33] analyzed contralateral SI and bilateral SII responses following median nerve stimulation. In MS patients, the mean latencies of all contralateral SI responses were prolonged, the signal strength of the N20m

Table 12.4 MEG in patients with multiple sclerosis

No	Number of patients (gender, mean age, range)	Number of control subjects (gender, mean age, range)	MEG apparatus	MEG modality (evoked/ spontaneous)	Recording method	Analysis	Main results	Levels of evidence	Grades of recommendation	Authors	Year	Journal
1	21, right handed the relapsing remitting form of MS (RRMS) (EDSS<3.5) (F 16, M5; mean age 40)	14 right handed (F 7, M 7;38 ± 8)	28-channel system	Somatosensory evoked(thumb or little finger electrical stimulation)	Somatosensory-evoked fields were recorded from the right/ left central scalp by electrical stimulation of the thumb or little finger of the right/left hand (interstimulus interval of 641 ms).	Source activity representing the thumb and little finger was identified at around 24 ms poststimulus. The sensory cortical connectivity was estimated as the phase locking between these source activities in the gamma frequency range.	Altered pattern of the intracortical connectivity index was found in MS patients. The intracortical connectivity index in the primary sensory networks devoted to the thumb in the left hemisphere was higher than those related to the little finger in controls. This pattern vanished in MS patients.	IIb	B	Tecchio et al.	2008	Brain
2	21, the relapsing remitting form of MS (RRMS) (EDSS<3.5) (F 16, M 5; 39 ± 9)	Control 21 (F 16, M 5; 38 ± 12)	28-channel system	Somatosensory evoked (median nerve)	Somatosensory-evoked fields were recorded from the right/ left central scalp by electrical stimulation of the right/left median nerve at wrist (at around 2 Hz).	The strength and location for equivalent current dipoles of M20 and M30 were analyzed.	The latency and signal strength for M20 were not affected in MS patients. The analysis of M30 response in MS patients showed the prolonged latency, decreased signal strength, and asymmetry of right and left M30 dipole locations.	IIb	B	Dell'Acqua et al.	2010	Hum Brain Mapp

3	23, definite MS (F 18, M 5; 38.8 ± 8.1)	23 (F 18, M 5; 37.3 ± 10.6)	204 Ch, (102 sites, planar gradiometer) Neuromag (Elekta Neuromag)	Somatosensory evoked (median nerve)	Somatosensory-evoked fields were recorded by electrical stimulation of the right/left median nerve at wrist with interstimulus intervals ranged from 2.5 to 3.5 s (mean interval: 3 s).	The latency and strength for equivalent current dipoles of SI response and SII response were analyzed. The phase locking values of the induced gamma-band activity were also analyzed.	In MS patients, the mean latencies of all of the contralateral SI responses were prolonged in MS patients. By contrast, there were no differences in the latencies of bilateral SII responses. The signal strength of the N20m response was decreased, and induced-gamma activity of SI was relatively reduced in MS patients. The phase locking values of the induced gamma-band activity between SI and SII increased during the time interval of 30–100 ms poststimulus in controls; such an increase in phase locking values was diminished in the MS patients.	IIb	B	Hagiwara et al.	2010	Neuroimage

(continued)

Table 12.4 (continued)

No	Number of patients (gender, mean age, range)	Number of control subjects (gender, mean age, range)	MEG apparatus	MEG modality (evoked/ spontaneous)	Recording method	Analysis	Main results	Levels of evidence	Grades of recommendation	Authors	Year	Journal
4	34, definite MS (F 17, M17; 41.4 ± 8.0)	28 (F 14, M 14; 39.8 ± 10.5)	151Ch, radial gradiometer system (CTF)	Spontaneous	Spontaneous MEGs of 30 s were digitally filtered into standard frequency bands, and sensor-space (artifact-free 137 Ch data) analyses were performed for bilateral frontal, central, parietal, temporal, occipital regions.	The results were band-pass filtered into delta (0.5–4 Hz), theta (4–8 Hz), low alpha (8–10 Hz), upper alpha (10–13 Hz), beta (13–30 Hz), and gamma (30–48 Hz) bands. Functional connectivity between MEG sensors was assessed as calculating synchronization likelihood (SL). The resulting weight matrix was used to compute eigenvector centrality (EC).	In controls, from the θ to the γ band, the pattern of distribution showed highest EC values that were found over both parietal areas and medium high EC values over both temporal regions (a left dominance). In MS patients, EC values in theta band were found higher over both parietal areas; EC values in upper alpha and beta bands were lower over left temporal regions and so were in gamma band over right parietal regions.	IIb	B	Hardmeier et al.	2012	Plos One

5	34, definite MS (F 17, M17; 41.4 ± 8.0)	28 (F 14, M 14; 39.8 ± 10.5)	151Ch, radial gradiometer system (CTF)	Spontaneous	Spontaneous MEGs of 30 s were digitally filtered into standard frequency bands, and sensor-space (artifact-free 137 Ch data) analyses were performed for bilateral frontal, central, parietal, temporal, occipital regions.	MEGs were band-pass filtered into delta (0.5–4 Hz), theta (4–8 Hz), low alpha (8–10 Hz), upper alpha (10–13 Hz), beta (13–30 Hz), and gamma (30–45 Hz) bands. Functional connectivity was assessed by calculation of synchronization likelihood (SL) and graph theoretical analysis.	Increased synchronization in the theta, low alpha, and beta bands and decreased synchronization in the upper alpha band were found in MS patients. For the graph theoretical analysis, the lower alpha band displayed increased clustering coefficient and path length values, indicating a change toward a more regular network topology in MS.	IIb	B	Schoonheim et al.	2013	Hum Brain Mapp
6	21, definite MS (41.9 ± 7.7), selected from 34 definite MS (F 17, M17; 41.4 ± 8.0) [34]	17 (39.8 ± 9.8), selected from 28 (F 14, M 14; 39.8 ± 10.5) [34]	151Ch, radial gradiometer system (CTF)	Spontaneous	>25 eye-closed MEG epochs of 6.6 s were analyzed in conjunction with subject's MRI using a beamforming approach (synthetic aperture magnetometry, SAM). Source-space analyses were performed for 78 cortical regions.	Beamforming analysis was performed for delta (0.5–4 Hz), theta (4–8 Hz), alpha1 (8-10 Hz), alpha2 (10–13 Hz), beta (13–30 Hz), and gamma (30–48 Hz) bands. The phase lag index (PLI) was obtained to calculate the asymmetry of the distribution of phase differences between two time series.	Mean PLI was lower in the alpha2 band and higher in the beta band in MS patients. Lower functional connectivity in the alpha2 band was found in the default mode network (DMN) and the visual processing network. Higher functional connectivity in the beta band was found in the DMN and the temporo-parietal network.	III	B	Tewarie et al.	2013	PLoS One

(continued)

Table 12.4 (continued)

No	Number of patients (gender, mean age, range)	Number of control subjects (gender, mean age, range)	MEG apparatus	MEG modality (evoked/ spontaneous)	Recording method	Analysis	Main results	Levels of evidence	Grades of recommendation	Authors	Year	Journal
7	21, definite MS (41.9 ± 7.7), selected from 34 definite MS (F 17, M17; 41.4 ± 8.0) [34]	17 (39.8 ± 9.8), selected from 28 (F 14, M 14; 39.8 ± 10.5) [34]	151Ch, radial gradiometer system (CTF)	Spontaneous	>25 eye-closed MEG epochs of 6.6 s were analyzed in conjunction with subject's MRI using a beamforming approach (Synthetic Aperture Magnetometry, SAM). Source-space analyses were performed for 78 cortical regions.	Beamforming analysis was performed for delta (0.5–4 Hz), theta (4–8 Hz), alpha 1 (8–10 Hz), alpha 2 (10–13 Hz), beta (13–30 Hz), and gamma (30–45 Hz) bands. The phase lag index (PLI) was obtained. Minimum spanning tree (MST) network analyses were added.	MSTs differ between MS patients and controls in the alpha2, beta, and theta bands. The MSTs in the alpha2 band of MS patients were characterized by a larger eccentricity and lower leaf fraction and "tree hierarchy." These changes suggestive of a loss of hierarchical structure were associated with poorer cognitive performance.	III	B	Tewarie et al.	2014	Neuroimage

8	102 (68 relapsing–remitting MS/22 secondary-progressive MS /12 primary-progressive MS) (F 65, M37; 54.23 ± 9.76)	42 (F 26, M16; 51.1 ± 5.98)	306Ch, (102 sites, planar gradiometer) Neuromag (Elekta Neuromag)	Spontaneous	Eye-closed MEG epochs of approximate 5 min were analyzed in conjunction with subject's MRI. Using a scalar beamformer implementation (Elekta Neuromag), source-space analyses were performed for 78 cortical regions.	Beamforming analysis was performed for delta (0.5–4 Hz), theta (4–8 Hz), alpha1 (8–10 Hz), alpha2 (10–13 Hz), beta (13–30 Hz), and gamma (30–45 Hz) bands. The phase lag index (PLI) was obtained. Minimum spanning tree (MST) network analyses were added.	In MS patients, higher PLI values were present in the delta band in many cortical areas, except for right temporal and occipital areas, and in the theta band in many cortical areas. Lower PLI values were found in the alpha2 band in the occipital, temporal, and parietal areas. MST analyses revealed that MST topology was only different in the alpha2 band for MS patients, reflecting a lower leaf fraction, lower degree divergence, and lower tree hierarchy in the alpha2 frequency band for MS patients.	IIb	B	Tewarie et al.	2014	Human Brain Mapp

(continued)

Table 12.4 (continued)

No	Number of patients (gender, mean age, range)	Number of control subjects (gender, mean age, range)	MEG apparatus	MEG modality (evoked/ spontaneous)	Recording method	Analysis	Main results	Levels of evidence	Grades of recommendation	Authors	Year	Journal
9	86 (71 relapsing–remitting MS/5 secondary-progressive MS /10 primary-progressive MS) (41.6 ± 8.8)	21 (42.5 ± 10.3)	306Ch, (102 sites, planar gradiometer) Neuromag (Elekta Neuromag)	Spontaneous	Eye-closed MEG epochs of approximate 5 min were analyzed in conjunction with subject's MRI. Using a scalar beamformer implementation (Elekta Neuromag), source-space analyses were performed for 78 cortical regions.	Beamforming analysis was performed for delta (0.5–4 Hz), theta (4–8 Hz), alpha1 (8–10 Hz), alpha2 (10–13 Hz), beta (13–30 Hz), and gamma (30–45 Hz) bands. The phase lag index (PLI) was obtained. Minimum spanning tree (MST) network analyses were added.	As compared to controls, the cortical functional connectivity was higher in theta band and lower in gamma band in MS patients. MST analyses revealed lower leaf fraction for the delta, theta, and alpha2 bands in MS patients: A lower degree divergence also was found in the delta, theta, alpha1, and alpha2.	IIb	B	Tewarie et al.	2015	Human Brain Mapp

response was decreased, and induced SI gamma activity was relatively reduced [33]. Although the latencies of bilateral SII responses were within the normal range, phase locking in the induced gamma-band activity between SI and SII during the time interval of 30–100 ms poststimulus was diminished in MS patients, indicating impaired cortical somatosensory network activity.

The remaining six of the nine MEG studies in MS were published by the VU University Medical Center in Amsterdam [34–39]. Hardmeier et al. [34] performed sensor-space analyses of frequency power spectrum and functional connectivity in spontaneous MEG. Functional connectivity between MEG sensors was assessed by calculating the synchronization likelihood (SL), and the resulting weight matrix was used to compute eigenvector centrality (EC) [34]. Eigenvector centrality values in the theta (θ) band were higher over both parietal areas in MS patients compared to controls. Further, EC values in the upper alpha (α) (8–10 Hz) and beta (β) (13–30 Hz) bands over left temporal regions, and the gamma (γ) (30–48 Hz) band over right parietal regions, were lower in MS patients compared to controls [34]. In a further study of the same MS patients and controls, Schoonheim et al. [35] performed graph theoretical analysis to assess functional connectivity. Sensor-space analyses of the frequency power spectrum demonstrated increased synchronization in the theta (θ) (4–8 Hz),low alpha (α) (8–10 Hz), and beta (β) (10–13 Hz) bands and decreased synchronization in the upper alpha (α) (10–13 Hz) band in MS patients compared to controls [35]. In the graph theoretical analysis, the lower alpha (α) (8–10 Hz) band demonstrated increased clustering coefficient and path length values, indicating a change toward a more regular network topology in MS patients. Tewarie et al. published four studies of spontaneous MEG in MS patients focusing on network functional connectivity [36–39]. Using a beamforming approach (synthetic aperture magnetometry, SAM), source-space analyses were performed in 78 cortical regions. First, the phase lag index (PLI) was determined to calculate the asymmetry of the distribution of phase differences between the two time series [36]. Lower functional connectivity (lower PLI) was observed in the α2 band in the default mode network (DMN), and the visual processing network and higher functional connectivity (higher PLI) were observed in the beta (β) (13–30 Hz) band in the DMN and the temporoparietal network in MS patients [36]. The authors posited that altered functional connectivity may underlie the clinical and cognitive dysfunction in MS. In the second paper, minimum spanning tree (MST) network analyses were performed [37]. MSTs were found to differ between MS patients and controls in the alpha2 (α2) (10–13 Hz), beta (β) (13–30 Hz), and theta (θ) (4–8 Hz) bands [37]. The MSTs in the alpha2 (α2) (10–13 Hz) band of MS patients were characterized by a larger eccentricity and lower leaf fraction and "tree hierarchy" [37]. These changes indicated a loss of hierarchical structure and were associated with poor cognitive performance. Similar findings were reported by a further study with a large number of MS patients and controls [38, 39] (Table 12.4). In MS patients, higher PLI values were present in the delta (δ) (0.5–4 Hz) band in many cortical areas, except for the right temporal and occipital areas, and in the theta (θ) (4–8 Hz) band in many cortical areas [38]. Lower PLI values were observed in the alpha2 (α2) (10–13 Hz) band in occipital, temporal, and parietal areas. MST analyses demonstrated different MST topology only in the alpha2 (α2) band in MS patients, reflecting a lower leaf fraction, lower degree of divergence, and lower tree hierarchy in the alpha2 (α2) frequency band of MS patients [38]. A lower degree of divergence also was observed in all frequency bands, except the gamma (γ) (30–48 Hz) band, in MS patients [39].

12.5 General Remarks

The present review of studies examining the clinical application of MEG in neurodegenerative diseases such as ALS, PD, and MS reveals the future potential and limitations of MEG as a diagnostic tool or neurophysiological marker. The simultaneous recording of MEG and STN-LFP in PD patients who underwent neurosurgery for STN-DBS provided an opportunity to explore functional network activity between the STN and distinct cortical areas. These studies in PD patients, in conjunction with experimental studies of MPTP parkinsonian primates, have provided invaluable data allowing the testing of emerging hypotheses regarding the pathogenesis of hypokinesia or bradykinesia in PD [22–25], the "abnormal firing pattern model," and novel insights into the pathogenesis of rest tremor in PD [20, 21, 26]. In addition, regardless of diagnosis such as PD and MS, patients with cognitive decline or impairment demonstrated altered or disruptive network functional connectivity during spontaneous MEG [15–18, 36–39]. Thus, functional connectivity analyses using spontaneous MEG may provide data with utility in informing the diagnosis of PD-related dementia or the presence of frontotemporal dementia during the early stages of ALS. Further, spontaneous MEG may also have clinically utility in diagnosing and predicting cognitive decline in patients suffering from other neurodegenerative diseases including multisystem atrophy, spinocerebellar degeneration, and progressive supranuclear palsy. However, network functional connectivity analyses of spontaneous MEG in PD patients or MS patients were repeatedly performed in the same institute with MEG examinations performed on a limited number of patients and control subjects. Therefore, studies from other institutes or collaborations between many institutions with a large number of participants are required to evaluate the validity of functional connectivity analyses and confirm the relationship between abnormal functional connectivity analysis results of spontaneous MEG and cognitive decline or deficits, thereby enhancing the clinical utility of MEG examinations.

References

1. McCormick KA, Fleming B. Clinical practice guidelines. The Agency for Health Care Policy and Research fosters the development of evidence-based guidelines. Health Prog. 1992;73: 30–4.
2. Brown RH, Cannon SC, Rowland LP. Chapter 14: Diseases of the nerve and motor unit. In: Kandel ER, Schwartz JH, Jessell TM, Siegelbaum SA, Hudspeth AJ, editors. Principles of neural sciences. 5th ed. New York: McGraw-Hill Professional; 2012. p. 307–30.

3. Gordon PH. Amyotrophic lateral sclerosis: an update for 2013 clinical features, pathophysiology. Manag Ther Trials Aging Dis. 2013;4(5):295–310.
4. Pekkonen E, Osipova D, Laaksovirta H. Magnetoencephalographic evidence of abnormal auditory processing in amyotrophic lateral sclerosis with bulbar signs. Clin Neurophysiol. 2004; 115(2):309–15.
5. Boyajian RA, Amo C, Otis SM, Romine JS, Smith RA. Magnetic source imaging of cortical dysfunction in amyotrophic lateral sclerosis. Am J Phys Med Rehabil. 2008;87(6):427–37.
6. Teismann IK, Warnecke T, Suntrup S, Steinsträter O, Kronenberg L, Ringelstein EB, Dengler R, Petri S, Pantev C, Dziewas R. Cortical processing of swallowing in ALS patients with progressive dysphagia—a magnetoencephalographic study. PLoS One. 2011;6(5): e19987. doi:10.1371/journal.pone.0019987.
7. Wichmann T, DeLong MR. Chapter 43: The basal ganglia. In: Kandel ER, Schwartz JH, Jessell TM, Siegelbaum SA, Hudspeth AJ, editors. Principles of neural sciences. 5th ed. New York: McGraw-Hill Professional; 2012. p. 982–98.
8. Gale JT, Amirnovin R, Williams ZM, Flaherty AW, Eskandar EN. From symphony to cacophony: pathophysiology of the human basal ganglia in Parkinson disease. Neurosci Biobehav Rev. 2008;32(3):378–87.
9. Mink JW. The basal ganglia: focused selection and inhibition of competing motor programs. Prog Neurobiol. 1996;50(4):381–425.
10. Braak H, Ghebremedhin E, Rüb U, Bratzke H, Del Tredici K. Stages in the development of Parkinson's disease-related pathology. Cell Tissue Res. 2004;318(1):121–34.
11. Hawkes CH, Del Tredici K, Braak H. Parkinson's disease: a dual-hit hypothesis. Neuropathol Appl Neurobiol. 2007;33(6):599–614.
12. Stoffers D, Bosboom JL, Deijen JB, Wolters EC, Berendse HW, Stam CJ. Slowing of oscillatory brain activity is a stable characteristic of Parkinson's disease without dementia. Brain. 2007;130(Pt 7):1847–60.
13. Boesveldt S, Stam CJ, Knol DL, Verbunt JP, Berendse HW. Advanced time-series analysis of MEG data as a method to explore olfactory function in healthy controls and Parkinson's disease patients. Hum Brain Mapp. 2009;30(9):3020–30.
14. Gómez C, Olde Dubbelink KT, Stam CJ, Abásolo D, Berendse HW, Hornero R. Complexity analysis of resting-state MEG activity in early-stage Parkinson's disease patients. Ann Biomed Eng. 2011;39(12):2935–44.
15. Olde Dubbelink KT, Stoffers D, Deijen JB, Twisk JW, Stam CJ, Berendse HW. Cognitive decline in Parkinson's disease is associated with slowing of resting-state brain activity: a longitudinal study. Neurobiol Aging. 2013;34(2):408–18.
16. Olde Dubbelink KT, Stoffers D, Deijen JB, Twisk JW, Stam CJ, Hillebrand A, Berendse HW. Resting-state functional connectivity as a marker of disease progression in Parkinson's disease: a longitudinal MEG study. Neuroimage Clin. 2013;2:612–9.
17. Olde Dubbelink KT, Hillebrand A, Stoffers D, Deijen JB, Twisk JW, Stam CJ, Berendse HW. Disrupted brain network topology in Parkinson's disease: a longitudinal magnetoencephalography study. Brain. 2014;137(Pt 1):197–207.
18. Olde Dubbelink KT, Hillebrand A, Twisk JW, Deijen JB, Stoffers D, Schmand BA, Stam CJ, Berendse HW. Predicting dementia in Parkinson disease by combining neurophysiologic and cognitive markers. Neurology. 2014;82(3):263–70.
19. Vardy AN, van Wegen EE, Kwakkel G, Berendse HW, Beek PJ, Daffertshofer A. Slowing of M1 activity in Parkinson's disease during rest and movement – an MEG study. Clin Neurophysiol. 2011;122(4):789–95.
20. Pollok B, Makhloufi H, Butz M, Gross J, Timmermann L, Wojtecki L, Schnitzler A. Levodopa affects functional brain networks in Parkinsonian resting tremor. Mov Disord. 2009;24(1): 91–8.
21. Pollok B, Krause V, Martsch W, Wach C, Schnitzler A, Südmeyer M. Motor-cortical oscillations in early stages of Parkinson's disease. J Physiol. 2012;590(Pt 13):3203–12.

22. Litvak V, Jha A, Eusebio A, Oostenveld R, Foltynie T, Limousin P, Zrinzo L, Hariz MI, Friston K, Brown P. Resting oscillatory cortico-subthalamic connectivity in patients with Parkinson's disease. Brain. 2011;134(Pt2):359–74.
23. Litvak V, Eusebio A, Jha A, Oostenveld R, Barnes G, Foltynie T, Limousin P, Zrinzo L, Hariz MI, Friston K, Brown P. Movement-related changes in local and long-range synchronization in Parkinson's disease revealed by simultaneous magnetoencephalography and intracranial recordings. J Neurosci. 2012;32(31):10541–53.
24. Hirschmann J, Özkurt TE, Butz M, Homburger M, Elben S, Hartmann CJ, Vesper J, Wojtecki L, Schnitzler A. Distinct oscillatory STN-cortical loops revealed by simultaneous MEG and local field potential recordings in patients with Parkinson's disease. Neuroimage. 2011;55(3):1159–68.
25. Hirschmann J, Özkurt TE, Butz M, Homburger M, Elben S, Hartmann CJ, Vesper J, Wojtecki L, Schnitzler A. Differential modulation of STN-cortical and cortico-muscular coherence by movement and levodopa in Parkinson's disease. Neuroimage. 2013;68:203–13.
26. Hirschmann J, Hartmann CJ, Butz M, Hoogenboom N, Ozkurt TE, Elben S, Vesper J, Wojtecki L, Schnitzler A. A direct relationship between oscillatory subthalamic nucleus-cortex coupling and rest tremor in Parkinson's disease. Brain. 2013;136(Pt 12):3659–70.
27. Airaksinen K, Mäkelä JP, Taulu S, Ahonen A, Nurminen J, Schnitzler A, Pekkonen E. Effects of DBS on auditory and somatosensory processing in Parkinson's disease. Hum Brain Mapp. 2011;32(7):1091–9.
28. Airaksinen K, Butorina A, Pekkonen E, Nurminen J, Taulu S, Ahonen A, Schnitzler A, Mäkelä JP. Somatomotor mu rhythm amplitude correlates with rigidity during deep brain stimulation in Parkinsonian patients. Clin Neurophysiol. 2012;123(10):2010–7.
29. Airaksinen K, Mäkelä JP, Nurminen J, Luoma J, Taulu S, Ahonen A, Pekkonen E. Cortico-muscular coherence in advanced Parkinson's disease with deep brain stimulation. Clin Neurophysiol. 2014 (in press). doi:http://dx.doi.org/10.1016/j.clinph.2014.07.025.
30. Leary S, Giovannoni G, Howard R, Miller D, Thompson A. Chapter 10: Multiple sclerosis and demyelinating diseases. In: Clarke C, Howard R, Rossor M, Shorvon SD, editors. Neurology: a queen square textbook. Singapore: Willy-Blackwell; 2009. p. 411–47.
31. Tecchio F, Zito G, Zappasodi F, Dell' Acqua ML, Landi D, Nardo D, Lupoi D, Rossini PM, Filippi MM. Intra-cortical connectivity in multiple sclerosis: a neurophysiological approach. Brain. 2008;131(Pt 7):1783–92.
32. Dell'Acqua ML, Landi D, Zito G, Zappasodi F, Lupoi D, Rossini PM, Filippi MM, Tecchio F. Thalamocortical sensorimotor circuit in multiple sclerosis: an integrated structural and electrophysiological assessment. Hum Brain Mapp. 2010;31(10):1588–600.
33. Hagiwara K, Okamoto T, Shigeto H, Ogata K, Somehara Y, Matsushita T, Kira J, Tobimatsu S. Oscillatory gamma synchronization binds the primary and secondary somatosensory areas in humans. Neuroimage. 2010;51(1):412–20.
34. Schoonheim MM, Geurts JJ, Landi D, Douw L, van der Meer ML, Vrenken H, Polman CH, Barkhof F, Stam CJ. Functional connectivity changes in multiple sclerosis patients: a graph analytical study of MEG resting state data. Hum Brain Mapp. 2013;34(1):52–61.
35. Hardmeier M, Schoonheim MM, Geurts JJ, Hillebrand A, Polman CH, Barkhof F, Stam CJ. Cognitive dysfunction in early multiple sclerosis: altered centrality derived from resting-state functional connectivity using magnetoencephalography. PLoS One. 2012;7(7):e42087. doi:10.1371/journal. pone.0042087.
36. Tewarie P, Schoonheim MM, Stam CJ, van der Meer ML, van Dijk BW, Barkhof F, Polman CH, Hillebrand A. Cognitive and clinical dysfunction, altered MEG resting-state networks and thalamic atrophy in multiple sclerosis. PLoS One. 2013;8(7):e69318. doi:10.1371/journal.pone.0069318.
37. Tewarie P, Hillebrand A, Schoonheim MM, van Dijk BW, Geurts JJ, Barkhof F, Polman CH, Stam CJ. Functional brain network analysis using minimum spanning trees in Multiple Sclerosis: an MEG source-space study. Neuroimage. 2014;88:308–18.

38. Tewarie P, Steenwijk MD, Tijms BM, Daams M, Balk LJ, Stam CJ, Uitdehaag BM, Polman CH, Geurts JJ, Barkhof F, Pouwels PJ, Vrenken H, Hillebrand A. Disruption of structural and functional networks in long-standing multiple sclerosis. Hum Brain Mapp. 2014;35(12):5946–61. doi:10.1002/hbm.22596.
39. Tewarie P, Schoonheim MM, Schouten DI, Polman CH, Balk LJ, Uitdehaag BM, Geurts JJ, Hillebrand A, Barkhof F, Stam CJ. Functional brain networks: linking thalamic atrophy to clinical disability in multiple sclerosis, a multimodal fMRI and MEG study. Hum Brain Mapp. 2015;36(2):603–18.

Part VIII

Psychiatric Disorders

Autism Spectrum Disorder

13

Hidetoshi Takahashi, Yoko Kamio, and Shozo Tobimatsu

Abstract

Autism spectrum disorder (ASD) is a set of heterogeneous neurodevelopmental conditions characterized by early-onset difficulties in social communication and unusually restricted, repetitive behavior and interests. There has been substantial recent research on ASD. Magnetoencephalography (MEG) is used in ASD research for its noninvasive nature of recordings and for its excellent temporal and spatial resolution. The number of MEG-based ASD studies in children is increasing with larger study groups. Furthermore, the analyses are becoming more sophisticated. Research over the last several decades using MEG has identified consistent atypical electrophysiological signatures of ASD, indicating common neural circuit disruptions, such as reduced long-range resting-state neural connectivity. In addition, auditory processing MEG signatures, such as middle-latency response (M50/M100) and gamma-band oscillatory activity, hold particular promise in the study and treatment of ASD and as candidate biomarkers of ASD. With the development of MEG customized for a child's head, some studies even include children younger than 3 years of age. Thus, future studies investigating these MEG signatures across developmental stages

H. Takahashi, M.D., Ph.D. (✉)
Department of Child and Adolescent Mental Health, National Institute of Mental Health, National Center of Neurology and Psychiatry, 4-1-1 Ogawahigashicho, Kodaira, Tokyo 187-8553, Japan

Department of Advanced Neuroimaging, Integrative Brain Imaging Center, National Center of Neurology and Psychiatry, 4-1-1 Ogawahigashicho, Kodaira, Tokyo 187-8551, Japan
e-mail: htakahashi@ncnp.go.jp

Y. Kamio, M.D., Ph.D.
Department of Child and Adolescent Mental Health, National Institute of Mental Health, National Center of Neurology and Psychiatry, 4-1-1 Ogawahigashicho, Kodaira, Tokyo 187-8553, Japan

S. Tobimatsu, M.D., Ph.D.
Department of Clinical Neurophysiology, Neurological Institute, Graduate School of Medical Sciences, Kyushu University, 3-1-1 Maidashi, Higashiku, Fukuoka 812-8582, Japan

S. Tobimatsu, R. Kakigi (eds.), *Clinical Applications of Magnetoencephalography*,
DOI 10.1007/978-4-431-55729-6_13

are expected to reveal the underlying neurobiology of ASD and may uncover new avenues of treatments for ASD.

Keywords
Autism spectrum disorder • Magnetoencephalography • Neural connectivity • Sensory processing • Pervasive developmental disorders

13.1 Introduction

Autism spectrum disorder (ASD) is a set of heterogeneous neurodevelopmental conditions characterized by early-onset difficulties in social communication and unusually restricted, repetitive behavior and interests [1]. ASD is known to severely impact individuals throughout childhood and adolescence and can affect society itself [1].

Understanding of autism has evolved substantially over the past 70 years since it was first described by Kanner in 1943 [2]. There has been exponential growth in research since the mid-1990s and has involved scientists from a wide range of disciplines. Almost three times as many reports about autism were published between 2000 and 2012 as between 1940 and 1999 [3].

The prevalence of autism has been steadily increasing since the first epidemiological study [3, 4], which showed that 41 out of 10,000 individuals in the United Kingdom had autism. Currently, the median worldwide prevalence of autism is 0.62–0.70 % [5, 6], although estimates of 1–2 % have been reported in the latest large-scale surveys [3]. The rise of prevalence is found particularly in individuals without intellectual disability [7], partly owing to improved awareness and recognition, changes in diagnosis, and younger age of diagnosis [5, 6].

While there has been substantial recent focus on ASD in research, both the biological pathology and fully effective treatments for ASD have yet to be realized [8]. Findings of multiple weak or rare and often nonspecific genetic or environmental etiologies of ASD have made it difficult to identify the common neurobiological mechanisms underlying behavioral features that define ASD [9]. What seems to be of consensus is the hypothesis that both the phenotypic expression and the underlying etiology of ASD are highly heterogeneous [8].

Despite the genetic and phenotypic heterogeneity, research over the last several decades using magnetoencephalography (MEG) has identified consistent atypical electrophysiological signatures of ASD, indicating common neural circuit disruptions [9–13]. Electrophysiological signatures of ASD hold particular promise in the study and treatment of ASD because of the noninvasive nature of recordings for a population traditionally challenged by imaging procedure compliance [8]. Compared with other imaging techniques, MEG is favorable owing to the use of non-claustrophobia/anxiety-inducing setups. In contrast, MRI-based techniques require magnetic bores and generate loud machine noise. MEG has minimal sensory contact, as opposed to applying electroencephalogram (EEG) head caps

[8]. MEG has the important distinction of being exquisitely sensitive to both the temporal and spatial dimensions [8] and is also relatively immune from contaminating electromyogram artifact in high-frequency bands arising from scalp and facial muscle [14] and saccadic eye movements [15]. Thus, signatures derived from MEG are considered to be promising potential biomarkers for neurophysiological abnormalities that may underlie symptom domains of ASD [8].

In this chapter, we will first provide an overview of the major characteristics of ASD which MEG researchers of ASD should bear in mind. Then, we will examine the current state of MEG research in ASD.

13.2 Overview of the Characteristics of ASD

There are various important behavioral characteristics associated with ASD. The core features of ASD according to recent diagnostic criteria (fifth edition of the *Diagnostic and Statistical Manual of Mental Disorders* (DSM-5)) [1] are (1) persistent deficits in social communication and social interaction across multiple contexts and (2) restricted, repetitive patterns of behavior, interests, or activities. In *Diagnostic and Statistical Manual of Mental Disorders*, Fourth Edition (DSM-IV) [16], language delay was a defining feature of autism (autistic disorder), but is no longer included in DSM-5. In addition, attention-deficit/hyperactivity disorder was not diagnosed when it occurs in individuals with autism in DSM-IV, but it is diagnosed in DSM-V. Motor abnormalities and excellent attention to detail are also considered to be features associated with ASD [3]. Most of these features are probably the result of complex interactions between genetic and nongenetic risk factors.

Recently, the central significance of sensory symptoms in ASD have been increasingly recognized [17]. Abnormalities in sensory responsivity have been observed in ASD since Kanner's first characterization of autism [2]. Individuals with ASD have been found to have high rates of abnormalities of sensory functioning, varying from 69 to 95 % [17–21]. Abnormalities in all sensory modalities have been reported in ASD, and a broad range of disturbances may be observed [22]. Despite the high frequency of sensory symptoms in ASD, in the past, these symptoms have been viewed as peripheral to the disorder rather than as a core feature. Atypical sensory processing was not included as a diagnostic criteria of ASD in DSM-IV; however, it is now included in DSM-5. In addition, individuals with ASD have atypical profiles in various aspects of cognition. Cognition and neurobiology are related, and cognition provides a guide to simplify the various behavioral manifestations of autism and can help in investigating its underlying neurobiology. Atypical cognitive profiles of ASD include (1) impaired social cognition and social perception, (2) executive dysfunction, and (3) atypical bottom-up and top-down (local vs. global) perceptual and information processing [3]. Details of these cognitive functions are as follows:

1. Social cognition and social perception: Gaze and eye contact, emotion perception, face processing, biological motion perception, social attention and orienting, social motivation, social reward processing, nonverbal communication, imitation, affective empathy and sympathy, joint attention, pretend play, theory of mind or mental perspective taking, self-referential cognition, alexithymia (difficulty in understanding and describing own emotions), and metacognitive awareness
2. Executive function: Cognitive flexibility, planning, inhibitory control, attention shifting, monitoring, generativity, and working memory
3. Bottom-up and top-down (local vs. global) information processing: Global vs. local perceptual functioning (superior low-level sensory-perceptual processing), central coherence (global vs. local preference), and systemizing (drive to construct rule-based systems, ability to understand rule-based systems, knowledge of factual systems)

Comorbidity is quite frequently observed in ASD, and individuals with ASD often exhibit symptoms of other neurological or psychiatric conditions [23]. More than 70 % of individuals with autism have concurrent medical, developmental, or psychiatric conditions [3]. The high frequency of comorbidity could be a result of shared pathophysiology, secondary effects of growing up with autism, shared symptom domains and associated mechanisms, or overlapping diagnostic criteria. Common co-occurring conditions are as follows [3]:

- Developmental disorders: Intellectual disability, attention-deficit/hyperactivity disorder, motor abnormality, and language disorders
- Genetic syndromes: Fragile X syndrome, Rett syndrome, tuberous sclerosis complex, Down syndrome, and phenylketonuria
- General medical disorders: Epilepsy, gastrointestinal problems, and immune dysregulation
- Psychiatric disorders: Anxiety, depression, obsessive-compulsive disorder, psychotic disorders, oppositional defiant disorder, substance use disorders, eating disorders, and sleep disorders
- Personality disorders: Obsessive-compulsive personality disorder, schizoid personality disorder, avoidant personality disorder, schizotypal personality disorder, paranoid personality disorder, and borderline personality disorder
- Other behavioral problems: Aggressive behaviors, self-injurious behaviors, pica, and suicidal ideation or attempt

Sex differences are also a major factor in ASD. Autism affects more males than females. Early studies showed that autism affects 4–5 times more males than females [6], and recent large-scale population-based studies have shown that 2–3 times more males are affected than females [3]. Empirical data suggest high-functioning females are diagnosed later than males [24, 25], and females with autism might be under-recognized [26].

Because individuals with ASD often exhibit manifestations of other neurological or psychiatric conditions that make it difficult to establish a clear diagnosis [27], diagnostic assessment should be multidisciplinary and use a developmental framework of an interview with the parent or caregiver, interaction with the individual, collection of information about behavior in community settings, cognitive assessments, and a medical examination [28]. Ideally, a standardized, structured interview and/or observational measure, such as the Autism Diagnostic Interview-Revised; the Autism Diagnostic Observation Schedule (first or second edition); the Diagnostic Interview for Social and Communication Disorders; the Developmental, Dimensional, and Diagnostic Interview; or Childhood Autism Rating Scale (first or second edition), should be incorporated into the assessment process [3]. Adaptive skills should be checked with standardized instruments, such as Vineland Adaptive Behavior Scales.

13.3 MEG Research on ASD

Since the late 1990s, MEG has been used increasingly as an investigative tool to study the neurophysiological basis of ASD. A rapid increase in ASD research using MEG, especially after 2010, has been observed. We summarize these studies in Tables 13.1, 13.2, 13.3, and 13.4. The features of the recent MEG studies on ASD can be described as follows:

- Age of the subjects is getting younger. The number of ASD MEG studies, including children (mostly over 6–10 years), is increasing, and some studies involve even younger children (around 3 years old) because of the development of MEG customized for a child's head.
- Subjects in most of the recent studies were diagnosed by structural diagnostic criteria, such as DSM-IV, and diagnoses were supported by using a standardized structured interview and/or observational measure, such as the Autism Diagnostic Interview-Revised and the Autism Diagnostic Observation Schedule (first or second edition).
- The number of subjects in a study group is getting larger. Some studies even include more than 100 subjects.
- The analyses used are becoming more sophisticated than just a simple dipole analysis. Most of the recent ASD MEG studies used whole-head MEG and MRI data for more precise source localization and neural connectivity.

MEG is also a noninvasive technique for measuring brain activity, and its excellent temporal and spatial resolution makes it useful for various fields of ASD research. Because comorbid epilepsy is frequently found in ASD (8–30 % [3]), some studies investigated epileptic conditions in ASD [63, 64]. The high comorbidity of seizure disorders in ASD may be explained by hyperexcitability and an imbalance in GABA and glutamate [65]. MEG was used to record the electrophysiological response during stage III sleep in 50 children with regressive

Table 13.1 Summary of recent MEG-based resting-state findings in ASD

	Subjects	ASD	Control	Structural diagnostic criteria	Standardized structured interview and/or observation	MEG system	MRI	Task and measurements	Main results
Cornew et al. [29]	Children	27 ASD	23 TD	DSM-IV	Yes	275-channel		Resting-state eye-closed	Regionally specific elevations in delta, theta, alpha, and high-frequency power in ASD
Ghanbari et al. [30]	Children	26 ASD	22 TD	DSM-IV	Yes	275-channel		Resting state	Stronger short-range frontal connectivity and unclear long-range connectivity in ASD
Ghanbari et al. [31]	Children	26 ASD	22 TD	DSM-IV	Yes	275-channel		Resting-state eye-closed	Increased short-range connectivity in the frontal lobe in the delta band and in the temporal, parietal, and occipital lobes in the alpha band in ASD
									Decreased long-range connectivity in all the other bands in ASD
Ye et al. [32]	Adolescent	16 ASD	15 TD		Yes	151-channel	Yes	Maintain visual fixation	Hyperconnectivity among the frontal, temporal, and subcortical regions in beta and gamma frequency ranges in ASD
									Hypoconnectivity of parietal and occipital regions to widespread brain regions in theta and alpha bands and among orbitofrontal, subcortical, and temporal regions in the gamma band in ASD

Kitzbichler et al. [33]	Children	15 ASD	15 TD		Yes	306-channel	Yes	Maintain visual fixation	More integrated, more efficient, and more distributed networks in the α and γ bands in ASD
	adolescent								Less globally integrated, less locally efficient, and less clustered in the β band in ASD
Edgar et al. [34]	Children	41 ASD	47 TD	DSM-IV	Yes	306-channel	Yes	Resting-state eye-closed	Increased alpha activity in nonprimary alpha-generating areas in ASD
									A failure to shift calcarine region peak alpha frequency as a function of age in primary alpha-generating areas in ASD
Kikuchi et al. [35]	Children	35 ASD	35 TD	DSM-IV	Yes	151-channel Child sized		Concentrated on video programs	Aberrant rightward-lateralized connectivity via gamma oscillation in ASD
Kikuchi et al. [36]	Children	26 ASD	26 TD	DSM-IV	Yes	151-channel Child sized		Concentrated on video programs	Rightward lateralization of the neurophysiological connectivity between the parietal and temporal regions in ASD
Kikuchi et al. [37]	Children	50 ASD	50 TD	DSM-IV	Yes	151-channel Child sized		Concentrated on video programs	Reduced connectivity between the left anterior and right posterior brain areas, as demonstrated by 6 Hz oscillation patterns in ASD

(continued)

Table 13.1 (continued)

	Subjects	ASD	Control	Structural diagnostic criteria	Standardized structured interview and/or observation	MEG system	MRI	Task and measurements	Main results
Hiraishi et al. [38]	Children	38 ASD	38 TD	DSM-IV	Yes	151-channel Child sized		Concentrated on video programs	Reduced rightward-lateralized brain oscillations in the theta-1 frequency bands in ASD

ASD autism spectrum disorder, *DSM-IV* diagnostic and statistical manual of mental disorders, fourth edition, *TD* typical development

Table 13.2 Summary of recent MEG-based auditory processing findings in ASD, around M50 and M100

	Subjects	ASD	Control	Structural diagnostic criteria	Standardized structured interview and/or observation	MEG system	MRI	Task and measurements	Main results
Acoustic simple tone									
Oram Cardy et al. [39]	Children	14 ASD	16 TD		Yes	151-channel		1,000 Hz tones	Right hemispheric M50 latency as the best predictor of clinical LI
	Adolescent								
Schimidt et al. [40]	Children	8 ASD	7 TD		Yes	151-channel		Pairs of 1,000 Hz tones in two different interstimulus intervals	The anterior-posterior M100 asymmetry pattern of TD not found in ASD
	Adults		10 adults						
Roberts et al. [41]	Children	53 ASD	39 TD	DSM-IV	Yes	275-channel	Yes	Pairs of 1,000, 2,000 Hz tones	A delayed M50 response in ASD
									M50 latency negatively associated with age for both TD and ASD
Edgar et al. [12]	Children	105 ASD	36 TD	DSM-IV	Yes	275-channel		200, 300, 500, 1,000 Hz tones	Increased pre-stimulus activity and then early high-frequency abnormalities followed by low-frequency abnormalities in auditory superior temporal gyrus processes of ASD
Edgar et al. [42]	Children	101 ASD	35 TD	DSM-IV	Yes	275-channel		Pairs of 1,000, 2,000 Hz tones	Delayed left and right M50 responses in ASD without LI and ASD with LI

(continued)

Table 13.2 (continued)

	Subjects	ASD	Control	Structural diagnostic criteria	Standardized structured interview and/or observation	MEG system	MRI	Task and measurements	Main results
Yoshimura et al. [43]	Children	35 ASD	35 TD	DSM-IV	Yes	151-channel Child sized		Auditory Japanese syllable sound /ne/	Less leftward lateralization of the P50m dipole intensity in ASD
Matsuzaki et al. [44]	Children	18 ASD	12 TD	DSM-IV	Yes	160-channel	Yes	A tone pip presented with random interstimulus intervals	More delayed M50/M100 peak latencies and larger M50 dipole moments in ASD with hypersensitivity compared with ASD without hypersensitivity or the control
Matsuzaki et al. [45]	Children	12 ASD	15 TD	DSM-IV	Yes	160-channel	Yes	A tone pip presented with random interstimulus intervals	Increased M50 dipole moments and more prolonged response duration in ASD with auditory hypersensitivity compared with those for ASD without auditory hypersensitivity or controls

ASD autism spectrum disorder, *DSM-IV* diagnostic and statistical manual of mental disorders, fourth edition, *LI* language impairment, *TD* typical development

Table 13.3 Summary of recent MEG-based auditory processing findings in ASD. Mismatched magnetic fields and auditory steady-state response

	Subjects	ASD	Control	Structural diagnostic criteria	Standardized structured interview and/or observation	MEG system	MRI	Task and measurements	Main results
Acoustic steady-state response									
Wilson et al. [10]	Children Adolescent	10 ASD	10 TD	DSM-IV	Yes	37-channel		ASSR	Reduced left hemispheric 40 Hz power from 200 to 500 ms poststimulus onset in ASD
Rojas et al. [46]	Adults (parents)	16 parents of ASD, 11 adult ASD	16 healthy controls	DSM-IV	Yes	37-channel	Yes	ASSR	Higher induced gamma power at 40 Hz, reduced evoked gamma-band power, and lower phase-locking factor in ASD and the parent
Rojas et al. [13]	Adults (parents)	21 parents of ASD	20 healthy controls	DSM-IV	Yes	248-channel		ASSR	Lower phase-locked steady-state power in the left hemisphere, reduced total gamma-band power, lower gamma-band phase-locking factor in parents of ASD
Roberts et al. [47]	Children	26 ASD	17 TD	DSM-IV	Yes	275-channel		ASSR	Delayed right hemisphere average M100 latency in ASD
Gandal et al. [11]	Children	26 ASD	17 TD	DSM-IV	Yes	275-channel		ASSR	A similar 10 % latency delay in the N1/M100 evoked response and a reduction in frequency (30–50 Hz) phase-locking factor in ASD and valproic acid-exposed mice

(continued)

Table 13.3 (continued)

	Subjects	ASD	Control	Structural diagnostic criteria	Standardized structured interview and/or observation	MEG system	MRI	Task and measurements	Main results
Magnetic mismatch field									
Roberts et al. [48]	Children	51 ASD	27 TD	DSM-IV	Yes	275-channel		Tone or vowel deviant oddball task	Prolonged MMF latency in ASD, most pronounced in children with concomitant LI
Ingalhalikar et al. [49]	Children	93 ASD	42 TD	DSM-IV	Yes	275-channel	Yes	Tone or vowel deviant oddball task, ASSR	Largest contribution obtained by MMF latency to group discrimination, between ASD and TD and between ASD with LI and ASD without LI
Ingalhalikar et al. [50]		92 ASD	42 TD			275-channel	Yes	Tone or vowel deviant oddball task, ASSR	The fusion classifier performance superior to the classification achieved using single modalities as well as multimodal classifier using only complete data

ASD autism spectrum disorder, *ASSR* auditory steady-state response, *DSM-IV* diagnostic and statistical manual of mental disorders, fourth edition, *LI* language impairment, *MMF* magnetic mismatch field, *TD* typical development

Table 13.4 Summary of recent MEG-based cognitive processing findings in ASD

	Subjects	ASD	Control	Structural diagnostic criteria	Standardized structured interview and/or observation	MEG system	MRI	Task and measurements	Main results
Mišic' et al. [51]	Children	14 ASD	14 TD		Yes	151-channel	Yes	A set-shifting (color or the shape) task	Two distributed networks, operating at fast and slow time scales, respond completely differently to set shifting in ASD
Doesburg et al. [52]	Children	16 ASD	14 TD			151-channel	Yes	A set-shifting (color or the shape) task	Reduced theta-band phase synchronization in ASD during extradimensional set shifting
Bangel et al. [53]	Adults	14 ASD	14 TD	DSM-IV	Yes	151-channel	Yes	Dot number estimation task (globally integrated forms and randomly shaped stimuli)	Increased early task-dependent interregional phase synchrony in theta, alpha, and beta frequency bands in ASD
									Reduced long-range beta-band phase synchronization at 70–145 ms during presentation of globally coherent dot patterns encompassing numerous areas including occipital, parietal, temporal, and frontal lobe regions in ASD

(continued)

Table 13.4 (continued)

	Subjects	ASD	Control	Structural diagnostic criteria	Standardized structured interview and/or observation	MEG system	MRI	Task and measurements	Main results
Meaux et al. [54]	Adults	14 ASD	14 TD	DSM-IV	Yes	151-channel	Yes	Dot number estimation task (globally integrated forms and randomly shaped stimuli)	Smaller source amplitudes in visual areas, followed by temporal and then parietal regions in ASD
Leung et al. [55]	Adolescent	22 ASD	17 TD		Yes	151-channel	Yes	An implicit emotional processing task (happy, angry, and neutral faces)	Reduced beta-band interregional phase-locking during the perception of angry faces in a distributed network involving the right fusiform gyrus and insula of ASD
Leung et al. [56]	Adolescent	24 ASD	24 TD		Yes	151-channel	Yes	An implicit emotional processing task (happy, angry, and neutral faces)	Lower accuracy on the implicit emotional face processing task including angry faces in ASD
									Atypical neural activity in the insula, anterior and posterior cingulate, and temporal and orbitofrontal regions during angry and happy face processing in ASD

Vara et al. [57]	Adolescent	15 ASD	15 TD		Yes	151-channel	Yes	A go/no-go task	Recruitment first of the right middle frontal gyrus, followed by the left postcentral gyrus and finally the left middle frontal and right medial frontal gyri in ASD
									Recruitment first of the left middle frontal gyrus, followed by the left superior and inferior frontal gyri, then the right middle temporal gyrus, and finally the superior and precentral gyri and right inferior lobule in TD
Khan et al. [58]	Children	17 ASD	20 TD	DSM-IV	Yes	306-channel	Yes	Pictures of neutral, fearful, and angry faces and houses	Reduced local functional connectivity indicated by phase-amplitude coupling within the fusiform face area between alpha and gamma bands in ASD
	Adolescent								Reduced long-range functional connectivity quantified by alpha-band coherence in ASD

(continued)

Table 13.4 (continued)

	Subjects	ASD	Control	Structural diagnostic criteria	Standardized structured interview and/or observation	MEG system	MRI	Task and measurements	Main results
Kenet et al. [59]	Adults	11 ASD	11 TD	DSM-IV	Yes	306-channel	Yes	Saccadic paradigm	Reduced functional connectivity between the frontal eye field and dorsal anterior cingulate cortex during the saccadic paradigm and temporal and spectral specificity by demonstrating reduced coherence specifically during the preparatory period, driven by reduced synchronization specifically in the alpha band in ASD
Moseley et al. [60]	Adults	14 ASD	17 TD	DSM-IV		306-channel	Yes	Passive word reading task	Reduced activity of the lexical temporal pathway, and additional dorsal parietal processing route, for reading familiar, simple words, in ASD

Buard et al. [61]	Adults (parents)	16 parents of ASD, 12 adult ASD	35 healthy controls	DSM-IV	Yes	248-channel		Black and white line art images	Reduced evoked high-gamma activity in the right STG and reduced high-beta/low-gamma evoked power in the left IFG
									Reduced phase-locked beta band especially in OCC and higher beta- and gamma-band functional connectivity between the IFG and FG and between STG and OCC in the left hemisphere in ASD
									Higher high-gamma-band power in the left STG, increased high-beta/low-gamma evoked power in the left FG in parents of ASD
Mcfadden et al. [62]	Adults	23 parents of ASD	28 healthy controls	DSM-IV	Yes	248-channel		A continuous word recognition task	Increased evoked gamma and beta activity and greater left lateralization in some regions of interest in parents of ASD

ASD autism spectrum disorder, *DSM-IV* diagnostic and statistical manual of mental disorders, fourth edition, *FG* frontal gyrus, *IFG* inferior frontal gyrus, *LI* language impairment, *OCC* occipital lobes, *STG* superior temporal gyrus, *TD* typical development

ASD (15 with a clinical seizure disorder) [63] or spontaneous neural activity in 36 children with ASD without a diagnosis of seizure disorders [64]. In these studies, abnormal discharges were identified in 82 % or 86 % of all children with ASD using MEG, while only 68 % or 3 % used simultaneous EEG, respectively. These studies provided evidence that subclinical seizure disorder can be detected in many children with ASD using MEG and appear to have greater sensitivity than EEG [66]. This may be an advantage of MEG in researching abnormal brain activity in ASD.

On the other hand, most of the recent ASD MEG research has been directed at characterizing and understanding various aspects of neural processing that are affected in individuals with ASD, such as resting state, auditory processing, and other cognitive conditions (Tables 13.1, 13.2, 13.3, and 13.4.). Below, we review the major findings from recent MEG-based ASD studies.

13.3.1 Resting State

Neural activity during resting state has recently gained interest in MEG ASD research (Table 13.1). Although some studies focus on atypical lateralization of ASD, most of the recent studies investigating resting state of ASD using MEG focus on atypical neural connectivity of ASD.

The idea that autism is characterized by atypical neural connectivity, rather than by a discrete set of atypical brain regions, has been raised in various fields of research, such as electrophysiology, structural neuroimaging, functional neuroimaging, and molecular genetics [3]. The dominant theory of neural connectivity in ASD used to be long-range hypoconnectivity between the frontal cortex and other brain regions and local over-connectivity of the frontal cortex [67]. However, ideas vary about the precise way in which connectivity is atypical in subsequent studies.

There seems to be some consensus regarding reduced long-range resting-state connectivity of ASD in MEG-based ASD studies. Altered resting-state connectivity across various brain regions has been preliminarily demonstrated in adolescents and young adults with ASD ($n = 8$), with connectivity patterns in ASD participants that displayed significantly reduced interdependence strength, both within bilateral frontal and temporal sensors, as well as between temporal sensors and the remaining recording sites [68]. Most of the subsequent studies using larger samples reported reduced long-range connectivity in ASD at resting state (Table 13.1). Aberrant long-range connectivity in children with ASD was observed between the ages of 3 and 7 years ($n = 70$) [31] by customized child-sized whole-head MEG, suggesting that long-range hypoconnectivity is already present during early development. Thus, long-range hypoconnectivity appears to be a pervasive alteration of neural functioning in ASD. Longitudinal studies tracking neural connectivity in ASD across developmental stages are necessary to understand the maturational trajectories of hypoconnectivity, and in the future, hypoconnectivity patterns may provide insight into the heterogeneity of ASD and help in the development of individualized biomarkers for specific ASD subtypes.

However, there are varying results across ASD MEG studies regarding the question of whether short-range coherence is increased [30–32] or not [35, 37]. These findings from ASD MEG studies seem to indicate that abnormalities in connectivity in ASD are more complex than the pattern suggested by local hyperconnectivity and long-range hypoconnectivity [67]. The lack of uniformity among the different findings is likely attributable to differences in focus of the studies using various methods in recording resting state (eyes closed, maintaining visual fixation, or watching video programs), data analysis methods, and age of the subjects, but it may also result from the neurobiological heterogeneity of ASD. In addition, findings differ not only in the specific brain regions but also in the frequency bands of neural activity. Thus, patterns of connectivity should not only be examined by structural means but also by functional means. The field is in need of more MEG studies on neural connectivity in order to gain more insight into the characteristics of functional brain activation.

13.3.2 Sensory Symptoms in ASD

Individuals with ASD have been found to have high rates of abnormalities of sensory function [17]. In recent years, the central significance of sensory symptoms in ASD has been increasingly recognized.

Although the underlying neurobiology of sensory symptoms in ASD is unclear, several theories have been proposed linking possible etiologies of sensory dysfunction with known abnormalities in brain structure and function that are associated with ASD [8, 17]. Abnormalities in all sensory modalities have been reported in ASD [22], and atypical auditory processing has been well investigated in this area of research, including ASD MEG studies. Electrophysiological signatures of auditory processing examined by MEG are considered promising biomarkers not only for diagnostic and prognostic purposes but also for stratification and response indices for pharmaceutical development [8]. Additional research into the sensory symptoms associated with ASD has the potential to shed more light on the nature and pathophysiology of ASD and to help in the development of new, effective treatments. Below, we describe the recent developments in ASD MEG studies investigating auditory processing, including middle-latency response (M50/M100), gamma-band oscillatory activity, and mismatch magnetic fields.

13.3.2.1 Middle-Latency Response: M50/M100

One of the most replicable electrophysiological findings, even with a large sample [42], is the latency delay of middle-latency responses, M50/M100, in ASD as compared with typically developing children (Tables 13.2 and 13.3).

Besides the maturational changes (a pattern of latency shortening with increasing age) in both M100 (especially right hemisphere) and M50 (bilaterally resolved), a persistent shift or prolongation of latency exists in ASD compared with controls at any given age [8, 41, 69]. However, middle-latency-evoked potentials can differ depending on stimulus properties, such as frequency [70], intensity [71], and other features [72], and it is important to standardize all stimulus properties and delivery methodology [8].

Association of M50/M100 latency delays are frequently investigated with ultimate clinical language impairment. M50 and, to a lesser extent, M100 latencies are reported to predict oral language ability [39]. However, this was not replicated in all cohorts [47], and M100 delays were not evident in specific language impairments [73]. Although a direct mechanism of relationship between M50/M100 latency and ultimate high-level behavioral language performance is not currently present, it appears that the M50/M100 delays may indicate atypical auditory processing that might underlie clinical language impairments in ASD.

13.3.2.2 Gamma-Band Oscillatory Activity

Another area of focus in ASD MEG is gamma-band (30–80 Hz) oscillatory activity (Table 13.3).

Gamma-band oscillatory dysfunction may be related to certain core domains of symptomology, with the ultimately impacted system/domain being dependent on the time of critical dysfunction or region involved [9]. Auditory response-related gamma-band oscillatory activity has been repeatedly demonstrated to be altered in ASD [10–13, 46, 62]. Evoked (phase-locked) gamma power for both transient responses and simple sinusoidal tones driven by 40 Hz auditory steady-state stimuli was reduced in ASD [10, 13]. In addition, the transient and steady-state gamma responses both correlated to more basic functionality, such as communication and symptom severity rating of ASD [13]. Gamma dysfunction has also been identified in first-degree relatives of ASD patients [13, 46], suggesting that gamma-band oscillatory activity may be a candidate endophenotype of ASD [74]. On the other hand, gamma-band oscillatory dysfunction is not specific to ASD [75]. Gamma-band oscillatory alterations might occur in other disorders, such as bipolar disorders [76, 77] and schizophrenia [78, 79]. Thus, deeper investigations into gamma-band oscillatory activity in ASD and these psychiatric diseases may clarify the relationship between ASD and psychiatric comorbidities.

Gamma-band oscillatory activity is considered to be dependent on numerous constituents, including potassium channel subtypes, spike conductance trajectories, and a strong role for both glutamatergic and γ-aminobutyric acid (GABA)-ergic signaling [80]. With this link between GABA and the mechanism underpinning gamma-band oscillatory activity, the balance of excitation and inhibition has been hypothesized as a pathogenic mechanism for ASD [81]. Translational research has demonstrated ASD-related behavioral improvement due to pharmacological treatments that can mirror amelioration of gamma-band function in preclinical models [82], indicating that such an approach can help lead to better treatments in human patients.

Recently, we conducted a pilot study of evoked magnetic gamma-band oscillatory activity during auditory steady-state responses task in children with and without ASD (non-ASD). As recent evidence [83] seems to suggest, ASD should not be regarded as a discrete entity; instead its behavioral traits present a continuous distribution across the population. We split the median non-ASD group according to the quantitative autistic traits assessed with the Social Responsiveness Scale. Time-frequency analysis (wavelet analysis) was conducted, and increased gamma-band oscillation power dominant around the right auditory cortex was only found in non-ASD children with low autistic traits, and non-ASD children with high autistic traits that were similar to children with ASD did not show such laterality (Fig. 13.1). Our preliminary results suggest that auditory evoked gamma-band activity may be a useful biomarker related to autistic traits, not only for group comparisons of ASD and typically developing populations but also for relationships to ASD features among ASD and non-ASD or typical development samples. This might shed light on the underpinning neurobiology of ASD.

13.3.2.3 Mismatch Magnetic Fields

The auditory magnetic mismatch field (MMF), especially its latency delay, is another well-investigated electrophysiological signature of ASD (Table 13.3). MMF and its EEG counterpart, mismatch negativity, refer to the resultant component, which are derived from subtracting an average of standard (or frequent) stimulus component from that of a deviant (or rare) stimulus [84] and essentially representing the "difference wave" as an index of "preattentive change detection."

The MMF latency deficit was observed for sinusoidal tones [85] and for language stimuli [86]. The MMF latency delays scales within the ASD cohort with clinical measures of language ability that are reported to be sensitive and specific predictors of language impairment [48]. When compared with another MEG signature, such as M100 latency delay, or a structural imaging marker, MMF timing

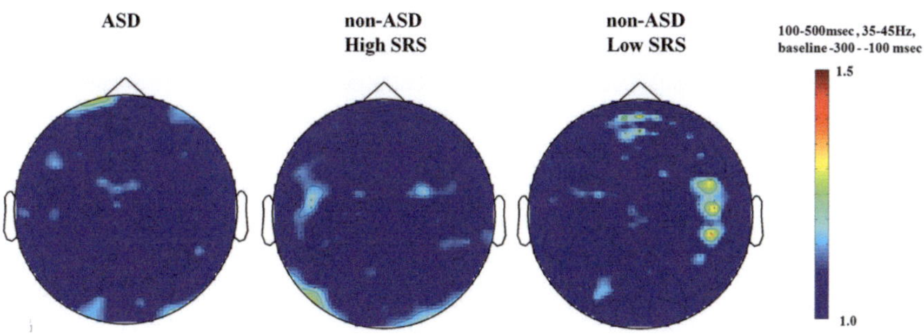

Fig. 13.1 Time-frequency analysis (2D topography of power spectrum) of evoked magnetic fields during auditory steady-state responses task in children with and without ASD. Representative examples: MEG recordings during auditory steady-state responses task were obtained for two children with ASD and ten without ASD (non-ASD) (age, 9–16; all right-handed). Subjects listened to 500 ms duration binaural click trains (200 trains) with a 25 ms inter-click interval (40 Hz) during 306-channel Elekta Neuromag MEG recordings. MaxFilter 2.2 software (Elekta Neuromag) was used offline to reduce movement-related artifacts. Time-frequency analysis (wavelet analysis) was conducted using FieldTrip toolbox (http://fieldtrip.fcdonders.nl/) of MATLAB. We also examined the quantitative autistic traits with the Social Responsiveness Scale (*SRS*). Increased gamma-band oscillation power dominant around right auditory cortex was only found in non-ASD children with lower than median SRS total score of non-ASD children

alterations have better accuracy for predicting both ASD and also language impairment within ASD [49].

However, the exact nature of the MMF latency delay is not stationary. For example, the MMF is sometimes missing in ASD populations [87]. Similar inconsistent results have also been exhibited in mismatch negativity latency [88]. In addition, MMF delays also exist in children who have language impairment but are not on the autism spectrum [73]. Thus, more detailed studies in the future that investigate the relationship of the MMF mechanism to ASD are necessary to derive conclusive evidence.

13.3.3 Cognitive Processing

Because individuals with ASD have atypical profiles in various aspects of cognition, many of the previous ASD MEG studies focused on neural activity during cognitive processing (Table 13.4). Compared with studies investigating resting-state or sensory (mostly auditory) processing, previous MEG contributions to ASD research on cognitive processing used to be limited, in that studies were rather fragmented in terms of experimental design and scientific objectives, sample sizes were often typically small, subject matching criteria such as age varied substantially between investigations, comparisons were often limited to ASD, and typically developing populations rather than pathologies exhibited some overlap with ASD.

As is the case in investigating resting-state neural activity, many of the recent studies focused on atypical neural connectivity of ASD. Similar to studies in resting state, findings from MEG studies on cognitive processing of ASD do not provide converging results that support the notion of long-range hypoconnectivity and local over-connectivity [67] but show instead that, in people with ASD, brain regions interact in a different way than they do in typically developing people. Some studies support an hypoconnectivity hypothesis; however, there have also been reports contradicting general hypoconnectivity in ASD. For example, reduced coherence between the fusiform face area and several other distant brain regions was demonstrated during the processing of emotional faces [58], suggesting that connectivity between functionally relevant brain areas may be disturbed in ASD in complex processes like facial perception. On the other hand, increased connectivity was found during a picture naming task in areas relevant to visual and language processing [61]. The specific pattern of connectivity alterations likely depends on the cognitive processes that were studied and the specific neural system involved, and the hyperconnectivity may be a correlate of compensating mechanisms resulting from connectivity deficits in other networks. As is the case in studies in resting state, the lack of uniformity in findings is likely attributable to the different focus of the studies using various methods, but may also result from the neurobiological heterogeneity of ASD, and findings differ not only in the specific brain regions but also in the frequency bands of neural activity. Thus, understanding altered brain activities during cognitive processing may provide insight into the neural basis of interindividual variability of symptoms in subjects with ASD, and future work may require investigating effective connectivity using imaging methods, such as MEG, with acute temporal resolution.

13.4 Conclusions

ASD is a set of heterogeneous neurodevelopmental conditions with various aspects of behavioral characteristics and cognition. Research over the last several decades using MEG has identified consistent atypical electrophysiological signatures of ASD, indicating common neural circuit disruptions, such as reduced long-range resting-state neural connectivity. MEG signatures of auditory processing, such as middle-latency response (M50/M100) and gamma-band oscillatory activity, hold particular promise for study and treatment as candidate biomarkers of ASD. With the development of MEG customized for a child's head, some studies even include children younger than 3 years of age. Our preliminary results not only suggest group comparisons of ASD and typically developing populations but also suggest a relationship to ASD features among ASD and non-ASD or typical development samples that may shed light on atypical neural activity engaged in ASD. Thus, future studies investigating these MEG signatures across developmental stages in the population would reveal the underlying neurobiology of ASD and open new avenues of effective treatments.

References

1. American Psychiatric Association. Diagnostic and statistical manual of mental disorders. 5th ed. Washington, DC: American psychiatric publishing; 2013.
2. Kanner L. Autistic disturbances of affective contact. Acta Paedopsychiatr. 1968;35(4):100–36.
3. Lai M-C, Lombardo MV, Baron-Cohen S. Autism. Lancet. 2014;383(9920):896–910. doi:10.1016/S0140-6736(13)61539-1.
4. Lotter V. Epidemiology of autistic conditions in young children. Soc Psychiatry. 1966;1(3):124–35. doi:10.1007/BF00584048.
5. Elsabbagh M, Divan G, Koh YJ, Kim YS, Kauchali S, Marcin C, Montiel-Nava C, Patel V, Paula CS, Wang C, Yasamy MT, Fombonne E. Global prevalence of autism and other pervasive developmental disorders. Autism Res : Off J Int Soc Autism Res. 2012;5(3):160–79. doi:10.1002/aur.239.
6. Fombonne E, Quirke S, Hagen A. Epidemiology of pervasive developmental disorders. In: Amaral DG, Dawson G, Geschwind DH, editors. Autism spectrum disorders. New York: Oxford University Press; 2011. p. 90–111.
7. Keyes KM, Susser E, Cheslack-Postava K, Fountain C, Liu K, Bearman PS. Cohort effects explain the increase in autism diagnosis among children born from 1992 to 2003 in California. Int J Epidemiol. 2012;41(2):495–503. doi:10.1093/ije/dyr193.
8. Port RG, Anwar AR, Ku M, Carlson GC, Siegel SJ, Roberts TP. Prospective MEG biomarkers in ASD: pre-clinical evidence and clinical promise of electrophysiological signatures. Yale J Biol Med. 2015;88(1):25–36.
9. Port RG, Gandal MJ, Roberts TP, Siegel SJ, Carlson GC. Convergence of circuit dysfunction in ASD: a common bridge between diverse genetic and environmental risk factors and common clinical electrophysiology. Front Cell Neurosci. 2014;8:414. doi:10.3389/fncel.2014.00414.

10. Wilson TW, Rojas DC, Reite ML, Teale PD, Rogers SJ. Children and adolescents with autism exhibit reduced MEG steady-state gamma responses. Biol Psychiatry. 2007;62(3):192–7. doi:10.1016/j.biopsych.2006.07.002.
11. Gandal MJ, Edgar JC, Ehrlichman RS, Mehta M, Roberts TP, Siegel SJ. Validating gamma oscillations and delayed auditory responses as translational biomarkers of autism. Biol Psychiatry. 2010;68(12):1100–6. doi:10.1016/j.biopsych.2010.09.031.
12. Edgar JC, Khan SY, Blaskey L, Chow VY, Rey M, Gaetz W, Cannon KM, Monroe JF, Cornew L, Qasmieh S, Liu S, Welsh JP, Levy SE, Roberts TP. Neuromagnetic oscillations predict evoked-response latency delays and core language deficits in autism spectrum disorders. J Autism Dev Disord. 2015;45(2):395–405. doi:10.1007/s10803-013-1904-x.
13. Rojas DC, Teale PD, Maharajh K, Kronberg E, Youngpeter K, Wilson LB, Wallace A, Hepburn S. Transient and steady-state auditory gamma-band responses in first-degree relatives of people with autism spectrum disorder. Mol Autism. 2011;2:11. doi:10.1186/2040-2392-2-11.
14. Muthukumaraswamy SD. High-frequency brain activity and muscle artifacts in MEG/EEG: a review and recommendations. Front Hum Neurosci. 2013;7:138. doi:10.3389/fnhum.2013.00138.
15. Yuval-Greenberg S, Tomer O, Keren AS, Nelken I, Deouell LY. Transient induced gamma-band response in EEG as a manifestation of miniature saccades. Neuron. 2008;58(3):429–41. doi:10.1016/j.neuron.2008.03.027.
16. American Psychiatric Association. Diagnostic and statistical manual of mental disorders. 4th ed. Washington, DC: American Psychiatric Publishing; 1994.
17. Hazen EP, Stornelli JL, O'Rourke JA, Koesterer K, McDougle CJ. Sensory symptoms in autism spectrum disorders. Harv Rev Psychiatry. 2014;22(2):112–24. doi:10.1097/01.hrp.0000445143.08773.58.
18. Baranek GT, David FJ, Poe MD, Stone WL, Watson LR. Sensory experiences questionnaire: discriminating sensory features in young children with autism, developmental delays, and typical development. J Child Psychol Psychiatry. 2006;47(6):591–601. doi:10.1111/j.1469-7610.2005.01546.x.
19. Baker AE, Lane A, Angley MT, Young RL. The relationship between sensory processing patterns and behavioural responsiveness in autistic disorder: a pilot study. J Autism Dev Disord. 2008;38(5):867–75. doi:10.1007/s10803-007-0459-0.
20. Klintwall L, Holm A, Eriksson M, Carlsson LH, Olsson MB, Hedvall A, Gillberg C, Fernell E. Sensory abnormalities in autism. A brief report. Res Dev Disabil. 2011;32(2):795–800. doi:10.1016/j.ridd.2010.10.021.
21. Tomchek SD, Dunn W. Sensory processing in children with and without autism: a comparative study using the short sensory profile. Am J Occup Ther: Off Publ Am Occup Ther Assoc. 2007;61(2):190–200.
22. Marco EJ, Hinkley LB, Hill SS, Nagarajan SS. Sensory processing in autism: a review of neurophysiologic findings. Pediatr Res. 2011;69(5 Pt 2):48R–54R. doi:10.1203/PDR.0b013e3182130c54.
23. Leyfer OT, Folstein SE, Bacalman S, Davis NO, Dinh E, Morgan J, Tager-Flusberg H, Lainhart JE. Comorbid psychiatric disorders in children with autism: interview development and rates of disorders. J Autism Dev Disord. 2006;36(7):849–61. doi:10.1007/s10803-006-0123-0.
24. Begeer S, Mandell D, Wijnker-Holmes B, Venderbosch S, Rem D, Stekelenburg F, Koot HM. Sex differences in the timing of identification among children and adults with autism spectrum disorders. J Autism Dev Disord. 2013;43(5):1151–6. doi:10.1007/s10803-012-1656-z.
25. Giarelli E, Wiggins LD, Rice CE, Levy SE, Kirby RS, Pinto-Martin J, Mandell D. Sex differences in the evaluation and diagnosis of autism spectrum disorders among children. Disabil Health J. 2010;3(2):107–16. doi:10.1016/j.dhjo.2009.07.001.

26. Baron-Cohen S, Lombardo MV, Auyeung B, Ashwin E, Chakrabarti B, Knickmeyer R. Why are autism spectrum conditions more prevalent in males? PLoS Biol. 2011;9(6):e1001081. doi:10.1371/journal.pbio.1001081.
27. Reiersen AM, Constantino JN, Volk HE, Todd RD. Autistic traits in a population-based ADHD twin sample. J Child Psychol Psychiatry. 2007;48(5):464–72. doi:10.1111/j.1469-7610.2006.01720.x.
28. Ozonoff S, Goodlin-Jones BL, Solomon M. Evidence-based assessment of autism spectrum disorders in children and adolescents. J Clin Child Adolesc Psychol: the official journal for the Society of Clinical Child and Adolescent Psychology, American Psychological Association, Division 53 2005;34(3):523–40. doi:10.1207/s15374424jccp3403_8.
29. Cornew L, Roberts TP, Blaskey L, Edgar JC. Resting-state oscillatory activity in autism spectrum disorders. J Autism Dev Disord. 2012;42(9):1884–94. doi:10.1007/s10803-011-1431-6.
30. Ghanbari Y, Bloy L, Batmanghelich K, Roberts TP, Verma R. Dominant component analysis of electrophysiological connectivity networks. Med Image Comput Comput Assist Interv: MICCAI International Conference on Medical Image Computing and Computer-Assisted Intervention 2012;15(Pt 3):231–8.
31. Ghanbari Y, Bloy L, Christopher Edgar J, Blaskey L, Verma R, Roberts TP. Joint analysis of band-specific functional connectivity and signal complexity in autism. J Autism Dev Disord. 2015;45(2):444–60. doi:10.1007/s10803-013-1915-7.
32. Ye AX, Leung RC, Schafer CB, Taylor MJ, Doesburg SM. Atypical resting synchrony in autism spectrum disorder. Hum Brain Mapp. 2014;35(12):6049–66. doi:10.1002/hbm.22604.
33. Kitzbichler MG, Khan S, Ganesan S, Vangel MG, Herbert MR, Hamalainen MS, Kenet T. Altered development and multifaceted band-specific abnormalities of resting state networks in autism. Biol Psychiatry. 2015;77(9):794–804. doi:10.1016/j.biopsych.2014.05.012.
34. Edgar JC, Heiken K, Chen YH, Herrington JD, Chow V, Liu S, Bloy L, Huang M, Pandey J, Cannon KM, Qasmieh S, Levy SE, Schultz RT, Roberts TP. Resting-state alpha in autism spectrum disorder and alpha associations with thalamic volume. J Autism Dev Disord. 2015;45 (3):795–804. doi:10.1007/s10803-014-2236-1.
35. Kikuchi M, Shitamichi K, Yoshimura Y, Ueno S, Hiraishi H, Hirosawa T, Munesue T, Nakatani H, Tsubokawa T, Haruta Y, Oi M, Niida Y, Remijn GB, Takahashi T, Suzuki M, Higashida H, Minabe Y. Altered brain connectivity in 3-to 7-year-old children with autism spectrum disorder. NeuroImage Clin. 2013;2:394–401. doi:10.1016/j.nicl.2013.03.003.
36. Kikuchi M, Yoshimura Y, Shitamichi K, Ueno S, Hirosawa T, Munesue T, Ono Y, Tsubokawa T, Haruta Y, Oi M, Niida Y, Remijn GB, Takahashi T, Suzuki M, Higashida H, Minabe Y. A custom magnetoencephalography device reveals brain connectivity and high reading/decoding ability in children with autism. Sci Rep. 2013;3:1139. doi:10.1038/srep01139.
37. Kikuchi M, Yoshimura Y, Hiraishi H, Munesue T, Hashimoto T, Tsubokawa T, Takahashi T, Suzuki M, Higashida H, Minabe Y. Reduced long-range functional connectivity in young children with autism spectrum disorder. Soc Cogn Affect Neurosci. 2015;10(2):248–54. doi:10.1093/scan/nsu049.
38. Hiraishi H, Kikuchi M, Yoshimura Y, Kitagawa S, Hasegawa C, Munesue T, Takesaki N, Ono Y, Takahashi T, Suzuki M, Higashida H, Asada M, Minabe Y. Unusual developmental pattern of brain lateralization in young boys with autism spectrum disorder: power analysis with child-sized magnetoencephalography. Psychiatry Clin Neurosci. 2015;69(3):153–60. doi:10.1111/pcn.12261.
39. Oram Cardy JE, Flagg EJ, Roberts W, Roberts TP. Auditory evoked fields predict language ability and impairment in children. Int J Psychophysiol: Off J Int Organ Psychophysiol. 2008;68(2):170–5. doi:10.1016/j.ijpsycho.2007.10.015.
40. Schmidt GL, Rey MM, Oram Cardy JE, Roberts TP. Absence of M100 source asymmetry in autism associated with language functioning. Neuroreport. 2009;20(11):1037–41. doi:10.1097/WNR.0b013e32832e0ca7.

41. Roberts TP, Lanza MR, Dell J, Qasmieh S, Hines K, Blaskey L, Zarnow DM, Levy SE, Edgar JC, Berman JI. Maturational differences in thalamocortical white matter microstructure and auditory evoked response latencies in autism spectrum disorders. Brain Res. 2013;1537:79–85. doi:10.1016/j.brainres.2013.09.011.
42. Edgar JC, Lanza MR, Daina AB, Monroe JF, Khan SY, Blaskey L, Cannon KM, Jenkins 3rd J, Qasmieh S, Levy SE, Roberts TP. Missing and delayed auditory responses in young and older children with autism spectrum disorders. Front Hum Neurosci. 2014;8:417. doi:10.3389/fnhum.2014.00417.
43. Yoshimura Y, Kikuchi M, Shitamichi K, Ueno S, Munesue T, Ono Y, Tsubokawa T, Haruta Y, Oi M, Niida Y, Remijn GB, Takahashi T, Suzuki M, Higashida H, Minabe Y. Atypical brain lateralisation in the auditory cortex and language performance in 3- to 7-year-old children with high-functioning autism spectrum disorder: a child-customised magnetoencephalography (MEG) study. Mol Autism. 2013;4(1):38. doi:10.1186/2040-2392-4-38.
44. Matsuzaki J, Kagitani-Shimono K, Goto T, Sanefuji W, Yamamoto T, Sakai S, Uchida H, Hirata M, Mohri I, Yorifuji S, Taniike M. Differential responses of primary auditory cortex in autistic spectrum disorder with auditory hypersensitivity. Neuroreport. 2012;23(2):113–8. doi:10.1097/WNR.0b013e32834ebf44.
45. Matsuzaki J, Kagitani-Shimono K, Sugata H, Hirata M, Hanaie R, Nagatani F, Tachibana M, Tominaga K, Mohri I, Taniike M. Progressively increased M50 responses to repeated sounds in autism spectrum disorder with auditory hypersensitivity: a magnetoencephalographic study. PLoS One. 2014;9(7):e102599. doi:10.1371/journal.pone.0102599.
46. Rojas DC, Maharajh K, Teale P, Rogers SJ. Reduced neural synchronization of gamma-band MEG oscillations in first-degree relatives of children with autism. BMC Psychiatry. 2008;8:66. doi:10.1186/1471-244X-8-66.
47. Roberts TP, Khan SY, Rey M, Monroe JF, Cannon K, Blaskey L, Woldoff S, Qasmieh S, Gandal M, Schmidt GL, Zarnow DM, Levy SE, Edgar JC. MEG detection of delayed auditory evoked responses in autism spectrum disorders: towards an imaging biomarker for autism. Autism Res: Off J Int Soc Autism Res. 2010;3(1):8–18. doi:10.1002/aur.111.
48. Roberts TP, Cannon KM, Tavabi K, Blaskey L, Khan SY, Monroe JF, Qasmieh S, Levy SE, Edgar JC. Auditory magnetic mismatch field latency: a biomarker for language impairment in autism. Biol Psychiatry. 2011;70(3):263–9. doi:10.1016/j.biopsych.2011.01.015.
49. Ingalhalikar M, Parker WA, Bloy L, Roberts TP, Verma R. Creating multimodal predictors using missing data: classifying and subtyping autism spectrum disorder. J Neurosci Methods. 2014;235:1–9. doi:10.1016/j.jneumeth.2014.06.030.
50. Ingalhalikar M, Parker WA, Bloy L, Roberts TP, Verma R. Using multiparametric data with missing features for learning patterns of pathology. Med Image Comput Comput Assist Interv MICCAI International Conference on Medical Image Computing and Computer-Assisted Intervention 2012;15(Pt 3):468–75.
51. Misic B, Doesburg SM, Fatima Z, Vidal J, Vakorin VA, Taylor MJ, McIntosh AR. Coordinated information generation and mental flexibility: large-scale network disruption in children with autism. Cereb Cortex. 2014. doi:10.1093/cercor/bhu082.
52. Doesburg SM, Vidal J, Taylor MJ. Reduced theta connectivity during set-shifting in children with autism. Front Hum Neurosci. 2013;7:785. doi:10.3389/fnhum.2013.00785.
53. Bangel KA, Batty M, Ye AX, Meaux E, Taylor MJ, Doesburg SM. Reduced beta band connectivity during number estimation in autism. NeuroImage Clin. 2014;6:202–13. doi:10.1016/j.nicl.2014.08.020.
54. Meaux E, Taylor MJ, Pang EW, Vara AS, Batty M. Neural substrates of numerosity estimation in autism. Hum Brain Mapp. 2014;35(9):4362–85. doi:10.1002/hbm.22480.
55. Leung RC, Ye AX, Wong SM, Taylor MJ, Doesburg SM. Reduced beta connectivity during emotional face processing in adolescents with autism. Mol Autism. 2014;5(1):51. doi:10.1186/2040-2392-5-51.

56. Leung RC, Pang EW, Cassel D, Brian JA, Smith ML, Taylor MJ. Early neural activation during facial affect processing in adolescents with Autism Spectrum Disorder. NeuroImage Clin. 2015;7:203–12. doi:10.1016/j.nicl.2014.11.009.
57. Vara AS, Pang EW, Doyle-Thomas KA, Vidal J, Taylor MJ, Anagnostou E. Is inhibitory control a 'no-go' in adolescents with autism spectrum disorder? Mol Autism. 2014;5(1):6. doi:10.1186/2040-2392-5-6.
58. Khan S, Gramfort A, Shetty NR, Kitzbichler MG, Ganesan S, Moran JM, Lee SM, Gabrieli JD, Tager-Flusberg HB, Joseph RM, Herbert MR, Hamalainen MS, Kenet T. Local and long-range functional connectivity is reduced in concert in autism spectrum disorders. Proc Natl Acad Sci U S A. 2013;110(8):3107–12. doi:10.1073/pnas.1214533110.
59. Kenet T, Orekhova EV, Bharadwaj H, Shetty NR, Israeli E, Lee AK, Agam Y, Elam M, Joseph RM, Hamalainen MS, Manoach DS. Disconnectivity of the cortical ocular motor control network in autism spectrum disorders. Neuroimage. 2012;61(4):1226–34. doi:10.1016/j.neuroimage.2012.03.010.
60. Moseley RL, Pulvermuller F, Mohr B, Lombardo MV, Baron-Cohen S, Shtyrov Y. Brain routes for reading in adults with and without autism: EMEG evidence. J Autism Dev Disord. 2014;44(1):137–53. doi:10.1007/s10803-013-1858-z.
61. Buard I, Rogers SJ, Hepburn S, Kronberg E, Rojas DC. Altered oscillation patterns and connectivity during picture naming in autism. Front Hum Neurosci. 2013;7:742. doi:10.3389/fnhum.2013.00742.
62. McFadden KL, Hepburn S, Winterrowd E, Schmidt GL, Rojas DC. Abnormalities in gamma-band responses to language stimuli in first-degree relatives of children with autism spectrum disorder: an MEG study. BMC Psychiatry. 2012;12:213. doi:10.1186/1471-244X-12-213.
63. Lewine JD, Andrews R, Chez M, Patil AA, Devinsky O, Smith M, Kanner A, Davis JT, Funke M, Jones G, Chong B, Provencal S, Weisend M, Lee RR, Orrison Jr WW. Magnetoencephalographic patterns of epileptiform activity in children with regressive autism spectrum disorders. Pediatrics. 1999;104(3 Pt 1):405–18.
64. Munoz-Yunta JA, Ortiz T, Palau-Baduell M, Martin-Munoz L, Salvado-Salvado B, Valls-Santasusana A, Perich-Alsina J, Cristobal I, Fernandez A, Maestu F, Dursteler C. Magnetoencephalographic pattern of epileptiform activity in children with early-onset autism spectrum disorders. Clin Neurophysiol: Off J Int Fed Clin Neurophysiol. 2008;119(3):626–34. doi:10.1016/j.clinph.2007.11.007.
65. Brooks-Kayal A. Epilepsy and autism spectrum disorders: are there common developmental mechanisms? Brain Dev. 2010;32(9):731–8. doi:10.1016/j.braindev.2010.04.010.
66. Braeutigam S. Magnetoencephalography: fundamentals and established and emerging clinical applications in radiology. ISRN Radiol. 2013;2013:529463. doi:10.5402/2013/529463.
67. Courchesne E, Pierce K. Why the frontal cortex in autism might be talking only to itself: local over-connectivity but long-distance disconnection. Curr Opin Neurobiol. 2005;15(2):225–30. doi:10.1016/j.conb.2005.03.001.
68. Tsiaras V, Simos PG, Rezaie R, Sheth BR, Garyfallidis E, Castillo EM, Papanicolaou AC. Extracting biomarkers of autism from MEG resting-state functional connectivity networks. Comput Biol Med. 2011;41(12):1166–77. doi:10.1016/j.compbiomed.2011.04.004.
69. Gage NM, Siegel B, Roberts TP. Cortical auditory system maturational abnormalities in children with autism disorder: an MEG investigation. Brain Res Dev Brain Res. 2003;144(2):201–9.
70. Roberts TP, Poeppel D. Latency of auditory evoked M100 as a function of tone frequency. Neuroreport. 1996;7(6):1138–40.
71. Stufflebeam SM, Poeppel D, Rowley HA, Roberts TP. Peri-threshold encoding of stimulus frequency and intensity in the M100 latency. Neuroreport. 1998;9(1):91–4.
72. Roberts TP, Ferrari P, Stufflebeam SM, Poeppel D. Latency of the auditory evoked neuromagnetic field components: stimulus dependence and insights toward perception. J Clin Neurophysiol: Off Publ Am Electroencephalographic Soc. 2000;17(2):114–29.

73. Roberts TP, Heiken K, Kahn SY, Qasmieh S, Blaskey L, Solot C, Parker WA, Verma R, Edgar JC. Delayed magnetic mismatch negativity field, but not auditory M100 response, in specific language impairment. Neuroreport. 2012;23(8):463–8. doi:10.1097/WNR.0b013e32835202b6.
74. Nagarajan S, Mahncke H, Salz T, Tallal P, Roberts T, Merzenich MM. Cortical auditory signal processing in poor readers. Proc Natl Acad Sci U S A. 1999;96(11):6483–8.
75. Herrmann CS, Demiralp T. Human EEG gamma oscillations in neuropsychiatric disorders. Clin Neurophysiol: Off J Int Fed Clin Neurophysiol. 2005;116(12):2719–33. doi:10.1016/j.clinph.2005.07.007.
76. Oda Y, Onitsuka T, Tsuchimoto R, Hirano S, Oribe N, Ueno T, Hirano Y, Nakamura I, Miura T, Kanba S. Gamma band neural synchronization deficits for auditory steady state responses in bipolar disorder patients. PLoS One. 2012;7(7):e39955. doi:10.1371/journal.pone.0039955.
77. Reite M, Teale P, Rojas DC, Reite E, Asherin R, Hernandez O. MEG auditory evoked fields suggest altered structural/functional asymmetry in primary but not secondary auditory cortex in bipolar disorder. Bipolar Disord. 2009;11(4):371–81. doi:10.1111/j.1399-5618.2009.00701.x.
78. Krishnan GP, Hetrick WP, Brenner CA, Shekhar A, Steffen AN, O'Donnell BF. Steady state and induced auditory gamma deficits in schizophrenia. Neuroimage. 2009;47(4):1711–9. doi:10.1016/j.neuroimage.2009.03.085.
79. Tsuchimoto R, Kanba S, Hirano S, Oribe N, Ueno T, Hirano Y, Nakamura I, Oda Y, Miura T, Onitsuka T. Reduced high and low frequency gamma synchronization in patients with chronic schizophrenia. Schizophr Res. 2011;133(1–3):99–105. doi:10.1016/j.schres.2011.07.020.
80. Buzsaki G, Wang XJ. Mechanisms of gamma oscillations. Annu Rev Neurosci. 2012;35:203–25. doi:10.1146/annurev-neuro-062111-150444.
81. Rubenstein JL, Merzenich MM. Model of autism: increased ratio of excitation/inhibition in key neural systems. Genes Brain Behav. 2003;2(5):255–67.
82. Gandal MJ, Sisti J, Klook K, Ortinski PI, Leitman V, Liang Y, Thieu T, Anderson R, Pierce RC, Jonak G, Gur RE, Carlson G, Siegel SJ. GABAB-mediated rescue of altered excitatory-inhibitory balance, gamma synchrony and behavioral deficits following constitutive NMDAR-hypofunction. Transl Psychiatry. 2012;2:e142. doi:10.1038/tp.2012.69.
83. Constantino JN, Todd RD. Autistic traits in the general population: a twin study. Arch Gen Psychiatry. 2003;60(5):524–30. doi:10.1001/archpsyc.60.5.524.
84. Naatanen R. The perception of speech sounds by the human brain as reflected by the mismatch negativity (MMN) and its magnetic equivalent (MMNm). Psychophysiology. 2001;38(1):1–21.
85. Oram Cardy JE, Flagg EJ, Roberts W, Roberts TP. Delayed mismatch field for speech and non-speech sounds in children with autism. Neuroreport. 2005;16(5):521–5.
86. Kasai K, Hashimoto O, Kawakubo Y, Yumoto M, Kamio S, Itoh K, Koshida I, Iwanami A, Nakagome K, Fukuda M, Yamasue H, Yamada H, Abe O, Aoki S, Kato N. Delayed automatic detection of change in speech sounds in adults with autism: a magnetoencephalographic study. Clin Neurophysiol: Off J Int Fed Clin Neurophysiol. 2005;116(7):1655–64. doi:10.1016/j.clinph.2005.03.007.
87. Tecchio F, Benassi F, Zappasodi F, Gialloreti LE, Palermo M, Seri S, Rossini PM. Auditory sensory processing in autism: a magnetoencephalographic study. Biol Psychiatry. 2003;54(6):647–54.
88. O'Connor K. Auditory processing in autism spectrum disorder: a review. Neurosci Biobehav Rev. 2012;36(2):836–54. doi:10.1016/j.neubiorev.2011.11.008.

Schizophrenia

14

Toshiaki Onitsuka and Shogo Hirano

Abstract

Despite the high prevalence of schizophrenia, which is a devastating mental disorder, the etiology and pathophysiology of this disease are still unclear. Magnetoencephalography (MEG) has a high spatial and temporal resolution and has been used to make important contributions to schizophrenia research. In this chapter, we review MEG studies of schizophrenia, with an emphasis on event-related responses and neural oscillations. Published MEG studies suggest that patients with schizophrenia have neurophysiological deficits from the early phase of sensory processing (i.e., M50, M100, mismatch negativity) in auditory perception. Moreover, schizophrenia patients may have altered neural oscillations, and abnormalities of auditory steady-state responses to 40 Hz click stimuli are repeatedly reported. Because this research can be conducted in living schizophrenia patients, these biological results are highly valuable for understanding the etiology and pathophysiology of the disorder. As advanced medical technology becomes increasingly globally available, the clinical application of MEG to schizophrenia treatment may be imminent.

Keywords

Schizophrenia • Electroencephalography (EEG) • Magnetoencephalography (MEG) • Event-related responses • Neural oscillations

T. Onitsuka (✉)
Department of Neuropsychiatry, Graduate School of Medical Sciences, Kyushu University, 3-1-1, Maidashi, Higashi-ku, Fukuoka 812-8582, Japan
e-mail: toshiaki@npsych.med.kyushu-u.ac.jp

S. Hirano
Department of Neuropsychiatry, Graduate School of Medical Sciences, Kyushu University, 3-1-1, Maidashi, Higashi-ku, Fukuoka 812-8582, Japan

Neural Dynamics Laboratory, Research Service, Veterans Affairs Boston Healthcare System and Department of Psychiatry, Harvard Medical School, Research 151C, 150 S. Huntington Ave, Boston, MA 02130, USA

S. Tobimatsu, R. Kakigi (eds.), *Clinical Applications of Magnetoencephalography*, DOI 10.1007/978-4-431-55729-6_14

14.1 Introduction

Schizophrenia is a devastating mental disorder and presenting a prevalence of about 1 % of the world population. The disorder is characterized by positive symptoms (e.g., delusion, auditory hallucination, and loose associations), negative symptoms (e.g., lack of emotional expression, lack of interest or enthusiasm), and cognitive dysfunctions. Diagnoses are usually made during adolescence, which is a crucial time for personality formation. These diagnoses may be chronic, causing enormous problems for social adjustment. Recent studies have suggested that schizophrenia is characterized as brain dysfunctions, some of which occurs before full symptom onset and some of which progresses in the 1–2 years after onset. Although many researchers have investigated this illness, termed schizophrenia by Eugen Bleuler [1], its etiology, and pathophysiology remain unclear.

Magnetoencephalography (MEG) is a noninvasive technique that can be used to detect magnetic fluctuations associated with electrical activities in the brain. Along with electroencephalography (EEG), which measures activity on the scale of millisecond, MEG has a high temporal resolution. MEG has several advantages over EEG; for instance, it can more easily penetrate the cerebrospinal fluid, certain membranes, the skull, and the scalp [2, 3]. Additionally, MEG is preferable to EEG in that it can be used to obtain more precise information about the spatial domain of neural activity. MEG has been used in many types of neuroscience research, including studies of psychiatric disorders such as schizophrenia [4]. In addition, recent developments in MEG signal processing technology have enabled increased visibility of neural activity. To date, few objective indices of schizophrenia pathophysiology have been confirmed, hindering research and therapeutic approaches. However, MEG has the potential to be a powerful tool in schizophrenia research. In this chapter, we present an overview of recent schizophrenia studies using MEG, especially those focused on event-related responses and neural oscillations.

14.2 Event-Related Responses

Event-related responses are characteristic electric or magnetic signal waves elicited by a stimulus or activity. They are usually measured by averaging a number of obtained signals elicited via the presentation of a repetitive sensory stimuli or performance of a cognitive task [5]. Event-related responses measured using EEG are called event-related potentials (ERPs) and those measured using MEG are called event-related fields (ERFs). ERP and ERF components are defined based on the polarity of wave peaks (i.e., negative or positive in EEG) and the latency from the time point at which a certain stimulus is presented to an individual. Generally, ERP and ERF components are described based on their polarity and latency (e.g., P50, N100, and P300). Event-related components can be measured with high reliability and reflect electrical brain activities coinciding with specific perception or cognition processes. ERP and ERF data have been applied to both academic and clinical fields. In studies of schizophrenia, ERFs have been used to characterize abnormal brain activity in visual, auditory, and somatosensory experimental conditions.

14.2.1 Auditory Event-Related Responses

14.2.1.1 M50

Measured via EEG, the P1-N1-P2 complex can be recorded to auditory stimuli. The typical auditory evoked magnetic responses are shown in Fig. 14.1. Specifically, the P1 component can be called the P50, because the positive peak of this component is located approximately 50 ms after the auditory stimulus is presented. The P50 component can be elicited by presenting a repetitive pair of auditory stimuli with a specific interstimulus interval. In this type of experiment, typical auditory stimuli are click sounds with short durations and the interstimulus interval is generally 500 ms. These stimuli are known to elicit a phenomenon in which the amplitude of the P50 evoked by the second auditory stimulus is smaller than the amplitude of the P50 evoked by the first stimulus [6, 7]. The first stimulus is thought to activate an inhibitory neural network, in which cholinergic neural circuits, involving the hippocampus, are thought to be implicated [8]. This P50 suppression, measured by paired auditory stimuli, is thought to be an index of "sensory gating," which is a process by which necessary and unnecessary information from both outer and inner sources is controlled. Patients with schizophrenia are known to have diminished P50 suppression [9], indicating that the symptoms of schizophrenia are associated with dysfunctional sensory gating, resulting in an information overload. Several MEG studies have also detected diminished P50 suppression in individuals with

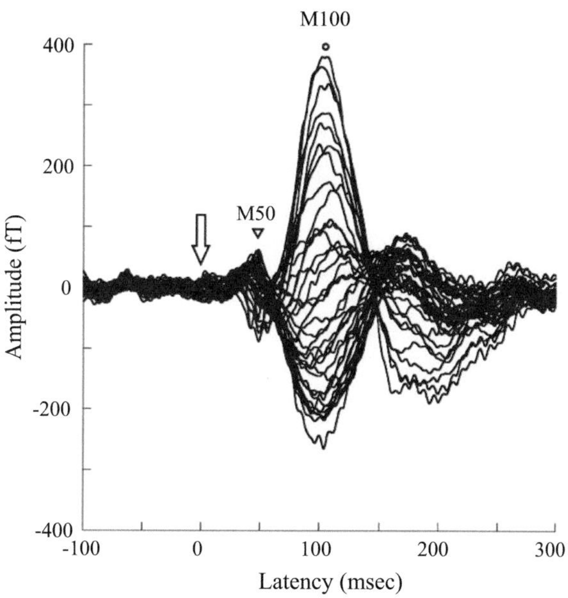

Fig. 14.1 The typical auditory evoked magnetic responses from one healthy subject. The MEG waveforms are obtained by the superimposition of 37 channels. The *triangle* indicates M50, and the *circle* indicates M100. The *arrows* signify the onset of the stimulus

schizophrenia. In MEG research, the magnetic counterpart of the P50 is often termed the M50 [10], thus, we will use this term as we review M50 studies of schizophrenia patients.

Thoma et al. [11] used simultaneous EEG and MEG to measure the P50 and the M50, respectively. The investigators presented auditory paired click stimuli to 20 individuals with schizophrenia and 15 healthy participants. They found diminished P50 suppression and diminished M50 suppression in the left hemisphere in the schizophrenia group. Additionally, they revealed that in the schizophrenia group, the M50 gating ratio, which is the ratio (S2/S1) contrasting the M50 amplitude evoked by the first stimulus (S1) and the M50 amplitude evoked by the second stimulus (S2), was negatively correlated with neuropsychological indexes of working memory, executive attention, and general memory in both hemispheres. Edgar et al. [12] used auditory paired click and electric stimuli and recorded the M50 and the M20 in patients with schizophrenia and healthy controls. The M20 is the magnetic counterpart of the N20, which is a component of somatosensory evoked potentials, and is thought to originate in the somatosensory cortex. The researchers found reduced M50 suppression in the left hemisphere in schizophrenia patients, but no significant difference in M50 suppression in the right hemisphere. They also found no M20 suppression in either group. As noted above, the presentation of paired click stimuli is a standard in measuring P50 and M50 suppression. For other stimuli, Hirano et al. [13] used auditory paired vowel stimuli to demonstrate reduced M50 suppression in schizophrenia patients. In addition, they showed that the M50 gating ratio in the left hemisphere was positively correlated with the severity of auditory hallucinations, while the M50 gating ratio in the right hemisphere was positively correlated with the severity of negative symptoms. The higher spatial resolution afforded by the use of MEG instead of EEG in these studies enabled the discovery of new information about hemispheric differences in sensory gating dysfunction in people with schizophrenia. Additionally, the gating ratio of the M50 is reported to have greater test–retest reliability compared with that of the P50, which is measured via EEG [14].

Thoma et al. [15] reported that lower levels of M50 gating were correlated with decreased auditory cortical thickness in both hemispheres in individuals with schizophrenia. This notion was reinforced by structural magnetic resonance imaging data. In a study combining M50 gating, neural oscillations, and functional magnetic resonance imaging (fMRI), Mathiak et al. [16] suggested that the alpha band (8–13 Hz) rhythm in the auditory cortex influences sensory gating. P50 and M50 gating dysfunction may have potential as a marker of risk of developing positive symptoms of schizophrenia (see the review by Potter et al. [17]). In the future, cross-modality studies like those mentioned above will be useful in sensory gating research and may provide clues regarding the source of abnormal sensory gating in people with schizophrenia.

14.2.1.2 M100

The auditory M100 is the magnetic counterpart of the N100, which is an auditory evoked potential component with a negative peak approximately 100 ms after the onset of an auditory stimulus. Hari et al. [18] used simultaneous EEG and MEG to record the N100 and the M100, respectively. They reported that the neural source of the M100 was the supratemporal auditory cortex when the interstimulus interval (ISI) was below 4 s. The auditory N100 component is elicited by any auditory stimulus without task demands, and the amplitude is influenced by several factors, including ISI, stimulus intensity, arousal level, and the attentional status of the participant. Several studies have measured the M100 component in schizophrenia patients. Hajek et al. [19, 20] reported that, unlike control participants, individuals with schizophrenia showed significantly different equivalent M100 dipole orientations. Smith et al. [21] measured the M100 elicited by an auditory paired click stimulus and reported that patients with schizophrenia showed smaller M100 amplitudes in response to the first click sound and diminished M100 suppression in response to the second click. The size of the auditory N100 amplitude, measured via EEG, is thought to be a marker of functional brain abnormalities related to a genetic predisposition to schizophrenia [22]. However, M100 studies of people with schizophrenia are sparse, and more data are required.

14.2.1.3 Mismatch Negativity

Näätänen et al. [23] investigated a negatively peaked wave by subtracting the evoked potential wave elicited by a low-frequency auditory stimulus from one elicited by a high-frequency auditory stimulus. Because this negative wave tends to be unaffected by the attentional status to stimuli of the participants, it is thought to reflect automatic brain activity—specifically, the mismatch between the memory trace made by the high-frequency (standard) stimulus and the low-frequency (deviant) stimulus. This negative wave is called mismatch negativity (MMN). Patients with schizophrenia have repeatedly been found to possess an attenuated MMN amplitude, which may indicate that they have a pathophysiological mechanism affecting automatic auditory detection [24]. Several studies [22, 25, 26] have used MEG to measure MMN in people with schizophrenia, in which the magnetic counterpart of MMN is often termed MMNm, and reported that these patients exhibit an attenuated MMNm amplitude. Although most MMNm studies have used combinations of several tone sounds that are different in terms of duration or pitch, Kasai et al. [27, 28] measured MMNm by presenting a combination of different vowels to patients with schizophrenia. They found that chronic administration of benzodiazepine did not affect MMNm in schizophrenic patients and suggested that the patients had abnormal automatic categorical differentiation of vowels. Yamasue et al. [29] also reported that changes in MMNm amplitude in the left hemisphere elicited by categorical differentiation of vowels were significantly correlated with left planum temporale gray matter volume in patients with schizophrenia. Interestingly, these results imply that language dysfunction in people with schizophrenia involves subconscious cognitive processing.

14.2.2 Other Event-Related Responses

Facial recognition is critical to human social interactions, and patients with schizophrenia are thought to have some facial processing deficits [30]. The N170, as measured via EEG, is a visual event-related potential component. The N170 has a negative peak that occurs approximately 170 ms after the visual stimuli onset. The N170 is considered to function as an index of the structural encoding of faces and the extraction of configural information [31]. Individuals with schizophrenia have been found to exhibit a diminished N170 in response to face stimuli [32]. However, findings of the M170, which is the magnetic counterpart of the N170, are still controversial in patients with schizophrenia [33, 34], probably because facial processing involves many sub-cognitive processes [35]. Rivolta et al. [36] conducted an extensive report regarding the components of visual magnetic responses M100, M170, and M250, which are prominent 100, 170, and 250 ms after stimulus onset, respectively. These components are associated with facial recognition in both drug-näive and single-episode patients with schizophrenia. Patients with schizophrenia showed no significant differences in terms of the M100 and M170 but had weaker M250 components and different spatial patterns in a source reconstruction of each component compared with controls.

In the somatosensory domain, as shown above, no significant differences have been found in somatosensory M20 suppression between individuals with schizophrenia and healthy controls [12]. Meanwhile, Huang et al. [37] used MEG to report that healthy controls exhibited an attention-mediated somatosensory response more than 40 ms after the presentation of an electrical median-nerve oddball task, while individuals with schizophrenia exhibited no such evoked response.

14.3 Neural Oscillations

In this subsection, the term "neural oscillation" is used to describe the phenomena in which electric or magnetic waves derived from the human brain change in terms of frequency or amplitude in temporal and/or spatial domains. These waves can correspond to diverse dynamic brain activities, involving perceptive, cognitive, motor, and/or spontaneous activities. Since Gray et al. [38] reported that a 40–60 Hz electrical oscillatory response in the cat visual cortex changed with the contents of the presented stimuli, the concept of neural oscillations has been broadly applied to various research topics, including mental disorders such as schizophrenia (see the review by Uhlhaas and Singer [39]). In the early stages of neural oscillation research, which largely focused on the 40 Hz "gamma" frequency band oscillation [40], the predominant oscillations in each frequency band were associated with distinct brain functions. For example, the gamma band (30–100 Hz) oscillation is thought to relate to perception [41], attention [42], memory [43], and consciousness [44]. Beta-band (13–30 Hz) oscillations have been associated with sensory gating [45] and inter-areal synchronization [46], theta (4–7 Hz) oscillations correspond to inter-areal synchronization and top-down processing [47], and alpha (8–12 Hz) oscillations are linked with attention [48] and consciousness [49]. Meanwhile, recent studies have indicated that each network is tremendously diverse and that the different processing levels in the brain play important roles in the generation and modulation of neural oscillations. For example, each processing level may be related with receptors of certain neurochemical transmitters [50], cortical microcircuits between pyramidal cells and interneurons [51], cortico–cortical connections [41], and inter-areal connections [52]. Neural oscillations represent a critical component in the connection between neurophysiology and neurochemistry data, and thus, are important for connecting pathophysiology with data from animal models of schizophrenia. Several MEG studies have focused on neural oscillation in patients with schizophrenia. In the following subsections, we review these oscillation studies.

14.3.1 Transient Neural Oscillations

Transient neural oscillations have transient temporal fluctuations in terms of power or phase at a certain frequency. Transient neural oscillations can be classified as either evoked oscillations or induced oscillations, depending on the timing of the temporal fluctuation and whether it is locked to an outer stimulus or the execution of a particular task. Evoked oscillations are closely synchronized with the timing of a stimulus presentation, while this is not the case for induced oscillations. Jitter in induced oscillations is thought to reflect higher cognitive processes involving memory [43] or the differentiation of contents of stimuli [53]. Evoked oscillations are also affected by the character of stimuli [54]. In the auditory domain, several MEG studies have reported that individuals with schizophrenia show diminished evoked oscillations in the gamma-frequency range in response to speech sounds. Additionally, individuals with schizophrenia exhibit different patterns between hemispheres [55, 56] and diminished gamma band oscillations evoked by pure tones [57] within a relatively early period (less than 150 ms) from stimulus onset. In the visual domain, Grützner et al. [58] reported that the schizophrenia patients showed diminished low (25–60 Hz) and high (60–120 Hz) gamma band oscillations in both the evoked and induced time windows (5–320 ms from stimulus onset) in response to Mooney faces. Their colleagues also showed that medication-näive schizophrenia patients exhibited diminished high-gamma band oscillations in the induced time window (105–220 ms from stimulus onset) in a face recognition study compared with healthy controls [59]. Herrmann et al. [60] described the match-and-utilization model for early (around 100 ms from stimulus onset) and late (around 300 ms from stimulus onset) transient gamma oscillations. They suggested that early gamma oscillations reflect the comparison of memory contents with stimulus-related information, and late gamma oscillations reflect the use of signals derived from this comparison. Indeed, the abnormal transient neural oscillations observed in schizophrenia patients might reflect a dysfunction in the temporal stream for information processing.

14.3.2 Spontaneous Neural Oscillations

Several studies have measured spontaneous MEG signals in schizophrenia patients without a cognitive task. These studies investigated temporal or spatial fluctuations of activity across several frequency bands. Cañive et al. [61, 62] reported that patients with schizophrenia exhibited signals with lower power and lower peak frequency in the alpha band compared with controls. This abnormal alpha pattern did not change after treatment with aripiprazole. Sperling et al. showed elevated dipole density in beta-band (12.5–30 Hz) activity over the left hemisphere in schizophrenia patients treated with clozapine [63], and a correlation between slow activity (2–6 Hz) and psychotic symptoms has been observed [64]. Fehr et al. [65] showed that enhanced delta (1.5–4 Hz) and theta (4–8 Hz) band activity and frontal delta and theta activity were associated with characteristics of the positive symptoms of schizophrenia. Rockstroh et al. [66] also reported that the schizophrenia patients exhibited prominent delta activity (0.5–4 Hz) in the frontal and central cortical regions. Rutter et al. [67] reported that the schizophrenia patients showed reduced gamma (30–80 Hz) activity in the posterior region of the medial parietal cortex. Combining several types of network analysis, various studies have attempted to reveal functional network pathophysiology in the resting state in people with schizophrenia [68, 69].

A number of studies have measured spontaneous MEG signals in relation to auditory hallucinations in people with schizophrenia. Van Lutterveld et al. [70] reported that auditory hallucinations were related to decreased beta-band power in the left temporal cortex, decreased alpha band power in the inferior frontal gyrus, and decreased theta-band power in the right hippocampus in schizophrenia patients. Other reports have shown that increased beta activity [71] and theta activity [72] in the left auditory cortex were associated with the appearance of auditory hallucinations in a schizophrenia patient. Considering the high spatial resolution of MEG compared with EEG, these studies imply that schizophrenia is associated with functional spatial disproportion, even in the resting state. Thus, this phenomenon might be a prospective state marker of schizophrenia.

14.3.3 Auditory Steady-State Response

A steady-state response is an evoked potential with constituent discrete frequency components that remain constant in amplitude and phase over an infinitely long period of time [73]. Steady-state responses are recorded when stimuli are presented periodically and demonstrate how the brain follows a stimulus or how the stimulus drives a response [74]. Steady-state paradigms can probe the ability of neuronal networks to generate and maintain oscillatory activity [39]. Auditory steady-state responses (ASSRs) are known to be most prominent when auditory stimuli are presented periodically at approximately 40 Hz to healthy subjects [74]. Since Kwon et al. [75] first used EEG to show diminished 40 Hz ASSRs in schizophrenia patients, this abnormality has been repeatedly reported in other EEG studies

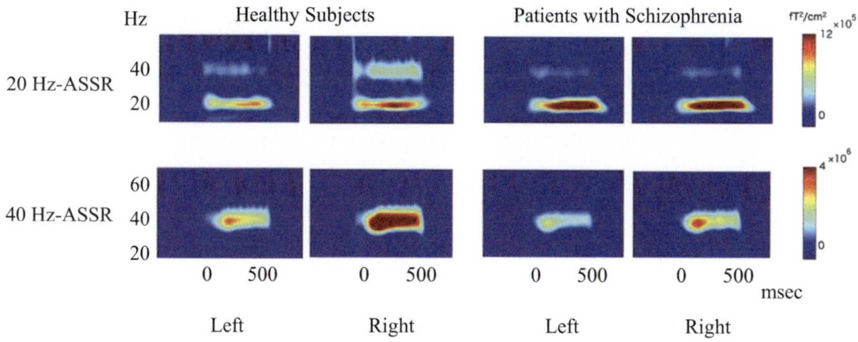

Fig. 14.2 Group-averaged time-frequency maps of MEG-ASSR power for each hemisphere. During the stimulation (from 0 to 500 ms), ASSR can be detected. The color scales signify ASSR power. Patients with schizophrenia showed significantly reduced 40 Hz-ASSR power, with no significant differences for 20 Hz-ASSR power

[76]. For instance, some ASSR studies with MEG have found bilateral diminished 40-Hz ASSRs [77], diminished bilateral 40- and 80-Hz ASSRs [78], diminished bilateral 5- and 80-Hz ASSRs, diminished 40-Hz ASSRs in the right auditory cortex [79], and diminished 40-Hz ASSRs in the left auditory cortex [80] of people with schizophrenia. Figure 14.2 shows group-averaged time-frequency maps of MEG-ASSR power to 20 Hz and 40 Hz click stimuli created by our data. Moreover, Tsuchimoto et al. [78] showed that 80-Hz ASSRs in the left hemisphere were significantly correlated with the severity of auditory hallucinations, and Edgar et al. [80] showed that the cortical thickness of the superior temporal gyrus was significantly correlated with 40-Hz ASSRs in the left hemisphere in healthy controls, but not schizophrenia patients. However, individuals with schizoaffective disorder show a different pattern of 40-Hz ASSRs compared with schizophrenia patients [57], and patients with bipolar disorder exhibit reduced bilateral 30-, 40-, and 80-Hz ASSRs [81]. Exactly why ASSRs are diminished in people with schizophrenia and other psychiatric disorders is not clear, but ASSRs are expected to be useful as biological markers of schizophrenia in translational studies with animal models of schizophrenia [82], as well as in mathematical model studies [83] with comparatively high test–retest reliability [84].

14.4 General Remarks

This section contained a review of MEG studies of patients with schizophrenia, with a focus on event-related responses and neural oscillations. Because the symptoms of schizophrenia can be investigated solely in living patients, biological information obtained from these individuals via EEG and MEG is indispensable to schizophrenia research. Compared with EEG, MEG is neither likely to be affected by volume conduction [2] nor tissue conductivity [3], and MEG enables a more direct estimation of the relationship between temporal and spatial brain current flow and associated brain activities. Moreover, new types of data processing software [85–87] allow users to project MEG data onto brain coordinates obtained via MRI. This enables the calculation and presentation of statistical data in the temporal and spatial domains, greatly facilitating functional brain research. Thus, MEG is expected to play a critical role in new developments in schizophrenia research. However, MEG is very sensitive to outer and inner environments, including artifacts. As a result, MEG experiments may necessitate the use of expensive apparatuses, including shielded rooms and the maintenance of superconducting quantum interference devices in ultralow temperature environments cooled by liquid helium. It is likely that these technical challenges prevent more frequent use of MEG in schizophrenia research. New advances in technology may increase the global application of MEG, perhaps leading to the clinical application of MEG in schizophrenia treatment.

References

1. Bleuler E. Dementia praecox or the group of schizophrenias. New York: International University Press; 1950.
2. van den Broek SP, Reinders F, Donderwinkel M, Peters MJ. Volume conduction effects in EEG and MEG. Electroencephalogr Clin Neurophysiol. 1998;106:522–34.
3. Wolters CH, Anwander A, Tricoche X, Weinstein D, Koch MA, MacLeod RS. Influence of tissue conductivity anisotropy on EEG/MEG field and return current computation in a realistic head model: a simulation and visualization study using high-resolution finite element modeling. NeuroImage. 2006;30:813–26.
4. Reite M, Teale P, Rojas DC. Magnetoencephalography: applications in psychiatry. Biol Psychiatry. 1999;45:1553–63.
5. Otten LJ, Rugg MD. Interpreting event-related brain potentials. In: Handy TC, editor. Event-related potentials: a methods handbook. Cambridge: The MIT Press; 2005. p. 3–16.
6. Adler LE, Pachtman E, Franks RD, Pecevich M, Waldo MC, Freedman R. Neurophysiological evidence for a defect in neuronal mechanisms involved in sensory gating in schizophrenia. Biol Psychiatry. 1982;17:639–54.
7. Freedman R, Adler LE, Waldo MC, Pachtman E, Franks RD. Neurophysiological evidence for a defect in inhibitory pathways in schizophrenia: comparison of medicated and drug-free patients. Biol Psychiatry. 1983;18:537–51.
8. Bickford-Wimer PC, Nagamoto H, Johnson R, Adler LE, Egan M, Rose GM, et al. Auditory sensory gating in hippocampal neurons: a model system in the rat. Biol Psychiatry. 1990;27:183–92.
9. Bramon E, Rabe-Hesketh S, Sham P, Murray RM, Frangou S. Meta-analysis of the P300 and P50 waveforms in schizophrenia. Schizophr Res. 2004;70:315–29.
10. Reite M, Teale P, Zimmerman J, Davis K, Whalen J. Source location of a 50 msec latency auditory evoked field component. Electroencephalogr Clin Neurophysiol. 1988;70:490–8.
11. Thoma RJ, Hanlon FM, Moses SN. Lateralization of auditory sensory gating and neuropsychological dysfunction in schizophrenia. Am J Psychiatry. 2003;160:1595–605.
12. Edgar JC, Miller GA, Moses SN, Thoma RJ, Huang MX, Hanlon FM, et al. Cross-modal generality of the gating deficit. Psychophysiology. 2005;42:318–27.
13. Hirano Y, Hirano S, Maekawa T, Obayashi C, Oribe N, Monji A, et al. Auditory gating deficit to human voices in schizophrenia: a MEG study. Schizophr Res. 2010;117:61–7. doi:10.1016/j.schres.2009.09.003.

14. Lu BY, Edgar JC, Jones AP, Smith AK, Huang MX, Miller GA, et al. Improved test–retest reliability of 50-ms paired-click auditory gating using magnetoencephalography source modeling. Psychophysiology. 2007;44:86–90.
15. Thoma RJ, Hanlon FM, Sanchez N, Weisend MP, Huang M, Jones A, et al. Auditory sensory gating deficit and cortical thickness in schizophrenia. Neurol Clin Neurophysiol. 2004;2004:1–7.
16. Mathiak K, Ackermann H, Rapp A, Mathiak KA, Shergill S, Riecker A, et al. Neuromagnetic oscillations and hemodynamic correlates of P50 suppression in schizophrenia. Psychiatry Res. 2011;194:95–104. doi:10.1016/j.pscychresns.2011.01.001.
17. Potter D, Summerfelt A, Gold J, Buchanan RW. Review of clinical correlates of P50 sensory gating abnormalities in patients with schizophrenia. Schizophr Bull. 2006;32:692–700.
18. Hari R, Kaila K, Katila T, Tuomisto T, Varpula T. Interstimulus interval dependence of the auditory vertex response and its magnetic counterpart: implications for their neural generation. Electroencephalogr Clin Neurophysiol. 1982;54:561–9.
19. Hajek M, Boehle C, Huonker R, Volz HP, Nowak H, Schrott PR, et al. Abnormalities of auditory evoked magnetic fields in the right hemisphere of schizophrenic females. Schizophr Res. 1997;24:329–32.
20. Hajek M, Huonker R, Boehle C, Volz HP, Nowak H, Sauer H. Abnormalities of auditory evoked magnetic fields and structural changes in the left hemisphere of male schizophrenics – a magnetoencephalographic-magnetic resonance imaging study. Biol Psychiatry. 1997;42:609–16.
21. Smith AK, Edgar JC, Huang M, Lu BY, Thoma RJ, Hanlon FM, et al. Cognitive abilities and 50- and 100-msec paired-click processes in schizophrenia. Am J Psychiatry. 2010;167:1264–75. doi:10.1176/appi.ajp.2010.09071059.
22. Ahveninen J, Jääskeläinen IP, Osipova D, Huttunen MO, Ilmoniemi RJ, Kaprio J, et al. Inherited auditory-cortical dysfunction in twin pairs discordant for schizophrenia. Biol Psychiatry. 2006;60:612–20.
23. Näätänen R, Gaillard AW, Mäntysalo S. Early selective-attention effect on evoked potential reinterpreted. Acta Psychol (Amst). 1978;42:313–29.
24. Näätänen R, Kähkönen S. Central auditory dysfunction in schizophrenia as revealed by the mismatch negativity (MMN) and its magnetic equivalent MMNm: a review. Int J Neuropsychopharmacol. 2009;12:125–35. doi:10.1017/S1461145708009322.
25. Pekkonen E, Katila H, Ahveninen J, Karhu J, Huotilainen M, Tiihonen J. Impaired temporal lobe processing of preattentive auditory discrimination in schizophrenia. Schizophr Bull. 2002;28:467–74.
26. Kircher TT, Rapp A, Grodd W, Buchkremer G, Weiskopf N, Lutzenberger W, et al. Mismatch negativity responses in schizophrenia: a combined fMRI and whole-head MEG study. Am J Psychiatry. 2004;161:294–304.
27. Kasai K, Yamada H, Kamio S, Nakagome K, Iwanami A, Fukuda M, et al. Neuromagnetic correlates of impaired automatic categorical perception of speech sounds in schizophrenia. Schizophr Res. 2002;59:159–72.
28. Kasai K, Yamada H, Kamio S, Nakagome K, Iwanami A, Fukuda M, et al. Do high or low doses of anxiolytics and hypnotics affect mismatch negativity in schizophrenic subjects? An EEG and MEG study. Clin Neurophysiol. 2002;113:141–50.
29. Yamasue H, Yamada H, Yumoto M, Kamio S, Kudo N, Uetsuki M, et al. Abnormal association between reduced magnetic mismatch field to speech sounds and smaller left planum temporale volume in schizophrenia. NeuroImage. 2004;22:720–7.
30. Onitsuka T, Shenton ME, Kasai K, Nestor PG, Toner SK, Kikinis R, et al. Fusiform gyrus volume reduction and facial recognition in chronic schizophrenia. Arch Gen Psychiatry. 2003;60:349–55.
31. Bentin S, Allison T, Puce A, Perez E, McCarthy G. Electrophysiological studies of face perception in humans. J Cogn Neurosci. 1996;8:551–65.

32. Onitsuka T, Niznikiewicz MA, Spencer KM, Frumin M, Kuroki N, Lucia LC, et al. Functional and structural deficits in brain regions subserving face perception in schizophrenia. Am J Psychiatry. 2006;163:455–62.
33. Liu J, Harris A, Kanwisher N. Stages of processing in face perception: an MEG study. Nat Neurosci. 2002;5:910–6.
34. Halgren E, Raij T, Marinkovic K, Jousmäki V, Hari R. Cognitive response profile of the human fusiform face area as determined by MEG. Cereb Cortex. 2000;10:69–81.
35. Watanabe S, Miki K, Kakigi R. Mechanisms of face perception in humans: a magneto- and electro-encephalographic study. Neuropathology. 2005;25:8–20.
36. Rivolta D, Castellanos NP, Stawowsky C, Helbling S, Wibral M, Grützner C, et al. Source-reconstruction of event-related fields reveals hyperfunction and hypofunction of cortical circuits in antipsychotic-näive, first-episode schizophrenia patients during mooney face processing. J Neurosci. 2014;34:5909–17. doi:10.1523/JNEUROSCI.3752-13.2014.
37. Huang MX, Lee RR, Gaa KM, Song T, Harrington DL, Loh C, et al. Somatosensory system deficits in schizophrenia revealed by MEG during a median-nerve oddball task. Brain Topogr. 2010;23:82–104. doi:10.1007/s10548-009-0122-5.
38. Gray CM, König P, Engel AK, Singer W. Oscillatory responses in cat visual cortex exhibit inter-columnar synchronization which reflects global stimulus properties. Nature. 1989;338:334–7.
39. Uhlhaas PJ, Singer W. Abnormal neural oscillations and synchrony in schizophrenia. Nat Rev Neurosci. 2010;11:100–13. doi:10.1038/nrn2774.
40. Tallon-Baudry C, Bertrand O. Oscillatory gamma activity in humans and its role in object representation. Trends Cogn Sci. 1999;3:151–62.
41. Gray CM, Singer W. Stimulus-specific neuronal oscillations in orientation columns of cat visual cortex. Proc Natl Acad Sci U S A. 1989;86:1698–702.
42. Fries P, Reynolds JH, Rorie AE, Desimone R. Modulation of oscillatory neuronal synchronization by selective visual attention. Science. 2001;291:1560–3.
43. Tallon-Baudry C, Bertrand O, Peronnet F, Pernier J. Induced gamma-band activity during the delay of a visual short-term memory task in humans. J Neurosci. 1998;18:4244–54.
44. Melloni L, Molina C, Pena M, Torres D, Singer W, Rodriguez E. Synchronization of neural activity across cortical areas correlates with conscious perception. J Neurosci. 2007;27:2858–65.
45. Hong LE, Buchanan RW, Thaker GK, Shepard PD, Summerfelt A. Beta (~16 Hz) frequency neural oscillations mediate auditory sensory gating in humans. Psychophysiology. 2008;45:197–204.
46. Uhlhaas PJ, Linden DE, Singer W, Haenschel C, Lindner M, Maurer K, et al. Dysfunctional long-range coordination of neural activity during Gestalt perception in schizophrenia. J Neurosci. 2006;26:8168–75.
47. von Stein A, Chiang C, König P. Top-down processing mediated by interareal synchronization. Proc Natl Acad Sci U S A. 2000;97:14748–53.
48. Thut G, Nietzel A, Brandt SA, Pascual-Leone A. α-band electroencephalographic activity over occipital cortex indexes visuospatial attention bias and predicts visual target detection. J Neurosci. 2006;26:9494–502.
49. Palva S, Linkenkaer-Hansen K, Näätänen R, Palva JM. Early neural correlates of conscious somatosensory perception. J Neurosci. 2005;25:5248–58.
50. Cunningham MO, Hunt J, Middleton S, LeBeau FEN, Gillies MJ, Gillies MG, et al. Region-specific reduction in entorhinal gamma oscillations and parvalbumin-immunoreactive neurons in animal models of psychiatric illness. J Neurosci. 2006;26:2767–76.
51. Sohal VS, Zhang F, Yizhar O, Deisseroth K. Parvalbumin neurons and gamma rhythms enhance cortical circuit performance. Nature. 2009;459:698–702. doi:10.1038/nature07991.
52. Engel AK, König P, Kreiter AK, Singer W. Interhemispheric synchronization of oscillatory neuronal responses in cat visual cortex. Science. 1991;252:1177–9.

53. Tallon-Baudry C, Bertrand O, Delpuech C, Pernier J. Stimulus specificity of phase-locked and non-phase-locked 40 Hz visual responses in human. J Neurosci. 1996;16:4240–9.
54. Palva S, Palva J, Shtyrov Y, Kujala T. Distinct gamma-band evoked responses to speech and non-speech sounds in humans. J Neurosci. 2002;22:RC211.
55. Hirano S, Hirano Y, Maekawa T, Obayashi C, Oribe N, Kuroki T, et al. Abnormal neural oscillatory activity to speech sounds in schizophrenia: a magnetoencephalography study. J Neurosci. 2008;28:4897–903. doi:10.1523/JNEUROSCI.5031-07.2008.
56. Oribe N, Onitsuka T, Hirano S, Hirano Y, Maekawa T, Obayashi C, et al. Differentiation between bipolar disorder and schizophrenia revealed by neural oscillation to speech sounds: an MEG study. Bipolar Disord. 2010;12:804–12. doi:10.1111/j.1399-5618.2010.00876.x.
57. Reite M, Teale P, Collins D, Rojas DC. Schizoaffective disorder – a possible MEG auditory evoked field biomarker. Psychiatry Res. 2010;182:284–6. doi:10.1016/j.pscychresns.2010.02.007.
58. Grützner C, Wibral M, Sun L, Rivolta D, Singer W, Maurer K, et al. Deficits in high- (>60 Hz) gamma-band oscillations during visual processing in schizophrenia. Front Hum Neurosci. 2013;7:88. doi:10.3389/fnhum.2013.00088.
59. Sun L, Castellanos N, Grützner C, Koethe D, Rivolta D, Wibral M, et al. Evidence for dysregulated high-frequency oscillations during sensory processing in medication-naïve, first episode schizophrenia. Schizophr Res. 2013;150:519–25. doi:10.1016/j.schres.2013.08.023.
60. Herrmann CS, Munk MHJ, Engel AK. Cognitive functions of gamma-band activity: memory match and utilization. Trends Cogn Sci. 2004;8:347–55.
61. Cañive JM, Lewine JD, Edgar JC, Davis JT, Torres F, Roberts B, et al. Magnetoencephalographic assessment of spontaneous brain activity in schizophrenia. Psychopharmacol Bull. 1996;32:741–50.
62. Cañive JM, Lewine JD, Edgar JC, Davis JT, Miller GA, Torres F, et al. Spontaneous brain magnetic activity in schizophrenia patients treated with aripiprazole. Psychopharmacol Bull. 1998;34:101–5.
63. Sperling W, Vieth J, Martus M, Demling J, Barocka A. Spontaneous slow and fast MEG activity in male schizophrenics treated with clozapine. Psychopharmacology (Berl). 1999;142:375–82.
64. Sperling W, Martus P, Kober H, Bleich S, Kornhuber J. Spontaneous, slow and fast magnetoencephalographic activity in patients with schizophrenia. Schizophr Res. 2002;58:189–99.
65. Fehr T, Kissler J, Moratti S, Wienbruch C. Source distribution of neuromagnetic slow waves and MEG-delta activity in schizophrenic patients. Biol Psychiatry. 2001;50:108–16.
66. Rockstroh BS, Wienbruch C, Ray WJ, Elbert T. Abnormal oscillatory brain dynamics in schizophrenia: a sign of deviant communication in neural network? BMC Psychiatry. 2007;7:44.
67. Rutter L, Carver FW, Holroyd T, Nadar SR, Mitchell-Francis J, Apud J, et al. Magnetoencephalographic gamma power reduction in patients with schizophrenia during resting condition. Hum Brain Mapp. 2009;30:3254–64. doi:10.1002/hbm.20746.
68. Rutter L, Nadar SR, Holroyd T, Carver FW, Apud J, Weinberger DR, et al. Graph theoretical analysis of resting magnetoencephalographic functional connectivity networks. Front Comput Neurosci. 2013;7:93. doi:10.3389/fncom.2013.00093.
69. Kim JS, Shin KS, Jung WH, Kim SN, Kwon JS, Chung CK. Power spectral aspects of the default mode network in schizophrenia: an MEG study. BMC Neurosci. 2014;15:104.
70. van Lutterveld R, Hillebrand A, Diederen KMJ, Daalman K, Kahn RS, et al. Oscillatory cortical network involved in auditory verbal hallucinations in schizophrenia. PLoS One. 2012;7:e41149. doi:10.1371/journal.pone.0041149.
71. Ropohl A, Sperling W, Elstner S, Tomandl B, Reulbach U, Kaltenhäuser M, et al. Cortical activity associated with auditory hallucinations. Neuroreport. 2004;15:523–6.

72. Ishii R, Shinosaki K, Ikejiri Y, Ukai S, Yamashita K, Iwase M, et al. Theta rhythm increases in left superior temporal cortex during auditory hallucinations in schizophrenia: a case report. Neuroreport. 2000;11:3283–7.
73. Regan D. Human brain electrophysiology: evoked potentials and evoked magnetic fields in science and medicine. Amsterdam: Elsevier; 1989.
74. Picton TW, John MS, Dimitrijevic A, Purcell D. Human auditory steady-state responses. Int J Audiol. 2003;42:177–219.
75. Kwon JS, O'Donnell BF, Wallenstein GV, Greene RW, Hirayasu Y, Nestor PG, et al. Gamma frequency-range abnormalities to auditory stimulation in schizophrenia. Arch Gen Psychiatry. 1999;56:1001–5.
76. O'Donnell BF, Vohs JL, Krishnan GP, Rass O, Hetrick WP, Morzorati SL. The auditory steady-state response (ASSR): a translational biomarker for schizophrenia. Suppl Clin Neurophysiol. 2013;62:101–12.
77. Teale P, Collins D, Maharajh K, Rojas DC, Kronberg E, Reite M. Cortical source estimates of gamma band amplitude and phase are different in schizophrenia. NeuroImage. 2008;42:1481–9. doi:10.1016/j.neuroimage.2008.06.020.
78. Tsuchimoto R, Kanba S, Hirano S, Oribe N, Ueno T, Hirano Y, et al. Reduced high and low frequency gamma synchronization in patients with chronic schizophrenia. Schizophr Res. 2011;133:99–105. doi:10.1016/j.schres.2011.07.020.
79. Hamm JP, Gilmore CS, Picchetti NAM, Sponheim SR, Clementz BA. Abnormalities of neuronal oscillations and temporal integration to low- and high-frequency auditory stimulation in schizophrenia. Biol Psychiatry. 2011;69:989–96. doi:10.1016/j.biopsych.2010.11.021.
80. Edgar JC, Chen Y-H, Lanza M, Howell B, Chow VY, Heiken K, et al. Cortical thickness as a contributor to abnormal oscillations in schizophrenia? NeuroImage Clin. 2014;4:122–9. doi:10.1016/j.nicl.2013.11.004.
81. Oda Y, Onitsuka T, Tsuchimoto R, Hirano S, Oribe N, Ueno T, et al. Gamma band neural synchronization deficits for auditory steady state responses in bipolar disorder patients. PLoS One. 2012;7:e39955. doi:10.1371/journal.pone.0039955.
82. Nakao K, Nakazawa K. Brain state-dependent abnormal LFP activity in the auditory cortex of a schizophrenia mouse model. Front Neurosci. 2014;8:168. doi:10.3389/fnins.2014.00168.
83. Vierling-Claassen D, Siekmeier P, Stufflebeam S, Kopell N. Modeling GABA alterations in schizophrenia: a link between impaired inhibition and altered gamma and beta range auditory entrainment. J Neurophysiol. 2008;99:2656–71. doi:10.1152/jn.00870.2007.
84. McFadden KL, Steinmetz SE, Carroll AM, Simon ST, Wallace A, Rojas DC. Test-retest reliability of the 40 Hz EEG auditory steady-state response. PLoS One. 2014;9:e85748. doi:10.1371/journal.pone.0085748.
85. Oostenveld R, Fries P, Maris E, Schoffelen JM. FieldTrip: open source software for advanced analysis of MEG, EEG, and invasive electrophysiological data. Comput Intell Neurosci. 2011;2011:156869. doi:10.1155/2011/156869.
86. Delorme A, Makeig S. EEGLAB: an open source toolbox for analysis of single-trial EEG dynamics including independent component analysis. J Neurosci Methods. 2004;134:9–21.
87. Gramfort A, Luessi M, Larson E, Engemann DA, Strohmeier D, Brodbeck C, et al. MNE software for processing MEG and EEG data. NeuroImage. 2014;86:446–60. doi:10.1016/j.neuroimage.2013.10.027.

Bipolar Disorder

15

Toshihiko Maekawa, Yuko Oda, and Shozo Tobimatsu

Abstract

Bipolar disorder (BD) is a common psychiatric condition that is associated with a diverse range of symptoms across several domains. While the core feature is disturbance in mood states, disruption of biological rhythms, energy behavior, and cognition is central to its impact on functioning in daily life. Diagnosis requires knowledge about an individual's episodes of distinctly elevated mood: mania for type I BD and hypomania for type II. Defining the nature of BD has always been something of a challenge for psychiatrists, and, thus, establishing a neurobiological model of the condition has been a matter of long debate. This overview of magnetoencephalographic (MEG) studies that investigated BD describes event-related responses related to auditory function such as M50, M100, M200, MMNm, and neural oscillations. Based on our review, auditory dysfunctions in BD are outlined, and interpretations are delivered in light of recent developments in the field of clinical neurophysiology. Thus, we lead clinicians and researchers through some of the more actively debated topics regarding BD research and its potential direction in the future.

Keywords

Bipolar disorder • M50 • M100 • M200 • Mismatch negativity • Neural oscillations

T. Maekawa (✉) • Y. Oda • S. Tobimatsu
Department of Clinical Neurophysiology, Graduate School of Medical Sciences, Kyushu University, 3-1-1 Maidashi, Higashi-ku, Fukuoka 812-8582, Japan
e-mail: t-mae@npsych.med.kyushu-u.ac.jp

S. Tobimatsu, R. Kakigi (eds.), *Clinical Applications of Magnetoencephalography*, DOI 10.1007/978-4-431-55729-6_15

15.1 Introduction

Bipolar disorder (BD), also known as manic depression, is a brain disorder that causes unusual shifts in mood, energy, activity levels, and the ability to carry out day-to-day tasks. Diagnosis requires knowledge of an individual's experiences during episodes of distinctly elevated mood: mania for type I BD and hypomania for type II. The majority of patients also suffer from periods of depression, with this being stipulated in the criteria for the type II disorder. Indeed, depression is increasingly being recognized as the greatest burden of the illness [1]. Classically described, those with BD are said to have discrete periods of illness from which they return to a normal state. Currently, people widely acknowledge that this pattern may not accurately describe the nature of the condition because dysfunction may be detectable even during euthymia. Further, many patients experience chronic illness, mixed presentations, rapid cycling, and/or suboptimal responses to treatment [2].

Defining the nature of BD has always been something of a challenge for psychiatrists. Thus, establishing a neurobiological model of the condition has been a matter of long debate. The episodic nature of the disorder and the ostensible polar extremes of mania and depression have historically resulted in a functional account being favored over an organic model, and disturbed activity within the brain regions has been postulated to return to normal with time or treatment. Abnormalities in neuronal transmission are thought to underlie the symptoms, often in a dichotomous manner that mirrors the polar nature of the illness. For example, mania has been explored in terms of increased dopaminergic activity, whereas depression has been discussed in terms of reduced dopaminergic activity [3]. Such a stance has merit, with much evidence supporting the assertions, but factors such as medication, psychosis, physical activity, and stress-axis activation (all of which can span episodes or emerge during both mania and depression) confound the investigation and interpretation of potential neurobiological abnormalities [4].

The brain can be viewed as the source or mediator of normal and abnormal mental states, but may also be considered as an organ vulnerable to the damaging effects of illness. Thus, investigating the BD brain may provide insights into the causes and consequences of the condition. This overview of magnetoencephalographic (MEG) studies related to BD relates abnormalities in event-related auditory responses of BD patients, including M50, M100, M200, MMNm, and neural oscillations. We have focused on the interpretation of results in light of recent developments from the field of clinical neurophysiology. With this narrative article, we will provide necessary information about actively debated topics in the field about which we think clinicians and researchers should know.

15.2 Altered Sensory Gating Function

Hippocampal structure and function are abnormal in BD [5]. The most conspicuous functional abnormalities associated with hippocampal damage are deficits in learning and memory [6]. The hippocampus is also critical for the gating of sensory responses to stimuli and is involved in prepulse inhibition (PPI) of the startle reflex, sometimes known as sensorimotor gating [7, 8]. Just as a large body of literature suggests that many patients with schizophrenia have impairments in sensory gating [9, 10], this deficit has also been reported in BD [11, 12]. Sensory gating is typically assessed by a paired-click paradigm in which two identical click stimuli are presented in succession, and the P50 wave of the EEG response following each click is measured. Normally, individuals show sensory gating as assessed by a reduced P50 response following the second click. Adler et al. [13] proposed that P50 gating is hippocampal dependent, and studies assessing N40, the rat P50 analog, have confirmed this [14, 15]. However, because P50 is traditionally measured at a single EEG electrode (Cz), little information is available about hemispheric gating differences and the role of the hippocampus. Presumably assessing M50 (the P50 counterpart in MEG) sensory gating is a more direct measure than general scalp ERP measurements in terms of the neural generators.

M50 responses are localized to posterior areas of the bilateral superior temporal gyrus (STG) [16–18]. Moreover, bilateral M50-STG sources have been reported to account for 97 % of the variance in P50 responses at the Cz electrode in healthy controls, which demonstrates that the P50 auditory evoked potential (AEP) can be attributed to bilateral M50 STG-source localizations [19]. These findings confirm the importance of evaluating hemispheres separately for sensory gating deficits. Wang et al. [20] recorded M50 from 20 BD type I patients and 20 healthy controls during a paired-click paradigm (Fig. 15.1). The M50 dipole source was localized to

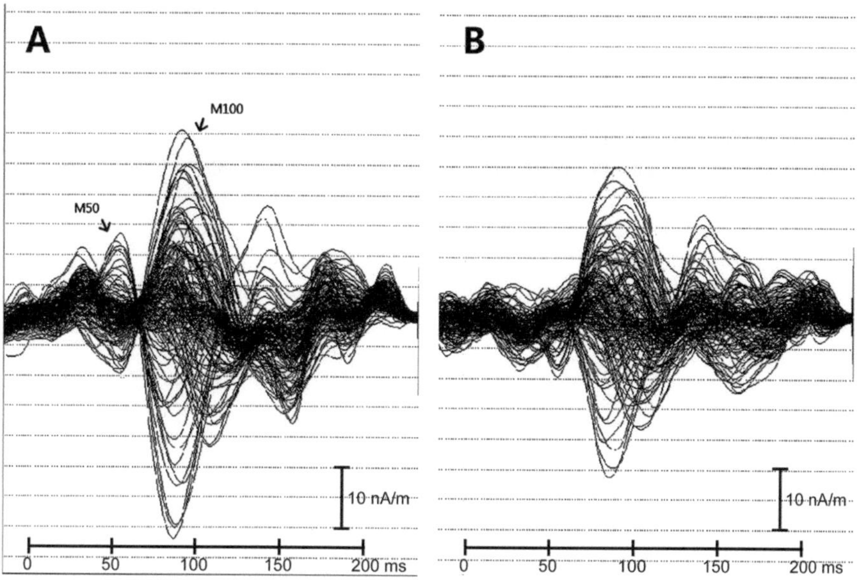

Fig. 15.1 Example of the 148 MEG-sensor overlaid waveforms in response to the first click for one BD type I patient. (**a**) Responses to the first stimulus. (**b**) Responses to the second stimulus. Note that the M50 component is strongly suppressed in response to the second stimulus (i.e., sensory gating) (Revised from [20])

the left and right posterior portion of the STG, and the authors could not find any significant difference in M50 strength between groups. However, bilateral M50-source strengths evoked by the second stimuli were significantly higher in the BD type I group than in the control group. Therefore, they concluded that the auditory gating deficit in type I BD was attributable to both hemispheres, suggesting that gating deficits are related to structural abnormalities in the STG.

15.3 Changes in Early Auditory Information Processing

The N100 (the EEG counterpart of M100) is the largest component among the AEPs, peaking 75–150 ms after stimulus onset and thus reflecting early auditory processing. N100 has been suggested as a physiological correlate of cognitive processes such as attention [21], memory [22], and stimulus classification [23]. It is generated in or near Heschl's gyrus [24] and the planum temporale [25, 26] of the STG. Its source strengths and latencies were not found to be different between BD patients and controls [27].

The P200 is a positive AEP elicited by auditory stimuli that is also most prominent over the vertex (Cz) and has a latency of approximately 150–250 ms. It represents attention-modulated stimulus detection or classification processes [28]. Moreover, the amplitude of the P200 decreases as attentiveness increases and increases during sleep [29]. Studies about auditory P200 potentials in patients with BD are very few and their results seem contradictory [30–33]. Although M200 deflection peaking at approximately 200 ms after stimulus onset is a prominent response relating to the STG [34], planum temporale, and area 22 (the auditory association cortex) [35], M200 response abnormalities in BD have not yet been investigated in detail. Wang et al. [36] measured M100 and M200 during an auditory oddball paradigm in 24 schizophrenia patients, 26 BD-I patients, and 31 healthy controls (Fig. 15.2). They found an asymmetric pattern of M100 and M200 auditory responses with more anterior sources in the right STG of healthy controls, while both schizophrenia and BD patients showed symmetric M100 and M200 source patterns. Additionally, both patient groups showed significantly reduced M100 and M200 source strength in both hemispheres. Therefore, they concluded that early auditory information-processing deficits may be similar in schizophrenia and BD and may be related to abnormalities of the STG.

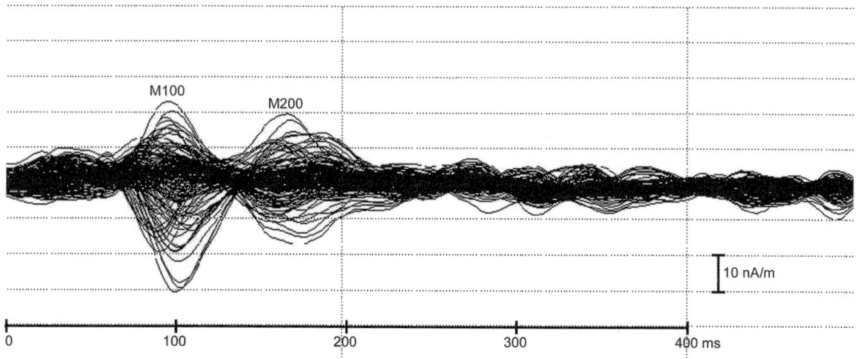

Fig. 15.2 An example of the 148 MEG-sensor overlaid waveforms in response to a 1-kHz tone in a healthy subject. M100 and M200 responses are clearly observed (Revised from [36])

15.4 Preattentive Dysfunction in Auditory Information Processing

Mismatch negativity (MMN) and its magnetic counterpart (the mismatch field; MMNm) are considered to reflect preattentive information processing and are elicited approximately 150–200 ms after inserting physically deviant auditory stimuli into a stream of identical and repeated stimulus sequences [37]. Although some evidence exists for attentional modulation of MMN [38, 39], it is elicited even when attention is directed away from sounds. Thus, interpreting MMN results is easier than interpreting P300 results when patient motivation cannot be controlled, as is the case in psychiatric patients. This automatic mismatch process may have an important role in initiating involuntary switches of attention to changes in auditory stimuli that occur outside the focus of attention [40].

While several studies have reported that MMN amplitude and latency in patients with BD are not significantly different from those in healthy controls [41–44], other studies have found MMN deficits in patients with BD [45–49]. Two MEG studies have investigated MMNm in patients with BD [50, 51]. Takei et al. reported that compared with healthy controls, the latency of pitch-MMNm was significantly delayed in the right hemisphere of patients with BD [50], and Shimano et al. [51] found that BD patients showed a significant bilateral reduction in the magnetic global field of the pitch-MMNm compared with healthy controls (Fig. 15.3). Moreover, the magnetic global field of pitch-MMNm in patients who had a history of being admitted to a hospital was significantly lower in amplitude compared with patients who had no such experience. Thus, patients with BD may have a preattentive dysfunction that can be indexed by abnormal pitch-MMNm.

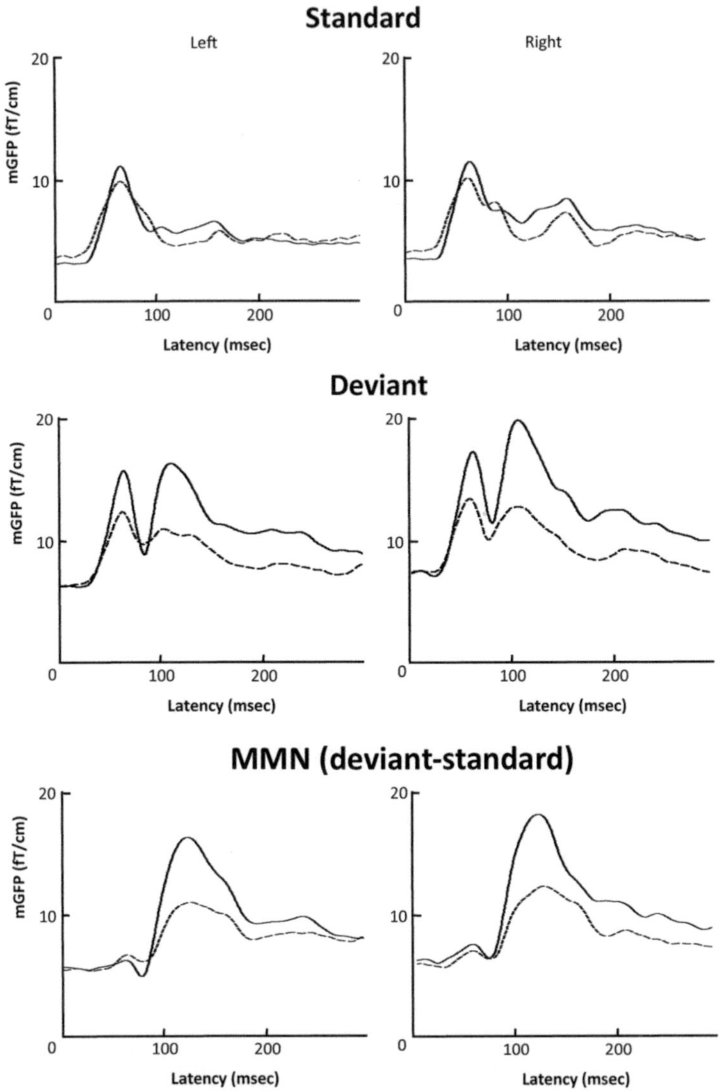

Fig. 15.3 Grand mean magnetic global field power (mGFP) waveforms of responses to standard stimuli, deviant stimuli, and mismatch fields (MMNm) in patients with BD (*dotted line*) and healthy controls (*solid line*). MMN mGFP in the BD group is significantly smaller than that of the healthy control. *Left* left hemisphere, *Right* right hemisphere (Reprinted from [51])

15.5 Abnormal Neural Oscillations

The term "neural (or brain) oscillations" refers to the rhythmic and/or repetitive electrical activity generated spontaneously and in response to stimuli by neural tissue in the central nervous system. The role of neural oscillations as functional building blocks in sensory-cognitive processes has gained support in recent decades. Compared with healthy brains, event-related oscillations in alpha-, beta-, gamma-, delta-, and theta-frequency windows are highly different throughout the cortex of pathologic brains, in particular, those from patients with cognitive impairments such as psychiatric disorders.

15.5.1 Neural Oscillations During Resting State

Chen et al. [52] analyzed MEG responses at rest (i.e., no specific task) to assess the synchronization of neural oscillations in frontal regions of BD patients and normal controls. Two-min recordings were made with a whole-head 306-channel MEG system from ten type I BD patients in euthymic phases and ten normal controls while they rested with their eyes closed. They found significant dynamic changes in BD patients that were characterized by increased synchronization of slow-frequency (delta) oscillation and decreased synchronization of fast-frequency (beta) oscillations. Furthermore, they found a positive correlation between beta-wave synchronization level and the number of perseveration errors made during the Wisconsin card-sorting task, a finding they inferred indicated a deficit of a pre-frontal function in BD patients. Therefore, they concluded that analyzing spontaneous MEG recordings taken at rest using nonlinear dynamic approaches might disclose subtle regional changes in the neural dynamics of BD patients.

15.5.2 Neural Oscillations to Speech Sounds

Neural oscillations (20–45 Hz) in response to speech sounds and pure tones have been recorded from BD, schizophrenia, and normal control groups (Fig. 15.4). BD patients were found to exhibit larger responses at bilateral temporal regions to speech sounds compared with either schizophrenia or control groups, but responses to pure tones did not differ across groups [53]. Whereas several EEG studies have shown decreased power of neural oscillations in BD patients compared with normal controls [54], increased neural oscillations in response to speech sounds are peculiar and may characterize the pathophysiological alterations found in BD.

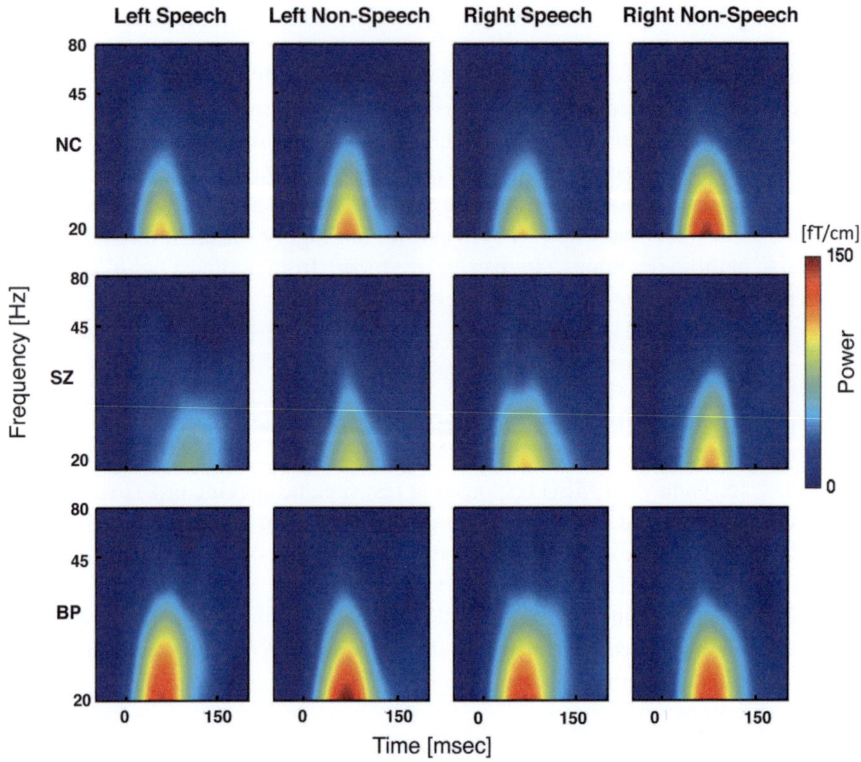

Fig. 15.4 Overall average time-frequency maps of the evoked neural oscillation (eNO) power to speech sounds and pure tone in patients with BD and schizophrenia and in normal controls. The color scale signifies the eNO power. The power from speech sounds in the BD group is clearly higher than that of the normal controls, whereas the power from pure tones is not significantly different between the two groups (Reprinted from [53])

15.5.3 Neural Oscillations Induced by Auditory Steady-State Stimulation

Periodic auditory click stimulation elicits an auditory steady-state response (ASSR) that synchronizes to both the phase and frequency of the click stimulus. Several MEG studies have reported that ASSR source generators are restricted to the primary auditory cortex [27, 55]. Hence, neural oscillatory function in the primary auditory cortex can be assessed using MEG-ASSR. The ASSR reveals information about neural activity with respect to phase synchronization and response magnitude and exhibits resonant frequencies in response to click trains at approximately 40 Hz [56].

Some have suggested that the ASSR reflects the efficacy of γ-amino butyric acid (GABA) inhibitory interneuronal activity, which controls the timing of pyramidal neuron firing in layers II/III of the cortex [57, 58]. Additionally, interactions between pyramidal neurons and inhibitory neurons have been found to produce emergent oscillations [59]. Because some studies imply GABAergic dysfunction plays a role in BD [60, 61], investigating ASSR is likely an important factor for understanding BD.

A MEG study of ASSR reported that compared with normal controls, BD patients showed a reduction in the right 40-Hz ASSR [27, 62]. Although ASSR laterality cannot be determined by EEG, its power reduction was consistent in EEG studies [32, 63]. Ulhaas et al. suggested that GABA is involved in generating the synchronization between beta and gamma oscillations [64]. Oda et al. [65] investigated the MEG-ASSR elicited by 20-, 30-, 40-, and 80-Hz click trains and symptom-ASSR associations in patients with BD. They found that BD patients exhibited bilaterally reduced mean ASSR power and phase-locking factor in response to 30-, 40-, and 80-Hz stimulation, with no significant reduction to 20-Hz stimulation (Fig. 15.5). Moreover, they found a significant negative correlation between right hemisphere 80-Hz ASSR-power values and structured interview guide for the Hamilton depression rating scale (SIGH-D) scores in patients with BD. Therefore, they concluded that BD is characterized by gamma-band ASSR deficits, which may be associated with dysfunctions of GABA inhibitory interneuronal activity.

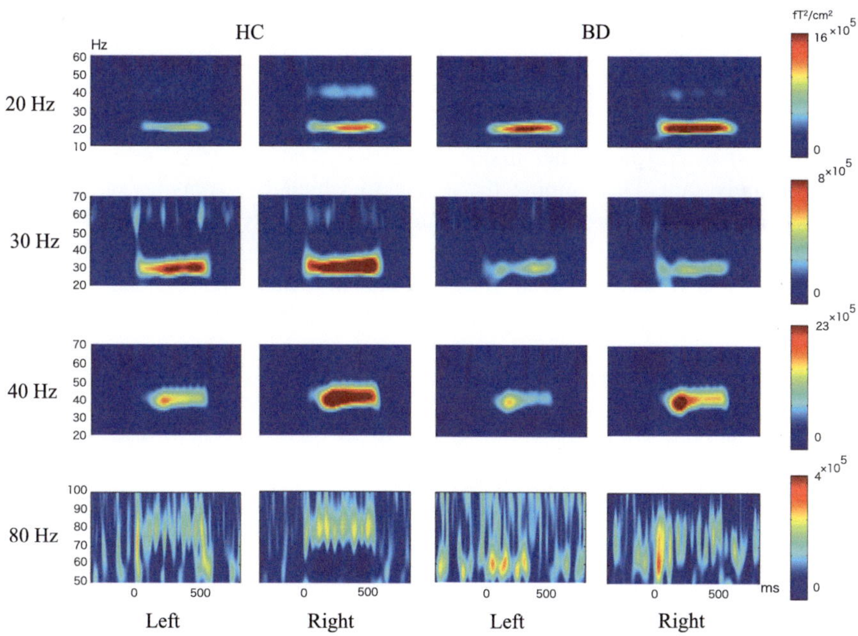

Fig. 15.5 Group averaging time-frequency maps of ASSR power for each hemisphere. The color scales signify ASSR power. BD patients exhibited bilaterally reduced mean ASSR power to 30-, 40-, and 80-Hz stimulation, with no significant reduction to 20-Hz stimulation. *HC* healthy controls, *BD* patients with bipolar disorder (Reprinted from [65])

15.6 General Remarks

To date, relatively few MEG studies of BD have been published. In this review, we have attempted to give an outline of recent findings and to clarify their neurophysiological and pathological interpretations. In particular, we focused on auditory tasks because as far as we know, only one report has used a visual task (face processing) [66]. Altered auditory processing in which the STG may play an important role is a common finding among many studies. The above inferences aside, a few limitations should be noted. First, the majority of studies recruited relatively modest numbers of participants per group (e.g., <30), thereby allowing only limited conclusions with regard to the wider population of individuals with BD. Similarly, few studies have compared findings from individuals with BD across different mood states, and few have been replicated. This is probably because of the different methodologies used, in addition to modest sample sizes. Clearly, more MEG studies with larger sample sizes that use similar techniques are needed in BD research [67].

Second, BD research has not taken full advantage of multimodal neuroimaging techniques to identify structure – function relationships in the brain [68]. In parallel, studies are beginning to identify relationships between genetic variants and function in adults and youths with BD [69]. Ultimately, an integrated system approach will help identify biomarkers that reflect neuropsychophysiological processes in individuals with mood, psychotic, and other psychiatric disorders that span genetic, molecular, neuroimaging, and behavioral levels of investigation [70]. Such studies will facilitate a more in-depth understanding of the neural mechanisms underlying BD.

Third, studies directly comparing different BD subtypes (e.g., BD type I and type II disorders) or BD and other major psychiatric disorders (e.g., schizophrenia) are extremely rare. Determining the extent to which BD subtypes or different psychiatric disorders share, or are distinguished by, underlying neural mechanisms is therefore quite difficult. Such studies have the potential to identify neural biomarkers that reflect neural mechanisms and thus aid diagnosis and treatment choice, particularly for disorders that are often difficult to distinguish using clinical assessment alone.

Fourth, studies in youths with BD and those at risk for BD are also an urgent matter, yet too few have been conducted. Longitudinal studies are clearly needed to examine developmental trajectories of structural and functional changes in the brains of individuals with BD and in youths at risk for future mood and psychotic disorders. Such studies will help identify abnormal developmental trajectories in the brain that are associated with BD as well as biomarkers that can help identify which at-risk youths are most likely to develop it. Unfortunately, a longitudinal MEG study for at-risk youths has yet to be conducted.

Fifth, another major criticism of BD studies is the potentially confounding effects of psychotropic medication. An increasing number of MRI studies of BD indicate that psychotropic medication either has a normalizing effect on neuroimaging measures or does not have a significant impact on these measures [71]; however, very few MEG studies have focused on this issue. Longitudinal studies examining individuals before and after medication can address this important point, as can large cross-sectional studies comparing medication-free individuals with those taking different types of medication.

15.7 Conclusion

We surveyed studies that recorded M50, M100, M200, and neural oscillatory responses evoked by auditory stimuli in patients with BD. Analysis showed STG abnormalities might occur in these patients. One exciting aspect of MEG is its ability to relatively directly investigate neural activity and oscillations within and across brain regions with very high spatiotemporal resolution. For clinical applications, the results of MEG studies in BD are good examples in the area of neuropsychiatric disorders. The demonstration of very high sensitivity and specificity for the characterization of BD suggests that MEG is a promising diagnostic tool in this field of research. Hopefully, correct classification of individuals with BD using an automated analysis makes it likely to become clinically useful in the future. Finally, the demonstration that MEG can be used as a biomarker of the potential efficacy of psychotropic drugs such as mood stabilizers and antiepileptic drugs is very exciting. We therefore conclude that MEG is ready to become a regular tool in BD research and with further tests may well be useful for clinical practice in treating patients with BD.

References

1. Kessler RC, Akiskal HS, Ames M, et al. Prevalence and effects of mood disorders on work performance in a nationally representative sample of U.S. workers. Am J Psychiatry. 2006; 163:1561–8.
2. Treuer T, Tohen M. Predicting the course and outcome of bipolar disorder: a review. Eur Psychiatry. 2010;25:328–33.
3. Silverstone T. Dopamine in manic depressive illness. A pharmacological synthesis. J Affect Disord. 1985;8:225–31.
4. Cousins DA, Butts K, Young AH. The role of dopamine in bipolar disorder. Bipolar Disord. 2009;11:787–806.
5. Otten M, Meeter M. Hippocampal structure and function in individuals with bipolar disorder: a systematic review. J Affect Disord. 2015;174:113–25.
6. Henson RN, Gagnepain P. Predictive, interactive multiple memory systems. Hippocampus. 2010;20:1315–26.
7. Swerdlow NR, Powell SB, Breier MR, Hinse SR, Light GA. Coupling gene expression in medial prefrontal cortex and nucleus accumbens after neonatal ventral hippocampal lesions accompanies deficits in sensorimotor gating and auditory processing in rats. Neuropharmacology. 2013;75:38–46.
8. Bak N, Glenthoj BY, Rostrup E, Larsson HB, Oranje B. Source localization of sensory gating: a combined EEG and fMRI study in healthy volunteers. Neuroimage. 2011;54:2711–8.
9. Vicek P, Bob P, Raboch J. Sensory disturbances, inhibitory deficits, and the P50 wave in schizophrenia. Neuropsychiatr Dis Treat. 2014;10:1309–15.

10. Dissanayake DW, Mason R, Marsden CA. Sensory gating, Cannabinoids and schizophrenia. Neuropharmacology. 2013;67:66–77.
11. Carroll CA, Vohs JL, O'donnell BF, Shekhar A, Hetrick WP. Sensorimotor gating in manic and mixed episode bipolar disorder. Bipolar Disord. 2007;9:221–9.
12. Gogos A, van den Buuse M, Rossell S. Gender differences in prepulse inhibition (PPI) in bipolar disorder: men have reduced PPI, women have increased PPI. Int Neuropsychopharmacol. 2009;12:1249–59.
13. Adler LE, Olincy A, Waldo M, Harris JG, Griffith J, Stevens K, et al. Schizophrenia, sensory gating and nicotinic receptors. Schizophr Bull. 1998;24:189–202.
14. Adler LE, Rose G, Freedman R. Neurophysiological studies of sensory gating in rats: effects of amphetamine, phencyclidine, and haloperidol. Biol Psychiatry. 1986;21:787–98.
15. Bickford-Wimer PC, Nagamoto H, Johnson R, Adler LE, Egan M, Rose GM, et al. Auditory sensory gating in hippocampal neurons: a model system in the rat. Biol Psychiatry. 1990;27: 183–92.
16. Josef Golubic S, Aine CJ, Stephan JM, Adair JC, Knoefel JE, et al. Modulatory role of the prefrontal generator within the auditory M50 network. Neuroimage. 2014;92:120–31.
17. Hunter M, Villarreal G, McHaffie GR, Jimenez B, Smith AK, et al. Lateralization abnormalities in auditory M50 sensory gating and cortical thickness of the superior temporal gyrus in post-traumatic stress disorder. Preliminary results. Psychiatry Res. 2011;191:138–44.
18. Roberts TP, Khan SY, Rey M, Monroe JF, Cannon K, et al. MEG detection of delayed auditory evoked responses inn autism spectrum disorders: towards an imaging biomarker for autism. Autism Res. 2010;3:8–18.
19. Huang M, Edgar J, Thoma R, Hanlon F, Moses S, Lee R, et al. Predicting EEG responses using MEG sources in superior temporal gyrus reveals source asynchrony in patients with schizophrenia. Clin Neurophysiol. 2003;114:835–50.
20. Wang Y, Feng Y, Jia Y, Wang W, Xie Y, et al. Auditory M50 and M100 sensory gating deficits in bipolar disorder: a MEG study. J Affect Disord. 2014;152–154:131–8.
21. Thronton ARD, Harmer M, Lavoie BA. Selective attention increases the temporal precision of the auditory N100 event-related potentials. Hear Res. 2007;230:73–9.
22. Sams M, Hari R, Rif J, Knuutila J. The human auditory sensory memory trace persists about 10 sec: neuromagnetic evidence. J Cogn Neurosci. 1993;5:363–70.
23. Pantev C, Hoke M, Lutkenhoner B, Lehnertz K. Tonotopic organization of the auditory cortex: pitch versus frequency representation. Science. 1989;246:486–8.
24. Reite M, Adams M, Simon J, Teale P, Sheeder J, et al. Auditory M100 component 1: relationship to Heschl's gyri. Brain Res Cogn Brain Res. 1994;2:12–20.
25. Hanlon FM, Miller GA, Thoma RJ, Irwin J, Jones A, Moses SN, et al. Distinct M50 and M100 auditory gating deficits in schizophrenia. Psychophysiology. 2005;42:417–27.
26. Teale P, Sheeder J, Rojas DC, Walker J, Reite M. Sequential source of the M100 exhibits interhemispheric asymmetry. Neuroreport. 1998;9:26–47.
27. Reite M, Teale P, Rojas DC. MEG auditory evoked fields suggest altered structural/functional asymmetry in primary but not secondary auditory cortex in bipolar disorder. Bipolar Disord. 2009;11:371–81.
28. Crowley KE, Colrain IM. A review of the evidence for P2 being an independent component process: age, sleep and modality. Clin Neurophysiol. 2004;115:732–44.
29. Cortoos A, De Valck E, Pattyn N, Mairesse O, Cluydts R. Excitatory versus inhibitory impairments in insomnia patients: an ERP study. Int J Psychophysiol. 2014;93:52–69.
30. Fridberg DJ, Hetrick WP, Brenner CA, Shekhar A, Steffen A, et al. Relationships between auditory event-related potentials and mood state, medication, and comorbid psychiatric illness in patients with bipolar disorder. Bipolar Disord. 2009;11:857–66.
31. Hamm JP, Ethridge IE, Shapiron JR, Pearison GD, Tamminga CA, et al. Family history of psychosis moderates early auditory cortical response abnormalities in non-psychotic bipolar disorder. Bipolar Disord. 2013;15:774–86.

32. O'Donnell B, Vohs J, Hetrick W, Caroll C, Shekhar A. Auditory event-related potential abnormalities inn bipolar disorder and schizophrenia. Int J Psychophysiol. 2004;53:45–55.
33. Tabras-Seisdedos R, Balanza-Martinez V, Pallardo Y, Salazar-Fraile J, Selva G, et al. Similar effect of family history of psychosis on Sylvian fissure size and auditory P200 amplitude in schizophrenia and bipolar disorder subjects. Psychiatry Res. 2001;108:29–38.
34. Hari R, Pelizzone M, Mäkelä J, Hällstöm, Leinonen L, et al. Neuromagnetic responses of the human auditory cortex to on-and offsets of noise bursts. Audiology. 1987;26:31–43.
35. Godey B, Schwartz D, de Graaf JB, Chauvel P, Liegeois-Chauvel C. Neuromagnetic source localization of auditory evoked fields and intracerebral evoked potentials: a comparison of data in the same patients. Clin Neurophysiol. 2001;112:1850–9.
36. Wang Y, Jia Y, Feng Y, Zhong S, Xie Y, et al. Overlapping auditory M100 and M200 abnormalities in schizophrenia and bipolar disorder: a MEG study. Schizophr Res. 2014; 160:201–7.
37. Näätänen R, Kujala T, Winkler I. Auditory processing dysfunction in schizophrenia as revealed by the mismatch negativity (MMN) and its magnetic equivalent MMNm: a review. Int J Neuropsychopharm. 2009;12:125–35.
38. Trejo LJ, Ryan Jones DL, Kramer AF. Attentional modulation of the mismatch negativity elicited by frequency differences between binaurally presented tone bursts. Psychophysiology. 1995;32:319–28.
39. Woldorff MG, Hillyard SA, Gallen CC, Hampson SR, Bloom FE. Magnetoencephalographic recordings demonstrate attentional modulation of mismatch-related neural activity in human auditory cortex. Psychophysiology. 1998;35:283–92.
40. Maekawa T, Hirano S, Onitsuka T. Auditory and visual mismatch negativity in psychiatric disorders: a review. Curr Psychiatry Rev. 2012;8:97–105.
41. Hall MH, Rijsdijk F, Kalidindi S, et al. Genetic overlap between bipolar illness and event-related potentials. Psychol Med. 2007;37:667–78.
42. Hall MH, Schultz K, Rijsdijk F, et al. Are auditory P300 and duration-MMN heritable and putative endophenotypes of psychotic bipolar disorder? A Maudsley Bipolar Twin and Family Study. Psychol Med. 2009;39:1277–87.
43. Catts SV, Shelley AM, Ward PB, Liebert B, McConaghy N, et al. Brain potential evidence for an auditory sensory memory deficit in schizophrenia. Am J Psychiatry. 1995;152:213–9.
44. Baldeweg T, Hirsch SR. Mismatch negativity indexes illness-specific impairments of cortical plasticity in schizophrenia: a comparison with bipolar disorder and Alzheimer's disease. Int J Psychophysiol. 2015;95:145–55.
45. Chitty KM, Logopoulos J, Kaur M, Hickie IB, Hermens DF. The N-methyl-D-aspartate receptor as a neurobiological intersection between bipolar disorder and alchol use: a longitudinal mismatch negativity study. Int J Neuropsychopharm. 2015;18(6). pii: pyu113.
46. Chitty KM, Logopoulos J, Hickie IB, Hermens DF. Investigating the role of glutathione in mismatch negativity: an insight into NMDA receptor disturbances in bipolar disorder. Clin Neurophysiol. 2015;126:1178–84.
47. Andersson S, Barder HE, Hellvin T, Løvdahl H, Malt UF. Neuropsychological and electrophysiological indices of neurocognitive dysfunction in bipolar II disorder. Bipolar Disord. 2008;10:888–99.
48. Kaur M, Battisti RA, Ward PB, Ahmed A, Hickie IB, et al. MMN/P3a deficits in first episode psychosis comparing schizophrenia-spectrum and affective-spectrum subgroups. Schizophr Res. 2011;130:203–9.
49. Jahshan C, Wynn JK, Mathis KI, Altshuler LL, Glahn DC, et al. Cross-diagnostic comparison of duration mismatch negativity and P3a in bipolar disorder and schizophrenia. Bipolar Disord. 2012;14:239–48.
50. Takei Y, Kumano S, Maki Y, Hattori S, Kawakubo Y, et al. Preattentive dysfunction in bipolar disorder: a MEG study using auditory mismatch negativity. Prog Neuropsychopharmacol Biol Psychiatry. 2010;34:903–12.
51. Shimano S, Onitsuka T, Oribe N, Maekawa T, Tsuchimoto R, et al. Preattentive dysfunction in patients with bipolar disorder as revealed by the pitch-mismatch negativity: a magnetoencephalography (MEG) study. Bipolar Disord. 2014;16:592–9.

52. Chen SS, Tu PC, Su TP, Hsieh JC, Lin YC, Chen LF. Impaired frontal synchronization of spontaneous magnetoencephalographic activity in patients with bipolar disorder. Neurosci Lett. 2008;445:174–8.
53. Oribe N, Onitsuka T, Hirano S, Hirano Y, Maekawa T, et al. Differentiation between bipolar disorder and schizophrenia revealed by neural oscillation to speech sounds: as MEG study. Bipolar Disord. 2010;12:804–12.
54. Başar E. Brain oscillations in neuropsychiatric disease. Dialogues Clin Neurosci. 2013;15: 291–300.
55. Gutschalk A, Mase R, Roth R, Rupp A, Hahnel S, et al. Deconvolution of 40 Hz steady-state fields reveals two overlapping source activities of the human auditory cortex. Clin Neurophysiol. 1999;110:856–68.
56. Picton TW, John MS, Dimitrijevic A, Purcell D. Human auditory steady-state responses. Int J Audiol. 2003;42:177–219.
57. Gonzalez-Burgos G, Lewis DA. GABA neurons and the mechanisms of network oscillations: implications for understanding cortical dysfunction in schizophrenia. Schizophr Bull. 2008;34: 944–61.
58. Brenner CA, Krishnan GP, Vohs JL, Ahn WY, Hetrick WP, et al. Steady state responses: electrophysiological assessment of sensory function in schizophrenia. Schizophr Bull. 2009; 35:1065–77.
59. Sohal VS. Insights into cortical oscillations arising from optogenetic studies. Biol Psychiatry. 2012;71:1039–45.
60. Emrich HM, von Zerssen D, Kissling W, Moller HJ, Windorfer A. Effect of sodium valproate on mania. The GABA-hypothesis of affective disorders. Arch Psychiatry Nervenkr. 1980;229: 1–16.
61. Konradi C, Zimmerman EI, Yang CK, Lohnmann KM, Gresch P, et al. Hippocampal interneurons in bipolar disorder. Arch Gen Psychiatry. 2011;68:340–50.
62. Maharajh K, Abrams DC, Rojas P. Auditory steady state and transient gamma band activity in bipolar disorder. Int Congr Ser. 2007;1300:707–10.
63. Rass O, Krishnan G, Brenner CA, Hetrick WP. Auditory steady state response in bipolar disorder: relation to clinical state, cognitive performance, medication status, and substance disorders. Bipolar Disord. 2010;12:793–803.
64. Uhlhaas PJ, Haeschel C, Nikolic D, Singer W. The role of oscillations and synchrony in cortical networks and their putative relevance for the pathophysiology of schizophrenia. Schizophr Bull. 2008;34:927–43.
65. Oda Y, Onitsuka T, Tsuchimoto R, Hirano S, Oribe N, et al. Gamma band neural synchronization deficits for auditory steady state responses in bipolar disorder patients. PLoS One. 2012; 7:e39955.
66. Liu TY, Hsieh JC, Chen YS, Tu PC, Su TP, et al. Different patterns of abnormal gamma oscillatory activity in unipolar and bipolar disorder patients during an implicit emotion task. Neuropsychologia. 2012;50:1514–20.
67. Poldrack RA, Gorgolewski KJ. Making big data open: data sharing in neuroimaging. Nat Neurosci. 2014;11:1510–7.
68. Phillips ML, Swartz HA. A critical appraisal of neuroimaging studies of bipolar disorder: toward a new conceptualization of underlying neural circuitry and a road map for future research. Am J Psychiatry. 2014;171:829–43.
69. Liu X, Akula N, Skup M, Brotman MA, Leibenluft E, et al. A genome-wide association study of amygdala activation in youths with and without bipolar disorder. J Am Acad Child Adolesc Psychiatry. 2010;49:33–41.
70. Singh I, Rose N. Biomarkers in psychiatry. Nature. 2009;460:202–7.
71. Hafeman DM, Chang KD, Garrett AS, Sanders EM, Phillips ML. Effects of medication on neuroimaging findings in bipolar disorder: an updated review. Bipolar Disord. 2012;14: 375–410.

Part IX

Cognition and Brain-Machine Interface

16 ECoG-Based BCI for BCI-MEG Research

Kyousuke Kamada

Abstract

A brain-computer interface (BCI) allows the user to control a device or software with brain activity. Many BCIs rely on dynamics of brain oscillations related to semantic and motor imaginary tasks in the electroencephalogram (EEG) and magnetoencephalography (MEG). Decoding brain activity of corresponding high-level tasks may lead to an independent and intuitively controlled BCI. Most of today's BCI research focuses on analyzing EEG and MEG which provide only limited signal-to-noise ratio. Derived electrocorticographic (ECoG) signals allow the investigation of spatially highly focused task-related activation within the high-gamma frequency band, making the discrimination of individual finger movements or complex grasping tasks possible. Common spatial patterns (CSPs) are commonly used for BCI systems and provide a powerful tool for feature optimization and dimensionality reduction. This work focused on the discrimination of (i) three complex hand movements as well as (ii) hand movement and idle state. Two subjects S1 and S2 performed single "open," "peace," and "fist" hand poses in multiple trials and (iii) one subject S3 controlled a humanoid in the remote place. Signals in the high-gamma frequency range between 100 and 500 Hz were spatially filtered based on a CSP algorithm for (i), (ii), and (iii). Additionally, a manual feature selection approach was tested for (i). A multi-class linear discriminant analysis (LDA) showed for (i) an error rate of 13.89 %/7.22 % and 18.42 %/1.17 % for S1 and S2 using manually/CSP selected features, where for (ii) a two-class LDA leads to a classification error of 13.39 % and 2.33 % for S1 and S2, respectively. For motor imaginary to remotely control the humanoid, (iii) S3 was able to perform such control over four sessions: three of those were performed consecutively over 1 day and the

K. Kamada, M.D., Ph.D. (✉)
Department of Neurosurgery, School of Medicine, Asahikawa Medical University, 2-1, Midorigaoka-Higashi, Asahikawa, Hokkaido 078-8510, Japan
e-mail: kamady-k@umin.ac.jp

S. Tobimatsu, R. Kakigi (eds.), *Clinical Applications of Magnetoencephalography*,
DOI 10.1007/978-4-431-55729-6_16

last one was performed 2 days afterward. After each session, the performance of the classifier improved, reaching about 90 % in the end. This experiment showed that ECoG-based motor imagery performed well despite a short training period, providing future possibility for MEG-based BCI with higher sensitivity than present MEG sensors.

Keywords
MEG • ECoG • BCI • Motor function • Brain signal decoding

16.1 Introduction

A brain-computer interface (BCI) is a direct communication pathway between the brain and an external device. BCIs are often directed at assisting, augmenting, or repairing human cognitive or sensorimotor functions. Most of them rely on processed EEG data containing event-related potentials (ERP) or oscillations showing event-related desynchronization/synchronization (ERD/ERS) [1, 2]. The EEG is widely spread in this area because of its low cost and easy setup, as well as its high temporal resolution [3]. However, the low spatial resolution and signal-to-noise ratio are limiting factors in many BCI applications [4].

MEG has been also used to record event-related responses with high time resolution, covering the whole head, and simply provides source localization of each response [5]. Both EEG and MEG are, however, sensitive to the topography of the brain – the details of the structure of its surface [6]. The brain surface is complexity folded, and whether the signal arises in the convex tops of the folds (gyri) or the concave depths of the folds (sulci) affects sensitivity to neuronal current sources (dipoles). MEG is particularly insensitive to radial dipoles, those directed toward or away from the scalp. It mainly sees tangential dipoles, which are parallel to the scalp. Electrocorticogram (ECoG) with subdural electrodes is more sensitive to both radial and tangential dipoles since electrodes are directly placed on the brain surface. In order to capture functional dynamics of neurophysiological phenomenon, this chapter explains electrophysiological characteristics of ECoG related to different motor tasks for brain signal decoding and BCI development. We believe that the ECoG study allows to demonstrate neurophysiological dynamics, overcoming the present MEG limitations.

We recently demonstrated ECoG profiles related to motor execution and visual recognition by normalizing ECoG electrodes of 20 patients on a template brain. This normalization technique clearly delineated a characteristic increase in band power for frequencies above 40 Hz, which is called high-gamma activity (HGA) [7].On the basis of the off-line analysis, we were able to discriminate HGA patterns related to different tasks (Fig. 16.1). This report adapted ECoG classification of motor execution and imagery for real-time processing. We focused on task-related HGA on specific cortical regions for brain signal decoding. In addition, we

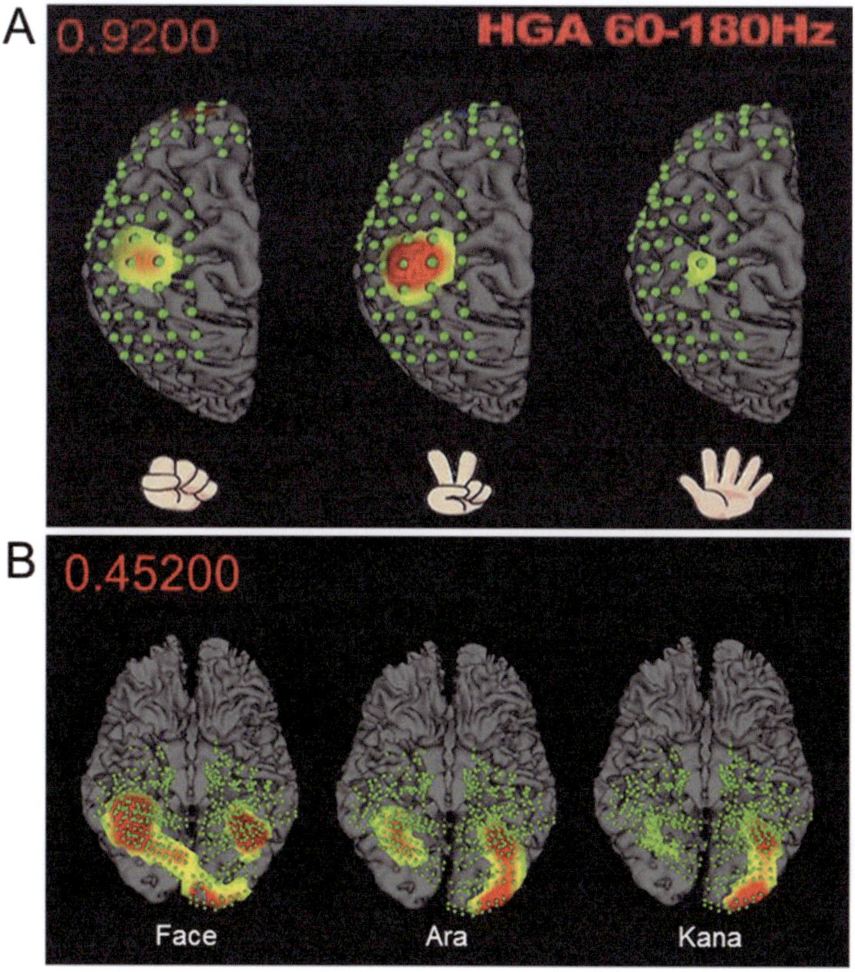

Fig. 16.1 Different dynamics of high-gamma activity (HGA) related to motor execution (**a**) and visual recognition (**b**). *Upper left* corner shows latency from visual ques. (**a**) A hand pose of peace sign demonstrates stronger and wider HGA than others. (**b**) Kana reading activated only the *left* temporal base. Face recognition showed *right* temporal activation with wider HGA than other tasks

demonstrated controlling a robot hand and a humanoid online to show future possibility for clinical practice.

16.2 Materials and Methods

We recorded ECoG in three patients with intractable epilepsy, who underwent implantation of subdural electrodes for diagnostic purposes at Asahikawa Medical University hospital between March 2013 and November 2014. During the ECoG recording, we instructed the patients to perform motor execution and imaginary tasks. Before epilepsy surgery, we confirmed the hemispheric dominance for language functions by Wada test or a combination of functional MRI and MEG as described elsewhere [8]. All the patients had subdural electrodes (Unique Medical, Tokyo) implanted over the primary motor and somatosensory cortices. The used platinum electrodes had an interelectrode distance of 5 mm and an exposure diameter of 1.5 mm, which indicated high resolution compared to the routine ECoG electrodes. Table 16.1 and Fig. 16.2 show more details of the patients.

This study was approved by the institutional review board of our institute. Written informed consent was obtained from each patient or their family after a detailed explanation of the ECoG and language evaluation.

16.2.1 General ECoG Recording

Each patient seated on a bed with a reclining backrest in a quiet, electrically shielded room. The ECoG data was recorded at bedside using a g.HIamp biosignal amplifier (Guger Technologies OG, Schiedlberg, Austria) with a sampling rate of 1200 Hz and a hardware band-pass filter range from 0.5 to 500 Hz.

16.2.2 Experimental Design

The subjects were asked to remain silent and follow the visual cues presented on a screen that was placed bedside to the patient. For stimulus presentation, a MATLAB/Simulink rapid prototyping environment was used.

Table 16.1 Demographic data of three patients

Subject	Gender	Age	Dominant hemisphere	Number of electrodes
S1	Female	35	Right	98
S2	Male	22	Left	60
S3	Female	54	Left	76

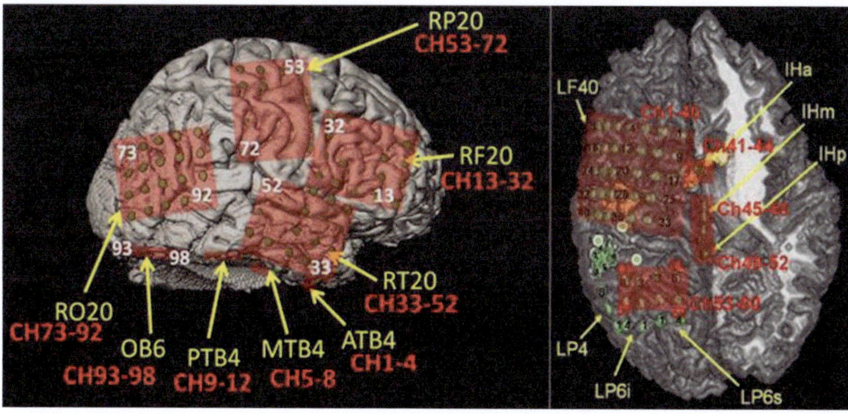

Fig. 16.2 Electrode configuration of subject S1 with coverage of the lateral side of the *right* hemisphere (*left*) and subject S2 with coverage of the *left* frontoparietal and the interhemispheric cortex (*right*)

16.2.3 Motor Execution/Imaginary Paradigms

The tasks consisted of different hand pose executions of (i) make a fist, (ii) show the peace gesture (V sign), and (iii) open the hand. In addition, motor imaginary tasks of hand opening and grasping were used for controlling the robot arm or humanoid. While three patients performed 90 trials in total, each consisting of a 1500 ms baseline and a 1500 ms task periods, during every single trial, exactly one pose was performed.

S3 remotely controls a robot arm in front of her and a humanoid robot in another city during the experiment. For the humanoid study, she is placed in front of a master screen that renders video fed back from humanoid's embedded camera. The humanoid is placed near a cupboard where an object is placed. A cue is shown to the patient, which has two states: (1) "Rest," a red hexagon with the Japanese character for rest and (2) "Move," a green circle with the Japanese character for move. For the patient, the mental strategy for "Rest" is to maintain a relaxed state in her mind, while the "Move" state corresponds to the imagination of a grasping movement with the left hand. Once the "Move" state has been recognized for 0.5 s, the humanoid grasps an object (soda can). The humanoid takes 2 s to perform the movement. It is executed smoothly while the patient is able to maintain her mental state to "Move." If the mental state is not maintained, the humanoid goes back to its starting position until the "Move" state is reactivated.

16.2.4 Data Analysis

16.2.4.1 ECoG Data Processing

All analysis of the ECoG data was performed using custom software written in MATLAB R2008b (MathWorks, Inc., Natick, MA). We chopped the continuous ECoG to create 2-s epochs (0.5 s before and 1.5 s after stimulus onset) with additional 0.125 s on both sides of each epoch to prevent "edge effect" artifact from clouding the results. Based on visual inspection of ECoG signals, trials with excessive epileptic activity were excluded from further analysis. For the motor task, the hand movement onset was determined via electromyogram (EMG) on the hand and stored as additional channel in the same recording.

16.2.4.2 Preprocessing

The data from subjects was band-pass filtered from 100 to 500 Hz (high-gamma range) with additional notch filters at all harmonics h_i of the 50 Hz power line frequency. All filters were designed as Butterworth filter of order 5 with a notch width of $h_i \pm 5$ Hz. Then the data was triggered 0.5 s before and 1.5 s after the stimulus onset of each task which resulted in an overall trial length of 2 s.

16.2.4.3 CSP Feature Extraction and Classification

Common spatial patterns (CSPs) are a standard method for ECoG data to extract optimal discriminant features in movement (or movement imagination) tasks [9]. To prevent influence of visual or auditory stimuli onto the classification, the CSPs were run exclusively on the electrode grids covering the motor cortex. The CSP weight matrix calculated with the optimal window size was then used to spatially filter the ECoG signals, and the four most discriminant feature channels were selected per decision pair (two largest eigenvalues from each side of the spectrum). Then the signal variances were computed and the resulting channels were normalized and had their logarithm taken for numeric stability. Based on those features, a two-class linear discriminant analysis (LDA) for the case "movement" vs. "idle" or a multi-class LDA (MLDA) for the hand pose decryption was calculated using 10×10-fold cross-validation [9, 10]. This was done for equally spaced time points (1/16 of sampling frequency) in the range of 0.5 s before and 1.5 s after stimulus onset, yielding a set of 32 different classifiers. For S3, the CSP method was adapted for ECoG to discriminate three motor execution tasks to control a robot arm and two motor mental states: "Rest" and "Move" for controlling the humanoid. In this study, the dimensions of the features' space are reduced and the classifier selection is automated. This allows online classification and performing chaining trials efficiently in order to improve the classifier's performances.

16.2.4.4 Manual Feature Extraction and Classification

The ERD/ERS was computed based on a 300 ms reference interval before the stimulus onset. Visual inspection of the time-frequency plots (0–200 Hz) showed that five channels were coding the individual hand poses for S1, while four channels were doing the same for S2 and S3. For each of the selected channels, the signal power within the 60–90 Hz, 110–140 Hz, and 160–190 Hz bands was estimated using a Butterworth filter of the fourth order, followed by squaring and averaging over consecutive samples of a 0.5 s window. All feature values were then logarithmically scaled. Based on the given band power features, a multi-class linear discriminant analysis (MLDA) was performed to compute a set of 32 linear classifiers, each represented the features for a given time point [11]. As in the CSP case, a 10 × 10 cross-validation was used to determine the classifier error rate. To determine the significance level of distinct distribution means between the two feature extraction methods, the nonparametric McNemar's test for paired nominal data was used with three different test statistics: exact binomial confidence interval [12], Yates continuity correction, and Edwards continuity correction.

16.3 Results

16.3.1 Discrimination of Hand Poses

Using the manual feature selection, the subjects S1 and S2 showed a minimal detection error for the three different hand poses of 13.89 % and 18.42 %, respectively. In contrast, the CSP-based features led to a minimal classification error of "fist," "peace," and "open" hand poses of 7.22 % and 1.17 % for S1 and S2, respectively. Figure 16.4 shows more details about the error rate of all the 32 different classifiers in the 2 s trial window for both subjects and methods. The shaded areas in Fig. 16.3 represent the 95 % (dark gray) and the 99 % (light gray) significance level, respectively, for the McNemar's test in all three test statistics.

16.3.2 Discrimination of Movement and Idle State

The discrimination of movement containing "fist," "peace," and "open" hand trials against the relaxed "idle" state showed a classification error of 13.39 % and 2.33 % for S1 and S2, respectively (Fig. 16.4).

16.3.3 Controlling the Robot Arm and the Humanoid

Mean training times to control the robot arm with motor execution and imaginary tasks were 58.2 and 25.4 min for all patients, respectively. In general, training session of EEG-based BCI needed more than 12 h [13]. ECoG demonstrated high signal-to-noise ratio and powerful potentials to create accurate decoders for developing BCI. Depending on motor execution, the maximum power and distribution were significantly different among hand poses that made decoding hand poses much

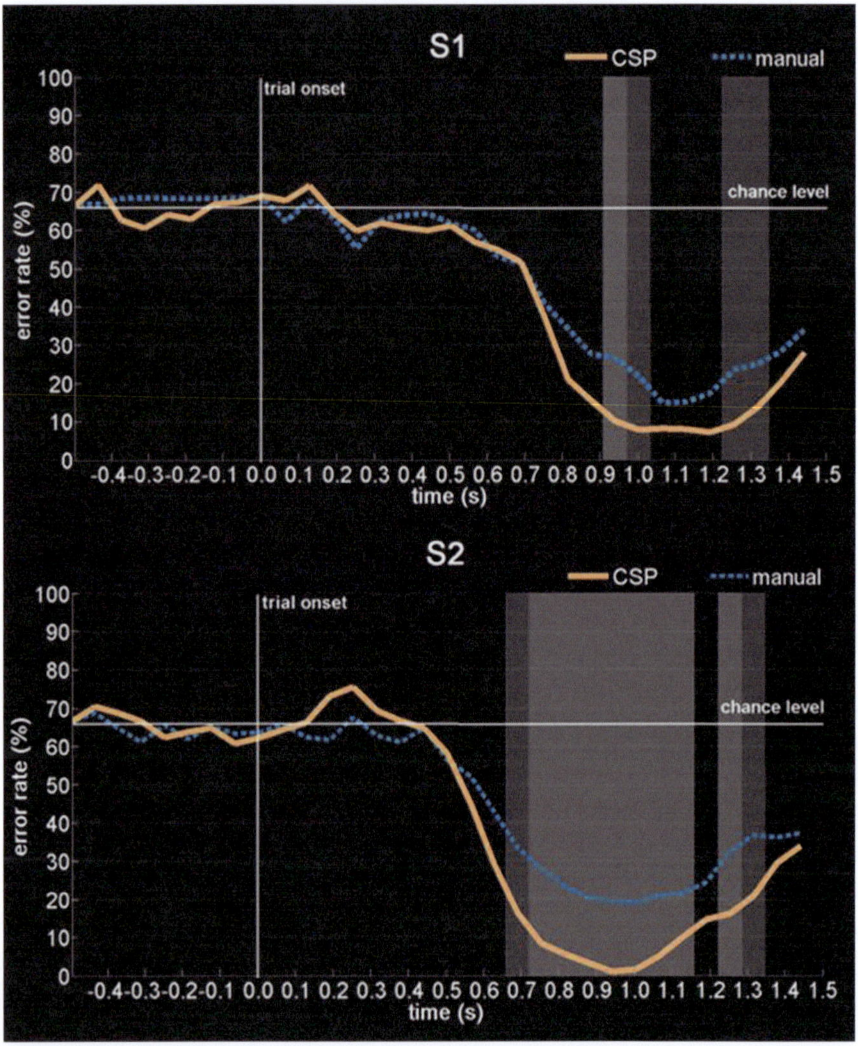

Fig. 16.3 Averaged three-class detection error for a single trial using manual (*blue dotted lines*) and CSP (*red solid lines*) feature extraction. The *vertical line* represents the time point of the visual stimulus that showed the subject which hand pose to perform. The *gray bars* represent the areas of significant differences between the two feature extraction method means; *dark gray* indicates $p < 0.05$ and *light gray* $p < 0.01$

easier (Fig. 16.5). For motor imaginary to control the robot arm with two poses such as fist and opening hand, training session took 35 min (Fig. 16.6). In addition to creating decoders and making short breaks, the whole setup took within 2 h. In the experiments of the motor execution and imaginary, decoding accuracies were 95.4 % and 97.7 %, respectively.

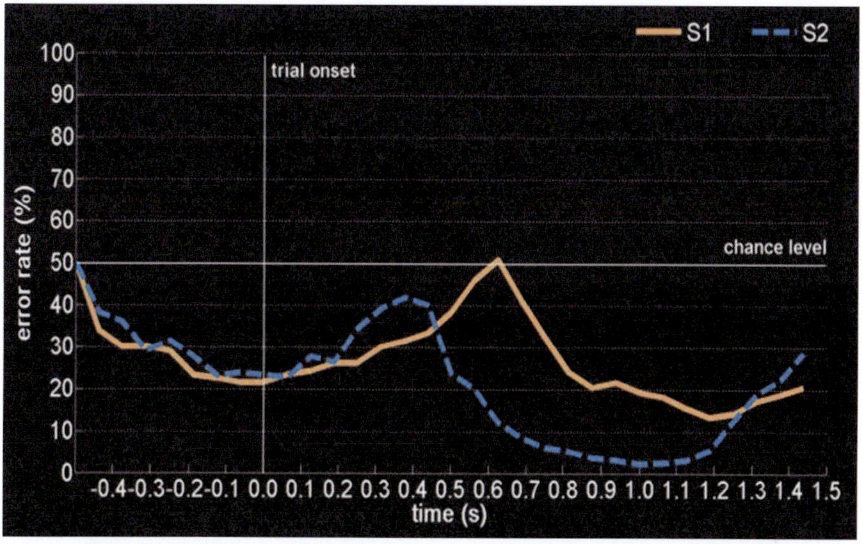

Fig. 16.4 Averaged two-class ("movement" vs. "idle") detection error for a single trial using the CSP feature extraction. The *vertical trial onset line* is identical with the onset of "movement" and "idle" blocks. The nonrandom classification behavior before 0.4 and 0.65 s, respectively, stems from previous movements contaminating the fake "idle" trials

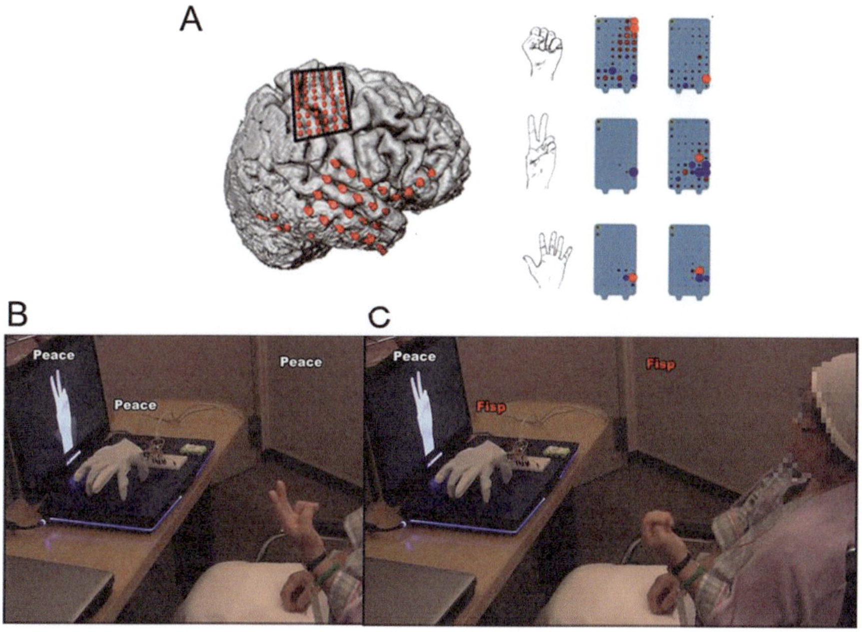

Fig. 16.5 (**a**) Electrode configuration of subject S3 with coverage of the lateral side of the *right* hemisphere (*left*) and maximum values of high-gamma activity depending on each hand pose. (**b**, **c**) ECoG decoding for hand poses of "peace or V sign" and "fist." Despite the same hand pose cue on the monitor, robot arm copied the real hand movement

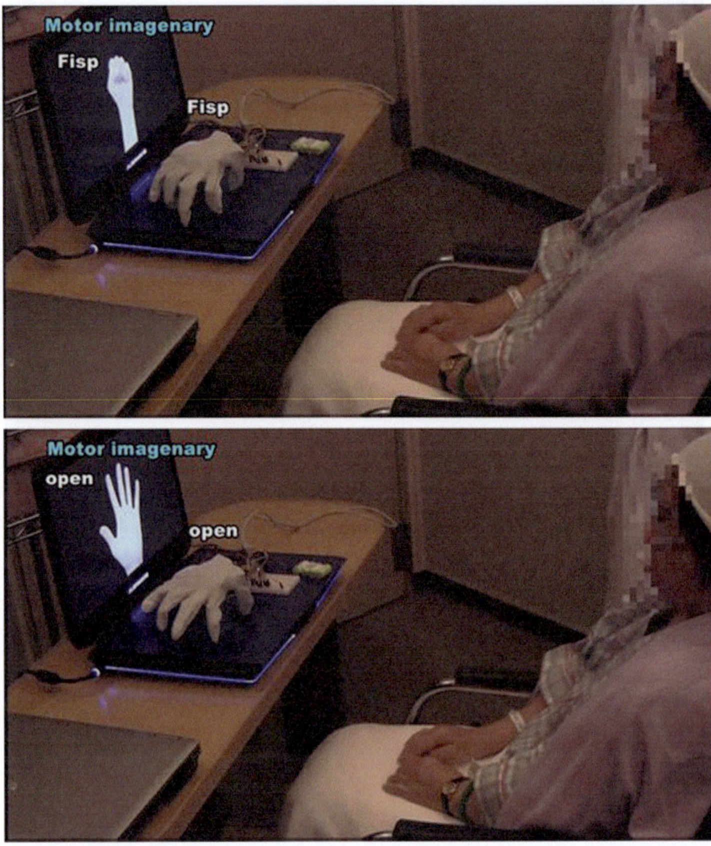

Fig. 16.6 Motor imaginary tasks with no muscle movement controlling a robot hand. CSP feature extraction well fits to decode the motor imaginary activity

For motor imaginary to remotely control the humanoid in the different institute, the patient was able to perform such control over four sessions: Three of those were performed consecutively over 1 day and the last one was performed 2 days afterward. After each session, the performance of the classifier improved, reaching about 90 % in the end. This experiment showed that ECoG-based motor imagery performed well despite a short training period compared to similar paradigms in EEG-based BCI [3]. The patient was also able to maintain a motor imagery state for an extended period of time (Fig. 16.7) [14].

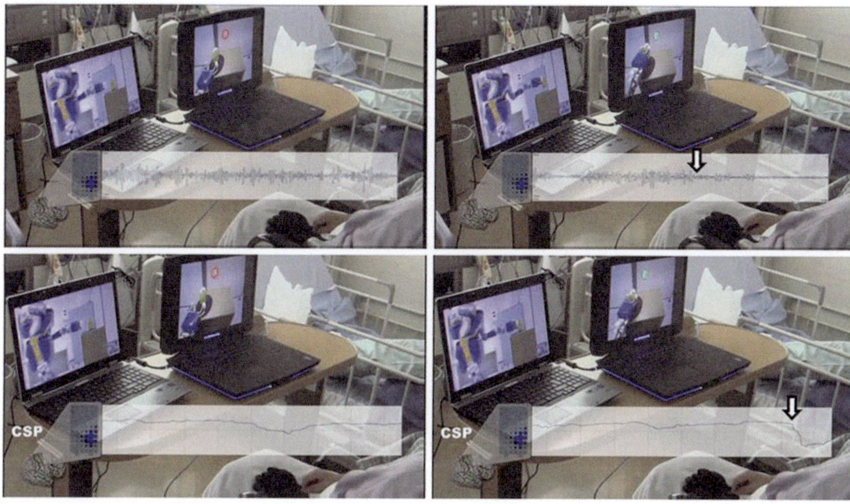

Fig. 16.7 Motor imaginary tasks with no hand movement controlling a humanoid to hold a can. Small windows in the *upper row* images show mean voltage changes of all targeted electrodes. Windows in the *lower row* demonstrate features of common spatial pattern of all targeted electrodes. *Red* and *green* spots on the monitor indicate resting and holding (active), respectively. *Arrows* indicated power suppression of electrocorticogram related to the motor imaginary task

16.4 General Remarks

BCI is being extensively studied for motor recovery after stroke and head injury. BCI for stroke motor recovery includes intensive BCI training linking brain activity related to the patient's intention to move the paretic limb with the contingent sensory feedback of the paretic limb movement guided by assistive devices. BCI studies in this area are mainly focused on EEG-BCI or MEG-BCI systems due to their high temporal resolution, which facilitates online contingency between the intention to move and sensory feedback of the intended movement [15]. EEG-BMI training was recently demonstrated in a controlled study to significantly improve motor performance in stroke patients with severe paresis [16]. Therefore, BCI contains clinical potentials for artificial prosthesis and rehabilitation. Despite invasiveness, ECoG detects the highest signal-to-noise ratio than EEG and MEG. Neural basis for BCI-induced restoration of motor function and perspectives for future BCI research for recovery of damaged motor functions have got strong attention in the clinical practice.

Figure 16.3 shows the error rates of the linear three-class classifiers along the different time points of trial duration. It can be seen that the CSP-based feature extraction expresses significantly lower error rates in detecting the correct hand poses compared to the manual feature selection. The error rates drop from 13.89 to 7.22 % for S1 and from 18.42 to 1.17 % for S2. This yields a maximal mean accuracy rate of 95.8 % by gaining on average 11.96 %. Due to the 0.5 s signal buffer for the band power computation and the reaction time of the subject, the movement type features were the most discriminant in a window between 0.8 and 1.2 s after trial onset (0.3–0.5 s after cue presentation). Furthermore, the features only remained stable for around 0.3 s, which leads to long phases of random state classification in an online experiment. Therefore, additional discrimination between the "movement" and "idle" state was used to minimize these false-positive assignments. Figure 16.4 shows that the respective error rates drop with the beginning of the movement around 0.5–0.8 s after the trial onset. The bowl-shaped deviation from the 50 % chance level for the classification error in the window from −0.5 to 0.65 s (S1) and −0.5 to 0.4 s (S2) stems from contamination with the end of the movement period of the previous trial. This is a direct consequence of the generation of the "idle" trials via trigger shifting. The contaminated window for S2 is shorter because of the longer intertrial period compared to S1. It is expected that the movement artifact would be more crucial in the MEG experiment, since the MEG sensors are fixed in the system. Although the MEG system has motion compensation [17] and other off-line motion correction programs, the artifact mostly affects decoding accuracy.

Visual inspection of the video material of the hand movements during the experiments reveals a possible reason for the weaker movement/idle discrimination of S1: It can clearly be seen that the flexion and extension of the fingers were executed much more powerful by subject S2. The presented experiments further show a strategy how to detect specific hand movements in an online environment. The presented BCI system allows the detection of movement in the first place, followed by a movement discrimination step. These two steps are computed in parallel (using their distinct CSP filters and classifiers). Compared to the manual feature extraction, the CSP filtering process stands out by the higher classification accuracy and the inherent dimensionality reduction, which decreases the computational effort tremendously and is an important factor for real-time computation within an online BCI system. In contrast to CSP-based classification algorithms for EEG data, the ECoG system stands out that similar accuracies can be achieved for more classes, shorter trial periods, and without the need of extra subject training. As a result, the presented ECoG-based BCI system provides an accurate and fast configuration for a corresponding online setup. In addition, the decoding principle of ECoG and MEG is mostly similar. The ECoG study in this chapter would be applicable for MEG-BCI research. In the future, we hope to lever our experience into future experiments of this type where the user would control the robot in more complex scenarios.

References

1. Ge S, Wang R, Yu D. Classification of four-class motor imagery employing single-channel electroencephalography. PLoS One. 2014;9(6):e98019.
2. Chaudhary U, Birbaumer N, Curado MR. Brain-Machine Interface (BMI) in paralysis. Ann Phys Rehabil Med. 2015;58(1):9–13.
3. Luaute J, Morlet D, Mattout J. BCI in patients with disorders of consciousness: clinical perspectives. Ann Phys Rehabil Med. 2015;58(1):29–34.
4. Hashimoto Y, Ushiba J. EEG-based classification of imaginary left and right foot movements using beta rebound. Clin Neurophysiol. 2013;124(11):2153–60.
5. Morel S, George N, Foucher A, Chammat M, Dubal S. ERP evidence for an early emotional bias towards happy faces in trait anxiety. Biol Psychol. 2014;99:183–92.
6. Ahn M, Ahn S, Hong JH, Cho H, Kim K, Kim BS, Chang JW, Jun SC. Gamma band activity associated with BCI performance: simultaneous MEG/EEG study. Front Hum Neurosci. 2013;7:848.
7. Kunii N, Kamada K, Ota T, Greenblatt RE, Kawai K, Saito N. The dynamics of language-related high-gamma activity assessed on a spatially-normalized brain. Clin Neurophysiol. 2013;124(1):91–100.
8. Kamada K, Sawamura Y, Takeuchi F, Kuriki S, Kawai K, Morita A, Todo T. Expressive and receptive language areas determined by a non-invasive reliable method using functional magnetic resonance imaging and magnetoencephalography. Neurosurgery. 2007;60(2):296–305; discussion 305–296.
9. Kapeller C, Schneider C, Kamada K, Ogawa H, Kunii N, Ortner R, Pruckl R, Guger C. Single trial detection of hand poses in human ECoG using CSP based feature extraction. Conf Proc IEEE Eng Med Biol Soc. 2014;2014:4599–602.
10. Prueckl R, Kapeller C, Potes C, Korostenskaja M, Schalk G, Lee KH, Guger C. CortiQ – clinical software for electrocorticographic real-time functional mapping of the eloquent cortex. Conf Proc IEEE Eng Med Biol Soc. 2013;2013:6365–8.
11. Kapeller C, Kamada K, Ogawa H, Prueckl R, Scharinger J, Guger C. An electrocorticographic BCI using code-based VEP for control in video applications: a single-subject study. Front Syst Neurosci. 2014;8:139.
12. Li P, Xu P, Zhang R, Guo L, Yao D. L1 norm based common spatial patterns decomposition for scalp EEG BCI. Biomed Eng Online. 2013;12:77.
13. Hinterberger T, Kubler A, Kaiser J, Neumann N, Birbaumer N. A brain-computer interface (BCI) for the locked-in: comparison of different EEG classifications for the thought translation device. Clin Neurophysiol. 2003;114(3):416–25.
14. Thomas E, Fruitet J, Clerc M. Combining ERD and ERS features to create a system-paced BCI. J Neurosci Methods. 2013;216(2):96–103.
15. Wang W, Sudre GP, Xu Y, Kass RE, Collinger JL, Degenhart AD, Bagic AI, Weber DJ. Decoding and cortical source localization for intended movement direction with MEG. J Neurophysiol. 2010;104(5):2451–61.
16. Chung E, Park SI, Jang YY, Lee BH. Effects of brain-computer interface-based functional electrical stimulation on balance and gait function in patients with stroke: preliminary results. J Phys Ther Sci. 2015;27(2):513–6.
17. Guo C, Li X, Taulu S, Wang W, Weber DJ. Real-time robust signal space separation for magnetoencephalography. IEEE Trans Biomed Eng. 2010;57(8):1856–66.

Oscillation and Cross-Frequency Coupling 17

Masayuki Hirata

Abstract

Cerebral oscillation is a neurophysiological phenomenon related to cerebral rhythmic activity that reflects neural activity. It changes depending on oscillatory frequency, reflecting various brain functions including sensory, motor, and language functions. These oscillatory changes are used as decoding features of brain-machine interfaces. Recently, it has become clear that these oscillations in the different frequency bands are synchronized with each other. This phenomenon is called cross-frequency coupling. We propose a hypothetical model: brain networks have a low-frequency oscillatory property and high γ activity in local circuits that are modulated by cross-frequency coupling between the phase of the low-frequency oscillation of the brain network and the amplitude of the high γ activity of local circuits.

Keywords

Cerebral oscillation • Cross-frequency coupling

17.1 Cerebral Oscillation

17.1.1 Cerebral Oscillatory Changes

Synchronous oscillations in specific frequency bands such as alpha waves are well known as basic brain rhythms. These basic rhythms change signal power due to brain activation. Event-related desynchronization (ERD) is an attenuation of the oscillation amplitude of a specific frequency band that occurs in relation to specific neural activities [1]. The opposite phenomenon, event-related synchronization

M. Hirata (✉)
Department of Neurosurgery, Osaka University Medical School, 565-0871 Suita, Osaka, Japan
e-mail: mhirata@nsurg.med.osaka-u.ac.jp

S. Tobimatsu, R. Kakigi (eds.), *Clinical Applications of Magnetoencephalography*,
DOI 10.1007/978-4-431-55729-6_17

(ERS), is an increase in that amplitude [2]. The best-known frequency band is α (8–13 Hz). So-called α waves are detected from the parieto-occipital area during relaxed wakefulness. The α wave increases when the eyes are closed and decreases when opened. This α blocking is one of the most typical examples of ERDs. In addition to the α band, frequency bands are divided depending on neurophysiological properties as follows: δ (1–4 Hz), θ (4–8 Hz), β (13–25 Hz), low γ (25–50 Hz), and high γ (>50 Hz) frequency bands.

Synchronous oscillations can be measured using EEG, MEG, and electrocorticography (ECoG) which is a method of recording neural activity from electrodes directly placed on the cortical surface. Of these three brain signals, ECoGs are the most precise with respect to both spatial and temporal resolution, and they provide a good and typical sample of time-frequency spectrograms for cerebral oscillatory changes including high-frequency components (Fig. 17.1). In Fig. 17.1, ECoGs were recorded during right-hand grasping from subdurally placed grid electrodes placed over the left sensorimotor areas in an intractable epilepsy patient. ERDs are observed in the α and β bands over the sensorimotor areas and are broadly distributed, whereas ERSs are observed in the high γ band in the sensorimotor areas and are more focally distributed [3, 4]. Regarding time domain, ERDs occur 500–1000 ms prior to muscle contraction and are sustained even after the end of muscle contraction, whereas ERSs are more restricted to the periods of muscle contraction. ERS in the high γ band is known to reflect functional localization of the brain better than ERD in the α and β bands. These oscillatory changes during movements are referred to as movement-related cerebral oscillatory changes. Cerebral oscillatory changes are observed not only during movements but also during language activities, sensory processing, and mental concentration.

ECoG provides us with precise information on neural activities directly from brain surface electrodes, but electrode placement requires brain surgery, whereas MEG is noninvasive as well as precise in functional localization. However,

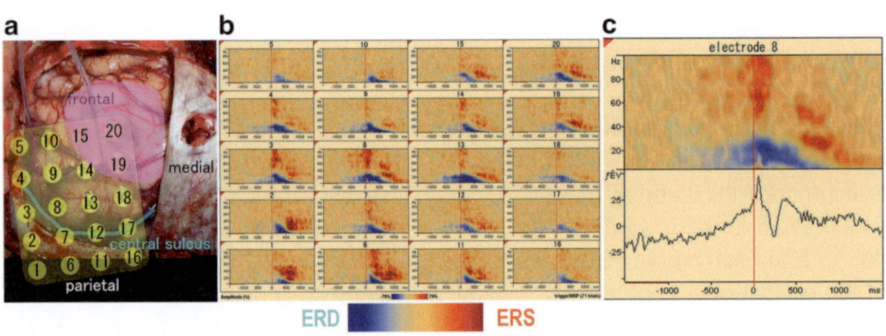

Fig. 17.1 Cerebral oscillatory changes recorded from subdural grid electrodes. (**a**) Electrode configuration on the brain. (**b**) Time-frequency spectrograms of the grid electrodes. ERS in the γ band (*red color*) is focally distributed, whereas ERD in the α and β bands (*blue color*) is broadly distributed. (**c**) A time-frequency spectrogram (*upper*) and a cortical potential graph (*lower*) of electrode 8

compared to ECoG, MEG does not consistently detect high γ band activity, unless time-locked stimuli or tasks are used in the case of the primary sensory or motor areas. Therefore, MEG is generally inferior to ECoG regarding functional localization.

17.1.2 Somatosensory Processing

Typical neuromagnetic somatosensory responses are observed when we stimulate major peripheral nerves of the body. We measured neuromagnetic responses using a whole-head type axial gradiometer equipped with 64 SQUID sensors. Figure 17.2a shows the averaged waveforms of the somatosensory evoked neuromagnetic fields recorded from all 64 SQUID sensors for 100 electrical stimuli to the right median nerve at the wrist of a healthy subject. Typical neuromagnetic components are clearly observed at 20 ms, 26 ms, 39 ms, and 51 ms after the stimuli. Isomagnetic field maps clearly show a pair of inflow and outflow neuromagnetic fields (Fig. 17.2b). A current dipole equivalent to the magnetic field at 20 ms after the stimuli is localized just in the contralateral postcentral gyrus (Fig. 17.2c). The postcentral gyrus is well known as the primary somatosensory area, where somatosensory processing such as touch and vibration sensation of the body is undertaken.

Beamformer analysis provides us with additional information about the spatiotemporal distribution of oscillatory activity of somatosensory processing. The ERS is focally localized in the contralateral postcentral gyrus in the high γ band, whereas the ERDs are more broadly localized over the bilateral postcentral gyri in the α band, lateralized to the contralateral side (Fig. 17.3) [5]. The high γ ERS is suggested to reflect activation of the primary somatosensory area, whereas α

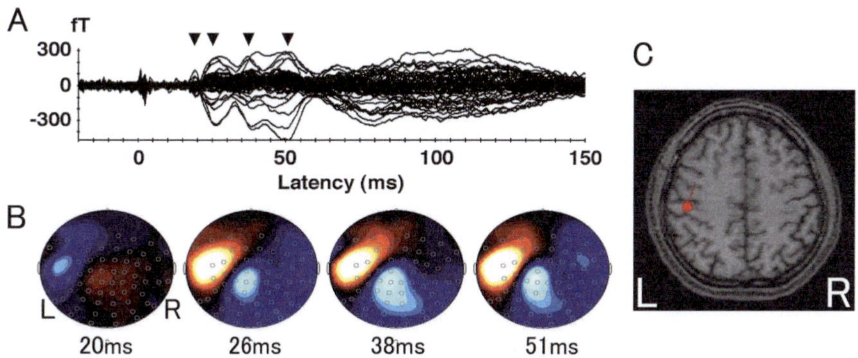

Fig. 17.2 Typical somatosensory evoked neuromagnetic fields (SEFs) induced by peripheral nerve stimulation. (**a**) Averaged waveforms of the SEFs for 100 electrical stimuli to the right median nerve. (**b**) Isomagnetic field maps showing a pair of inflow and outflow of neuromagnetic fields. (**c**) A current dipole equivalent to the magnetic field at 20 ms localized in the contralateral postcentral gyrus

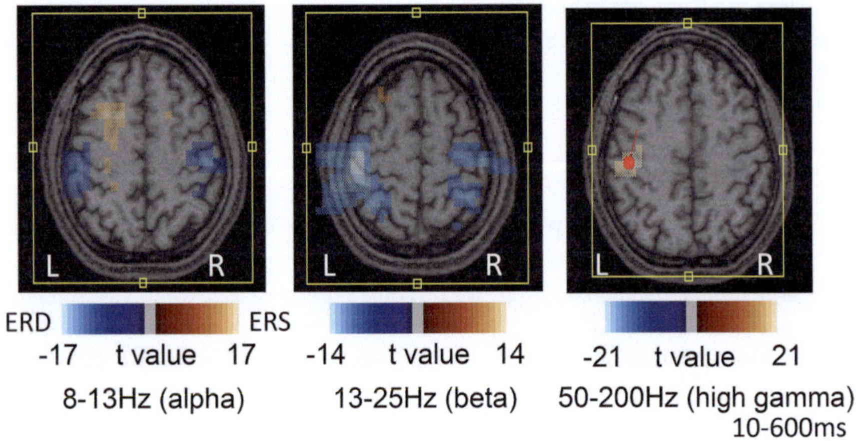

Fig. 17.3 Cerebral oscillatory changes during somatosensory stimulation revealed by beamforming MEG analyses. Distributed ERS in the contralateral somatosensory area in the high γ band (*right*) as well as the ERDs in the bilateral somatosensory areas in the α (*left*) and β (*center*) bands

ERDs are suggested to reflect the modulatory responses of somatosensory processing.

Clinically, high γ ERS induced by somatosensory stimulation is useful to identify the central sulcus as well as the dipole of neuromagnetic responses at 20 ms evoked by median nerve stimulation.

17.1.3 Motor Processing

Similar oscillatory changes are observed in the motor cortex during motor execution, as described above. These oscillatory changes during motor execution can also be observed using MEG. ERDs are observed in the α and β bands, broadly localized over the bilateral primary sensorimotor areas, lateralized to the contralateral side [6, 7]. Beta ERD is generally more anteriorly localized than α ERD, suggesting that β ERD reflects motor processing, while α ERD reflects somatosensory processing [7, 8]. In contrast, ERSs in the high γ band are focally localized on the contralateral precentral gyrus, well reflecting the activation of the contralateral primary motor area [9]. However, it is difficult to consistently detect high γ ERS using MEG. Therefore, β ERD is often used to evaluate the location of the primary motor area (Fig. 17.4). Although it is difficult to detect high γ ERS on a single trial basis with MEG, ECoG can detect high γ ERS on a single trial basis. Using this trial-by-trial high γ ERS as a decoding feature, hand movements can be differentiated, and a robotic arm can be controlled in real time [10, 11] (Fig. 17.5).

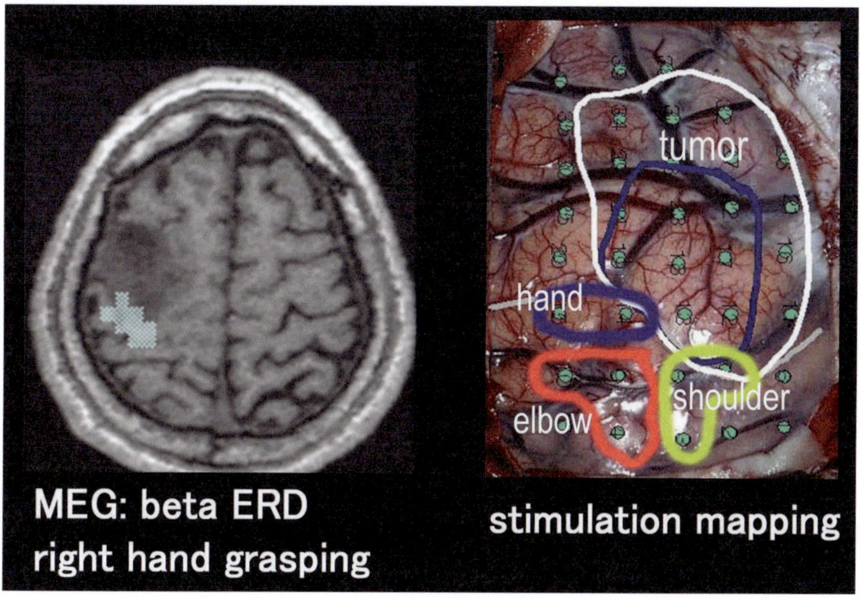

Fig. 17.4 Preoperative functional mapping based on the β ERD for evaluating the location of the primary motor area. Beta ERD was localized by MEG just posterior to a tumor in the precentral gyrus (*right*). The results of cortical electrical stimulation mapping were concordant with those of β ERD

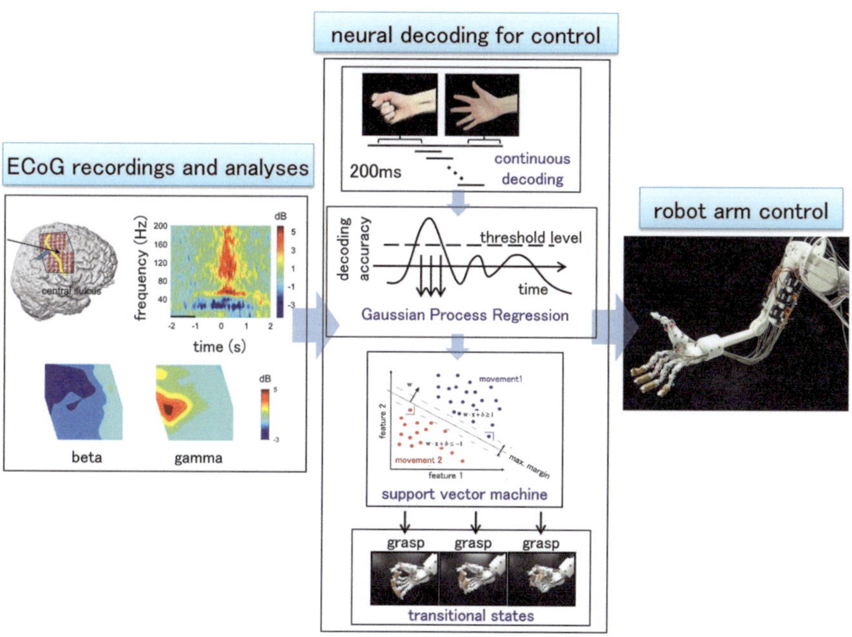

Fig. 17.5 Real-time robot control using trial-by-trial high γ ERS of ECoG. Using this trial-by-trial high γ ERS as a decoding feature, hand movements can be decoded, and a robotic arm can be controlled in real time

17.1.4 Language Processing

Compared to somatosensory processing and motor processing, neuromagnetic responses to language processing are more complex. We use a silent reading task to avoid noise contamination due to muscle contraction during phonation. A three-character hiragana word was presented on a display for 3 s. Subjects were instructed to silently read the words once, as soon as the words were presented. A total of 100 words were presented serially every 6 s. Figure 17.6a shows the averaged waveforms for neuromagnetic responses of all 64 SQUID sensors for visually presented hiragana words in a healthy subject. They have later latency components and isofield maps indicate more complex inflow and outflow distributions than those of somatosensory processing (Fig. 17.6b). Single dipole analyses managed to localize a dipole until 330 ms, but represented only part of the complex neuromagnetic fields. Multiple dipole analyses failed to localize stable dipoles with sufficient goodness of fit.

Beamformer analyses provide us with information on the complex spatial distribution of cerebral oscillatory changes during silent reading. Figure 17.7 shows the results of group analysis for 14 healthy right-handed subjects. It is

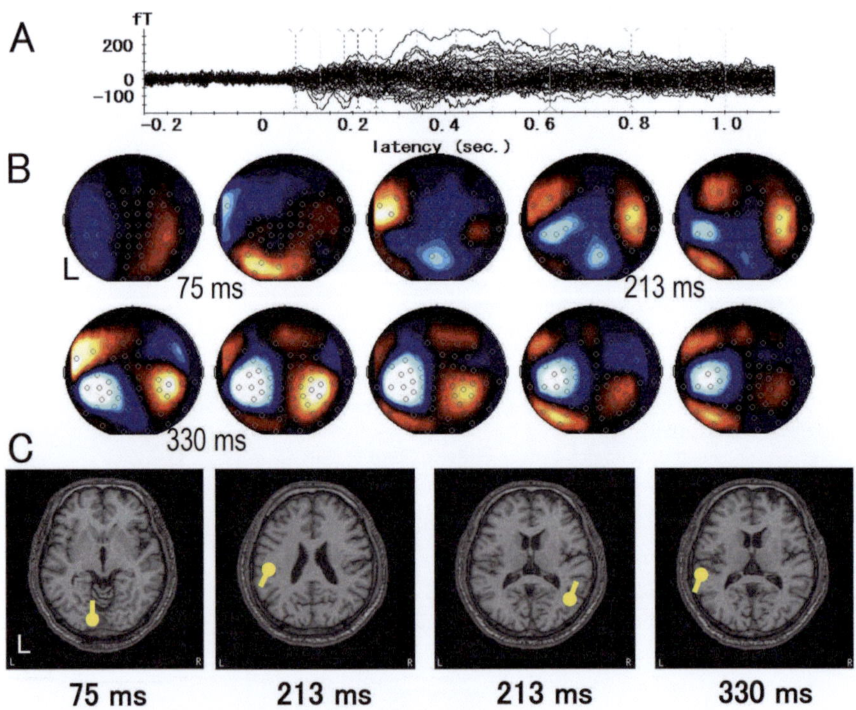

Fig. 17.6 Neuromagnetic responses during silent word reading. (**a**) Averaged waveforms of neuromagnetic responses for visually presented hiragana words in a healthy subject. (**b**) Isofield maps indicating complex inflow and outflow distributions. (**c**) Single dipole analyses

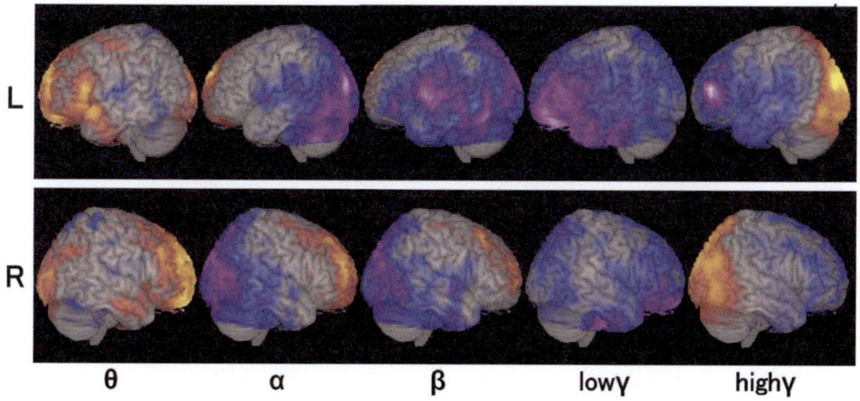

Fig. 17.7 Frequency-dependent spatial distribution of cerebral oscillatory changes during silent reading revealed by beamformer analyses. ERDs in the α band distributed in the posterior (receptive) language area, whereas ERDs in the low γ band distributed in the frontal (expressive) language area

noteworthy that the spatial distribution is dependent on the frequency bands of the oscillatory changes. ERDs in the α band are distributed in the posterior (receptive) language area, whereas ERDs in the low γ band are distributed in the frontal (expressive) language area [12]. In addition, we found that the left and right lateralization of ERD in the frontal area in the low γ band well corresponds to language dominance [13]. Using this property, we established a method to evaluate language dominance noninvasively with beamforming MEG analyses. Consistency is approximately 85 % with the standard but invasive method for evaluating language dominance (the Wada test), which requires the injection of anesthetic agents intra-arterially at the carotid artery [12]. Beamforming MEG analysis is considered to be an alternative to the Wada test in selected cases. Figure 17.8 shows an example of language dominance evaluated by this method.

Sliding time-window analyses of beamforming MEG revealed the temporal profiles of cerebral oscillatory changes during silent reading (Fig. 17.9) [14]. We found that the transient ERS first occurred in the occipital visual area and then in the temporo-occipital language areas and finally propagated to the frontal language areas. This transient θ (8–13 Hz) ERS was followed by α ERDs in the temporo-occipital language areas and low γ ERDs in the frontal language areas. High γ ERS was found in the occipital lobe, which reflects visual processing. It seems that transient θ ERS reflects serial processing, while α and low γ ERDs reflect parallel neural processing.

High γ ERSs related to language processing can be detected using ECoGs [15]. Figure 17.10 shows the relationship between the spatial distribution of high γ ERS induced by a verb generation task and the result of cortical electrical stimulation mapping. The localization of high γ ERS is well concordant with that of cortical mapping. Although it is generally difficult to detect high γ ERS related to language processing in individual subjects with MEG, high γ ERS can be detected

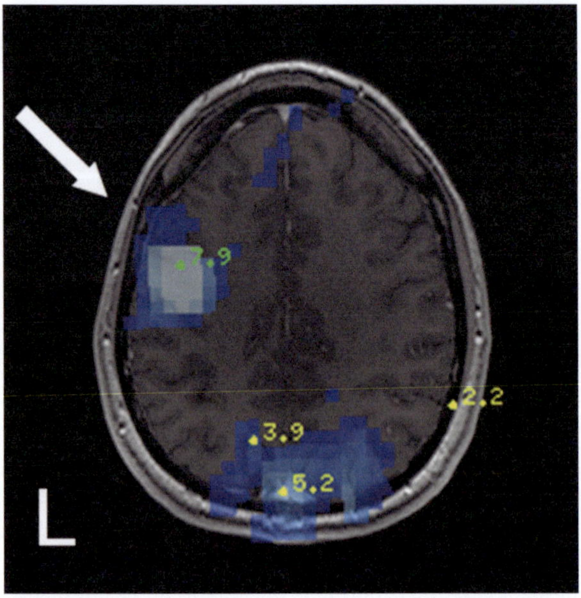

Fig. 17.8 Noninvasive preoperative evaluation of language dominance and localization. Low γ ERD indicating the frontal language area is localized in the left inferior and middle frontal gyri

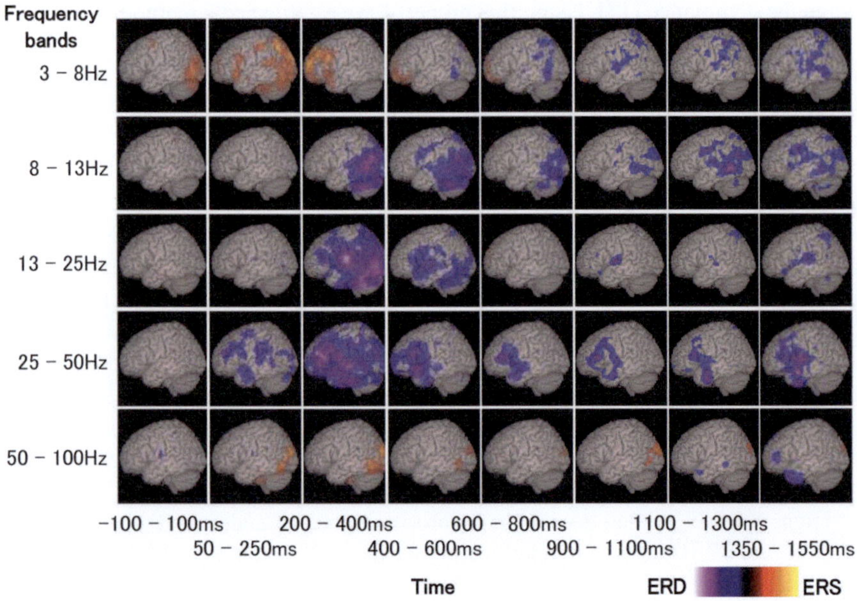

Fig. 17.9 Temporal profiles of cerebral oscillatory changes during silent reading. Transient θ ERS reflects serial processing, while sustained α and low γ ERDs reflect parallel neural processing

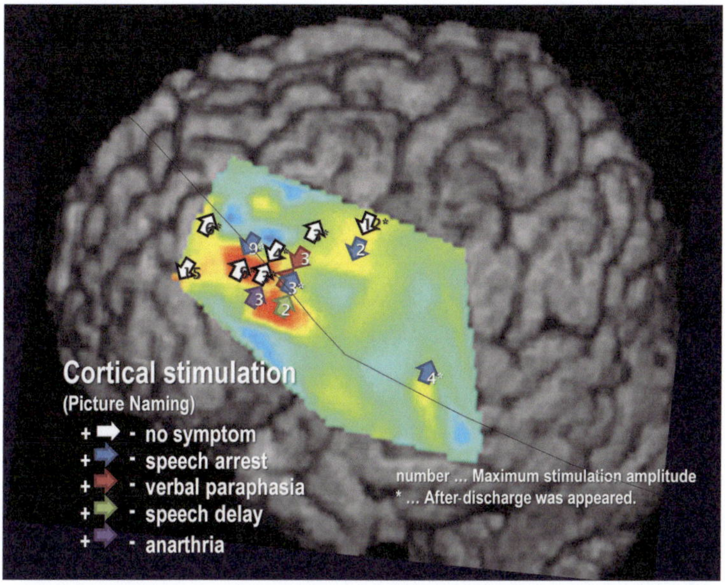

Fig. 17.10 High γ ERSs related to language processing detected using ECoGs. The localization of high γ ERS (*red color*) is well concordant with that of cortical mapping (*arrows*)

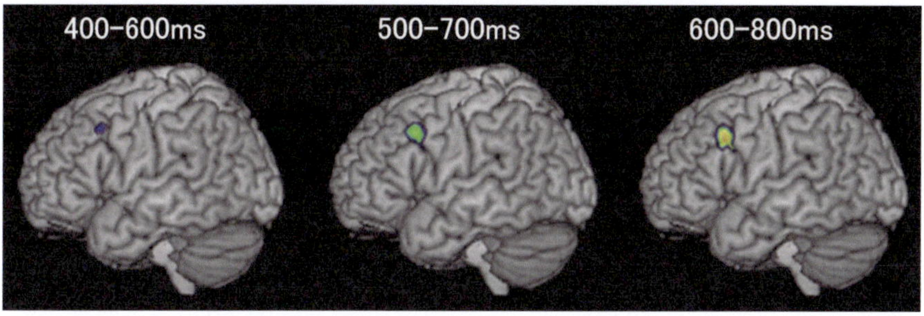

Fig. 17.11 High γ ERS related to language processing detected by MEG. Group analyses with targeted time windows detected high verb generation task induced γ ERS in the frontal language area

by MEG by combining group analyses, targeted time windows, and tasks with high task demand to the frontal language area such as a verb generation tasks (Fig. 17.11).

17.2 Cross-Frequency Coupling and Oscillatory Coupling Network Model

Cross-frequency coupling is a phenomenon where cerebral oscillations in the different frequency bands are synchronized (Fig. 17.12). Cross-frequency coupling has been reported in various cerebral areas [16, 17]. It is well known that the

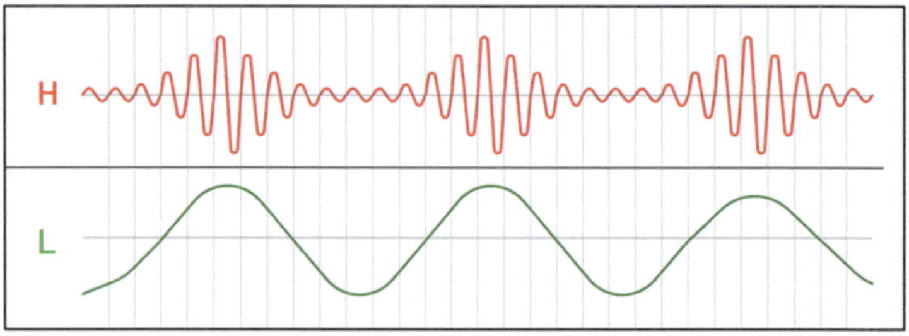

Fig. 17.12 Cross-frequency coupling. The amplitude of high-frequency activity (H) is synchronized with the phase of low-frequency activity (L)

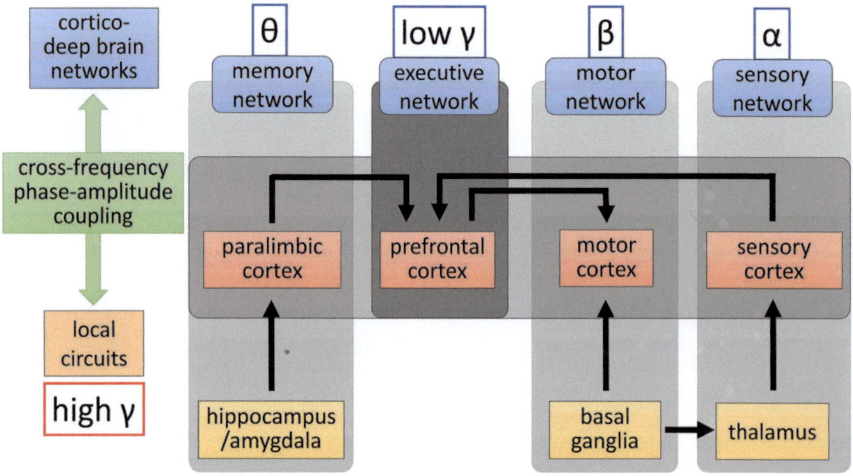

Fig. 17.13 Proposed oscillatory coupling network model. Sensory network dominated by α band activity, motor network dominated by α and β band activity, memory network dominated by θ band activity, and executive network dominated by low γ band activity. High γ band activity is a pivotal activity of local circuits modulated by the phase of the slow oscillatory activity including those of the θ, α, β, and low γ bands specific to each network

amplitude of the high γ band activity is modulated by the phase of the θ band activity. In general, slow oscillations in the θ, α, and β bands modulate fast oscillations in the high γ band [18]. We found that in using ECoGs, the amplitude in the high γ band is coupled with the phase in the α band more than 2 s before the onset of hand movements and decoupled just before movement onset [17]. This cross-frequency phase-amplitude coupling in the motor cortex might play an important role in control of movement initiation and postures.

Cerebral network models have been proposed regarding neural networks between the cortical areas and deep brain structures as well as cortico-cortical networks [19]. Based on the above-mentioned cerebral oscillation and cross-frequency coupling, we propose an oscillatory coupling network model of the brain (Fig. 17.13). Sensory networks consisting of thalamocortical loops are dominated by α band activity including α waves. Motor networks consisting of cortico-basal ganglia loops are dominated by α and β band activity including μ waves. Memory networks consisting of hippocampus/amygdala-cortical loops are dominated by θ band activity. Executive networks consisting of parieto-prefrontal circuits are dominated by low γ band activity. High γ band activity is a pivotal activity of local circuits in the cortical and deep brain structures consisting of these networks. Its activity may be modulated by the phase of slow oscillatory activity including those of the θ, α, β, and low γ bands specific to each network.

References

1. Pfurtscheller G, Aranibar A. Event-related cortical desynchronization detected by power measurements of scalp EEG. Electroencephalogr Clin Neurophysiol. 1977;2:817–26.
2. Pfurtscheller G. Event-related synchronization (ERS): an electrophysiological correlate of cortical areas at rest. Electroencephalogr Clin Neurophysiol. 1992;83:62–9.
3. Crone NE, Miglioretti DL, Gordon B, Sieracki JM, Wilson MT, Uematsu S, et al. Functional mapping of human sensorimotor cortex with electrocorticographic spectral analysis. I. Alpha and beta event-related desynchronization. Brain. 1998;121(Pt 12):2271–99.
4. Crone NE, Miglioretti DL, Gordon B, Lesser RP. Functional mapping of human sensorimotor cortex with electrocorticographic spectral analysis. II. Event-related synchronization in the gamma band. Brain. 1998;121(Pt 12):2301–15.
5. Hirata M, Kato A, Taniguchi M, Ninomiya H, Cheyne D, Robinson SE, et al. Frequency-dependent spatial distribution of human somatosensory evoked neuromagnetic fields. Neurosci Lett. 2002;318(2):73–6.
6. Jurkiewicz MT, Gaetz WC, Bostan AC, Cheyne D. Post-movement beta rebound is generated in motor cortex: evidence from neuromagnetic recordings. Neuroimage. 2006;32(3):1281–9. doi:10.1016/j.neuroimage.2006.06.005.
7. Taniguchi M, Kato A, Fujita N, Hirata M, Tanaka H, Kihara T, et al. Movement-related desynchronization of the cerebral cortex studied with spatially filtered magnetoencephalography. Neuroimage. 2000;12(3):298–306. doi:10.1006/nimg.2000.0611.
8. Salmelin R, Hari R. Spatiotemporal characteristics of sensorimotor neuromagnetic rhythms related to thumb movement. Neuroscience. 1994;60:537–50.
9. Cheyne D, Bells S, Ferrari P, Gaetz W, Bostan AC. Self-paced movements induce high-frequency gamma oscillations in primary motor cortex. Neuroimage. 2008;42(1):332–42. doi:10.1016/j.neuroimage.2008.04.178.
10. Yanagisawa T, Hirata M, Saitoh Y, Goto T, Kishima H, Fukuma R, et al. Real-time control of a prosthetic hand using human electrocorticography signals. J Neurosurg. 2011;114(6):1715–22. doi:10.3171/2011.1.jns101421.
11. Yanagisawa T, Hirata M, Saitoh Y, Kishima H, Matsushita K, Goto T, et al. Electrocorticographic control of a prosthetic arm in paralyzed patients. Ann Neurol. 2012;71(3):353–61. doi:10.1002/ana.22613.
12. Hirata M, Goto T, Barnes G, Umekawa Y, Yanagisawa T, Kato A, et al. Language dominance and mapping based on neuromagnetic oscillatory changes: comparison with invasive procedures. J Neurosurg. 2010;112(3):528–38. doi:10.3171/2009.7.JNS09239.
13. Hirata M, Kato A, Taniguchi M, Saitoh Y, Ninomiya H, Ihara A, et al. Determination of language dominance with synthetic aperture magnetometry: comparison with the Wada test. Neuroimage. 2004;23(1):46–53. doi:10.1016/j.neuroimage.2004.05.009.
14. Goto T, Hirata M, Umekawa Y, Yanagisawa T, Shayne M, Saitoh Y et al. Frequency-dependent spatiotemporal distribution of cerebral oscillatory changes during silent reading:

a magnetoencephalograhic group analysis. Neuroimage. 2011;54(1):560–7. doi:S1053-8119 (10)01105-5 [pii] 10.1016/j.neuroimage.2010.08.023 [doi].

15. Crone NE, Hao L, Hart Jr J, Boatman D, Lesser RP, Irizarry R, et al. Electrocorticographic gamma activity during word production in spoken and sign language. Neurology. 2001; 57(11):2045–53.
16. Canolty RT, Edwards E, Dalal SS, Soltani M, Nagarajan SS, Kirsch HE, et al. High gamma power is phase-locked to theta oscillations in human neocortex. Science (New York, NY). 2006; 313(5793):1626–8. doi:10.1126/science.1128115.
17. Yanagisawa T, Yamashita O, Hirata M, Kishima H, Saitoh Y, Goto T, et al. Regulation of motor representation by phase-amplitude coupling in the sensorimotor cortex. J Neurosci. 2012;32(44):15467–75. doi:10.1523/jneurosci.2929-12.2012.
18. Buzsaki G, Wang XJ. Mechanisms of gamma oscillations. Annu Rev Neurosci. 2012;35: 203–25. doi:10.1146/annurev-neuro-062111-150444.
19. Krack P, Hariz MI, Baunez C, Guridi J, Obeso JA. Deep brain stimulation: from neurology to psychiatry? Trends Neurosci. 2010;33(10):474–84. doi:10.1016/j.tins.2010.07.002.

MIX
Papier aus verantwortungsvollen Quellen
Paper from responsible sources
FSC® C105338

If you have any concerns about our products,
you can contact us on
ProductSafety@springernature.com

In case Publisher is established outside the EU,
the EU authorized representative is:
Springer Nature Customer Service Center GmbH
Europaplatz 3, 69115 Heidelberg, Germany

Printed by Libri Plureos GmbH
in Hamburg, Germany